Statistics and Computing

Series Editors:
W. Eddy
W. Härdle
S. Sheather
L. Tierney

Statistics and Computing

W. Härdle   S. Klinke   B.A. Turlach

XploRe:
An Interactive Statistical
Computing Environment

With 143 Figures, 32 in Color

Springer-Verlag

New York Berlin Heidelberg London Paris
Tokyo Hong Kong Barcelona Budapest

W. Härdle
Institut für Statistik und Ökonometrie
Humboldt-Universität zu Berlin
D-10178 Berlin
Germany

S. Klinke
CORE and Institut de Statistique
Université Catholique de Louvain
B-1348 Louvain-la-Neuve
Belgium

B.A. Turlach
CORE and Institut de Statistique
Université Catholique de Louvain
B-1348 Louvain-la-Neuve
Belgium
and
Statistics
Centre for Mathematics and
  Its Applications
The Australian National University
Canberra ACT 0200
Australia

Series Editors:

W. Eddy	W. Härdle	S. Sheather	L. Tierney
Department of	Institut für Statistik und	Australian Graduate	School of Statistics
Statistics	Ökonometrie	School of Management	University of
Carnegie Mellon	Humboldt-Universität zu	Kensington	Minnesota
University	Berlin	New South Wales 2033	Minneapolis, MN 55455
Pittsburgh, PA 15213	D-10178 Berlin	Australia	USA
USA	Germany		

On the cover: Based on a figure showing the local polynomial estimation of the derivative curve for motorcycle data. See Figure 5.11 in the text.

Library of Congress Cataloging-in-Publication Data
Härdle, Wolfgang.
    XploRe: an interactive statistical computing environment/W.
  Härdle, S. Klinke, B.A. Turlach.
      p.   cm. — (Statistics and computing)
  Includes bibliographical references and indexes.
  ISBN 0-387-94429-X
   1. XploRe. 2. Mathematical statistics — Data processing.
  I. Klinke, S. II. Turlach, B.A. III. Title. IV. Series.
  QA276.4.H365   1995
  519.5'0285'5369 — dc20   94-41488

Printed on acid-free paper.

Production managed by Francine McNeill; manufacturing supervised by Jacqui Ashri.
Photocomposed copy prepared using the editors' LaTeX files.
Color separations and printing of the cover and insert by New England Book Components, Hingham, MA.
Printed and bound by Braun-Brumfield, Inc., Ann Arbor, MI.
Printed in the United States of America.

9 8 7 6 5 4 3 2 1

ISBN 0-387-94429-X Springer-Verlag New York Berlin Heidelberg

Preface

This book describes an interactive statistical computing environment called XploRe.[1] As the name suggests, support for exploratory statistical analysis is given by a variety of computational tools. XploRe is a matrix-oriented statistical language with a comprehensive set of basic statistical operations that provides highly interactive graphics, as well as a programming environment for user-written macros; it offers hard-wired smoothing procedures for effective high-dimensional data analysis. Its highly dynamic graphic capabilities make it possible to construct student-level front ends for teaching basic elements of statistics. Hot keys make it an easy-to-use computing environment for statistical analysis.

The primary objective of this book is to show how the XploRe system can be used as an effective computing environment for a large number of statistical tasks. The computing tasks we consider range from basic data matrix manipulations to interactive customizing of graphs and dynamic fitting of high-dimensional statistical models. The XploRe language is similar to other statistical languages and offers an interactive help system that can be extended to user-written algorithms. The language is intuitive and readers with access to other systems can, without major difficulty, reproduce the examples presented here and use them as a basis for further investigation.

The book can be used as a basis for a variety of courses in statistical and econometrical model building, computational statistics, exploratory data analysis, and computer-aided teaching. The authors have used it in courses on applied multivariate statistics, computer-aided statistics, and econometrics. A course emphasizing the use of programming paradigms could be based on the material in Chapter 3. A course on interactive graphics with an emphasis on smoothing methods could use primarily the material in Chapters 5, 6, 7, and 15. A course on semiparametric modeling could use the examples in Chapters 9, 11, and 12. XploRe may also be used in undergraduate teaching when only the teachware module with student front end is implemented.

---

[1] XploRe is a trademark of W. Härdle.

Prerequisites

The statistical background assumed for most of the book is undergraduate knowledge. Methods like basic linear regression analysis should be known, but can also be learnt and taught with the XploRe teachware module. A few sections assume somewhat more knowledge; these sections indicate the background that they require.

No prior knowledge of any statistical computing system is assumed. Some experience with systems like S,[2] GAUSS,[3] or MATLAB[4] may be helpful but is neither required nor essential. XploRe runs on IBM[5]-compatible machines (386 or higher). Network settings, for example, for teaching in computer classrooms have been successfully implemented. A UNIX[6] version tested for Sun[7] workstations is also available.

Software Availability

As with all books on computing, the reader will benefit most from reading the book while trying the experiments with the software. At the time of this writing, there is a free copy available (Version 3.0 with memory limited to 640 KB) via FTP from "amadeus.wiwi.hu-berlin.de" (141.20.100.2). Version 3.2 of XploRe corresponds to the description in this book. Most of the macros and data sets used in this book will be also available on this server.

Numerical Precision

All calculations are done in double precision. The C++ source was compiled with the ZORTECH[8] C++ 3.0 compiler.

Reviews

The XploRe software was reviewed by Stone (1990), Ng and Sickles (1990), Schimek (1991), Lee (1992), Schimek and Kubik (1992), Hilbe (1993), Schimek and Schmaranz (1993), and Hilbe (1994a).

---

[2]S and S-Plus are a trademark of StatSci.
[3]GAUSS is a trademark of Aptech Systems, Inc.
[4]MATLAB is a trademark of Mathsoft.
[5]IBM is a trademark of International Business Machines Corporation.
[6]UNIX is a trademark UNIX Systems Laboratory, Inc.
[7]SUN is a trademark of Sun Microsystems, Inc.
[8]ZORTECH is a trademark of Zortech Systems.

Obtaining the XploRe Software

The XploRe software may be obtained by writing to

GfKI
Postfach 101248
D-68012 Mannheim
Germany
Fax: +49 621 812852

Berlin, Germany	Wolfgang Härdle
Berlin, Germany, and Louvain-la-Neuve, Belgium	Sigbert Klinke
Canberra, Australia, and Louvain-la-Neuve, Belgium	Berwin A. Turlach

Acknowledgments

Many of the functions and basic operations were motivated by similar functions in the S system (Becker, Chambers and Wilks, 1988). Several ideas were borrowed from the S book (Becker et al., 1988) and from the GAUSS system. The highly interactive graphic interface uses ideas from PRIM9 (as did many other systems before ours). Work on XploRe was supported in part by the Deutsche Forschungsgemeinschaft (Sonderforschungsbereiche 303 and 373), by CORE and the Institut de Statistique, Université Catholique de Louvain, Belgium, and by CentER, Katholieke Universiteit Brabant, Tilburg, Netherlands. The book was mainly written while the third author was at the Université Catholique de Louvain. It was completed during a sabbatical leave of the first author at INRA and INSEE, France.

Over the past seven years, XploRe has changed from a completely menu-driven special-purpose smoothing package to a real statistical computing environment. This convergence was possible because we had strong programming assistance by Peter Klinke and had colleagues around the world who shaped our ideas and made clear what an easy-to-use computing interface means. We would like to thank Irene Bertschek, Adrian Bowman, Ricardo Cao-Abad, Dianne Cook, Pierre Dehez, Alain Desdoigts, Irène Gijbels, Wenceslao González-Manteiga, Christian Gourieroux, Peter Hall, Joel Horowitz, Sylvie Huet, Mariette Huysentruit, Emmanuel Jolivet, Alan Kirman, Alois Kneip, Oliver Linton, Etienne Loute, Charles Manski, Steve Marron, Byeong Park, Juan Rodriguez Poo, Bernd Rönz, Gilbert Saporta, Michael Schimek, Sibylle Schmerbach, David Scott, Robin Sickles, Elisabeth de Turckheim, and Phillipe Vieu. Our friend Leopold Simar has criticized many missing features of earlier versions of XploRe. His criticisms were justified and we used his ideas to improve the system. We would like to thank him for his encouragement and support.

Finally, we would like to thank all the contributors to this book for their careful preparation of their manuscripts and the time and effort they have invested in creating a better XploRe system. Thanks go also to Nick Fischer, Michael Schimek, Sanford Weisberg, and the members of the Institut für Statistik und Ökonometrie for proof reading the manuscript at various stages. Donald Arseneau helped us with a subtle error in the Springer LATEX macros. Thanks go also to the editor, Martin Gilchrist of Springer-Verlag, for his helpful cooperation.

Berlin, Germany	Wolfgang Härdle
Berlin, Germany, and Louvain-la-Neuve, Belgium	Sigbert Klinke
Canberra, Australia, and Louvain-la-Neuve, Belgium	Berwin A. Turlach

Contents

Contributors

Marco Bianchi, Bank of England, London SE 1, UK.
  bianchi@core.ucl.ac.be

Raymond J. Carroll, Department of Statistics, Texas A&M University,
  College Station, TX 77843, USA.
  carroll@stat.tamu.edu

Rong Chen, Department of Statistics, Texas A&M University, College
  Station, TX 77843, USA.
  chen@stat.tamu.edu

Jianqing Fan, Department of Statistics, University of North Carolina,
  Chapel Hill, NC 27599-3260, USA.
  jfan@stat.unc.edu

Claudia Gajewski, Sonderforschungsbereich 373, Humboldt-Universität
  zu Berlin, D-10178 Berlin, Germany.
  claudia@wiwi.hu-berlin.de

Wolfgang Härdle, Institut für Statistik und Ökonometrie, Humboldt-
  Universität zu Berlin, D-10178 Berlin, Germany.
  haerdle@wiwi.hu-berlin.de

Christian Hafner, Graduiertenkolleg "Angewandte Mikroökonomik",
  Freie Universität Berlin, D-14195 Berlin, Germany.
  hafner@wiwi.hu-berlin.de

Joeseph M. Hilbe, Department of Sociology and Graduate College, Committee on Statistics, Arizona State University, Tempe, AZ 857210,
  USA.
  atjmh@asuvm.inre.asu.edu

Sigbert Klinke, Institut de Statistique, Université Catholique de Louvain, B-1348 Louvain-la-Neuve, Belgium; and Institut für Statistik
  und Ökonometrie, Humboldt-Universität zu Berlin, D-10178 Berlin,
  Germany.
  sigbert@wiwi.hu-berlin.de

Thomas Kötter, Institut für Statistik und Ökonometrie, Humboldt-Universität zu Berlin, D-10178 Berlin, Germany.
  thomas@wiwi.hu-berlin.de

Helmut Küchenhoff, Seminar für Ökonometrie und Statistik, Ludwig-Maximilian Universität, Akademiestraße 1, D-80799 München, Germany.
helmut.kuechenhoff@statistik.uni-muenchen.d400.de

Hans-Joachim Mucha, Weierstraß Institut für Angewandte Analysis und Stochastik, D-10117 Berlin, Germany.
mucha@iaas-berlin.d400.de

Marlene Müller, Institut für Statistik und Ökonometrie, Humboldt-Universität zu Berlin, D-10178 Berlin, Germany.
marlene@wiwi.hu-berlin.de

Jörg Polzehl, Konrad-Zuse-Institut für Informationstechnik Berlin, Heilbronner Str. 10, D-10711 Berlin, Germany.
polzehl@sc.zib-berlin.de

Isabel Proença, Instituto Superior de Economia e Gestao, Universidade Técnica de Lisboa, Rua Miguel Lupi 20, P-1200 Lisboa, Portugal.
isabelp@pascal.iseg.utl.pt

Christian Ritter, CORE and Institut de Statistique, Université Catholique de Louvain, B-1348 Louvain-la-Neuve, Belgium.
ritter@stat.ucl.ac.be

Berwin A. Turlach, CORE and Institut de Statistique, Université Catholique de Louvain, B-1348 Louvain-la-Neuve, Belgium; and Statistics, Centre for Mathematics and Its Applications, The Australian National University, Canberra ACT 0200, Australia.
berwin@alphasun.anu.edu.au

Axel Werwatz, Sonderforschungsbereich 373, Humboldt-Universität zu Berlin, D-10178 Berlin, Germany.
axel@wiwi.hu-berlin.de

Color Illustrations

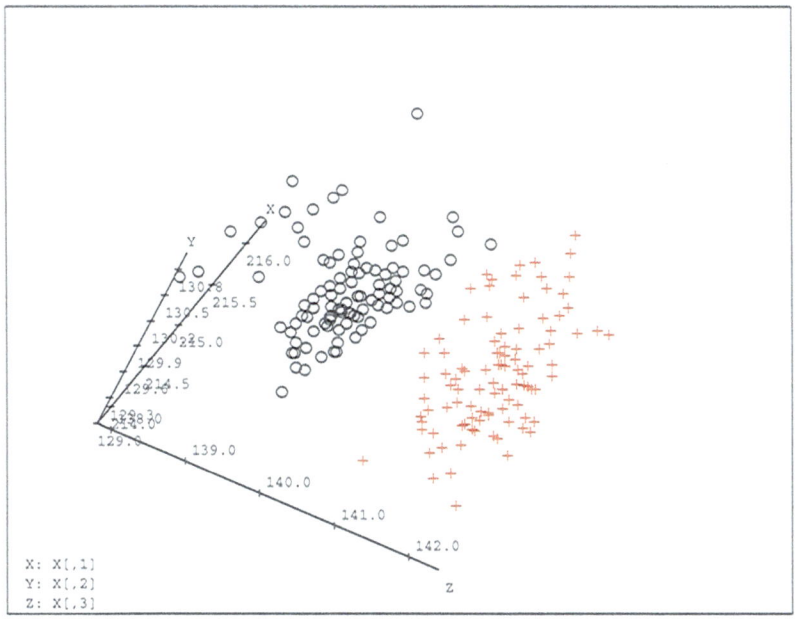

Figure 2.17

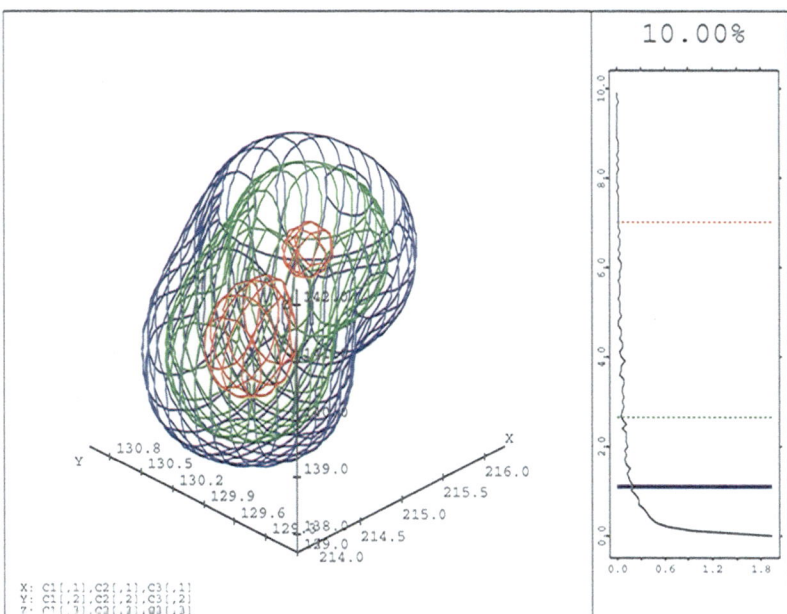

Figure 2.20

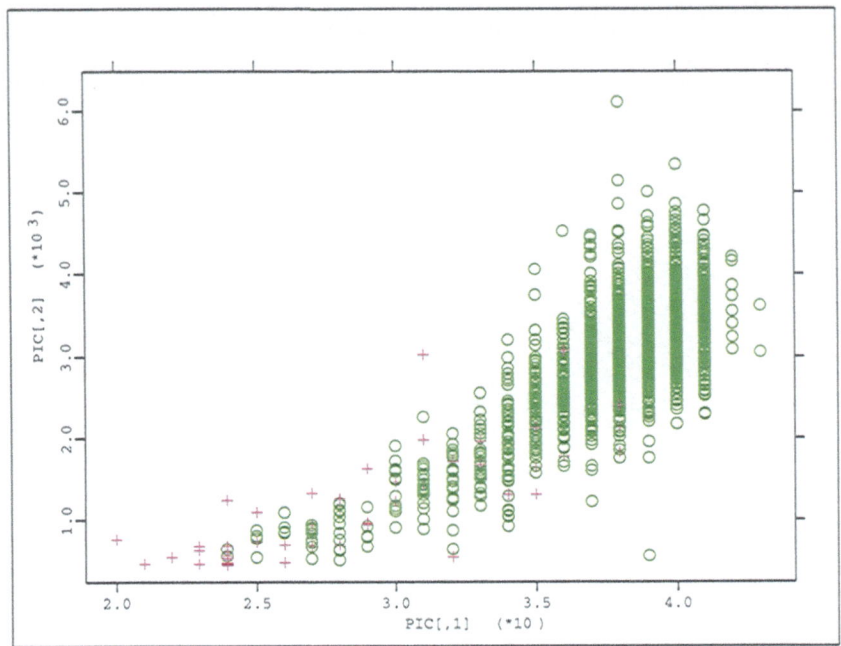

Figure 4.1

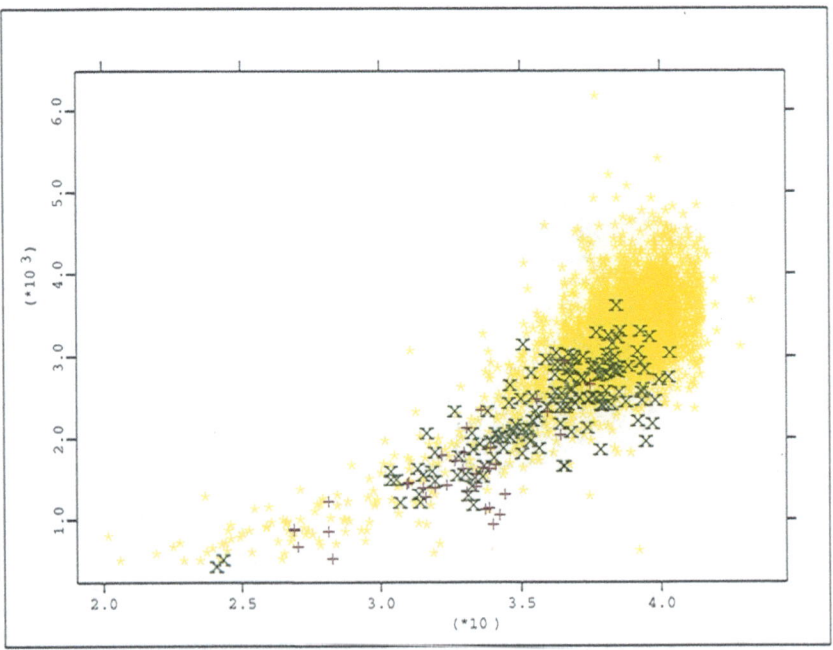

Figure 4.5

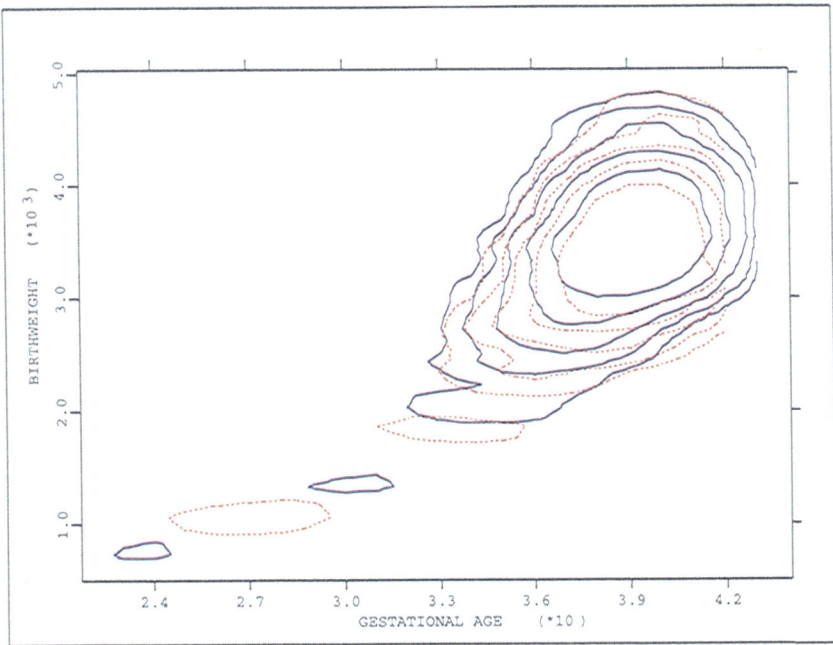

Figure 4.7

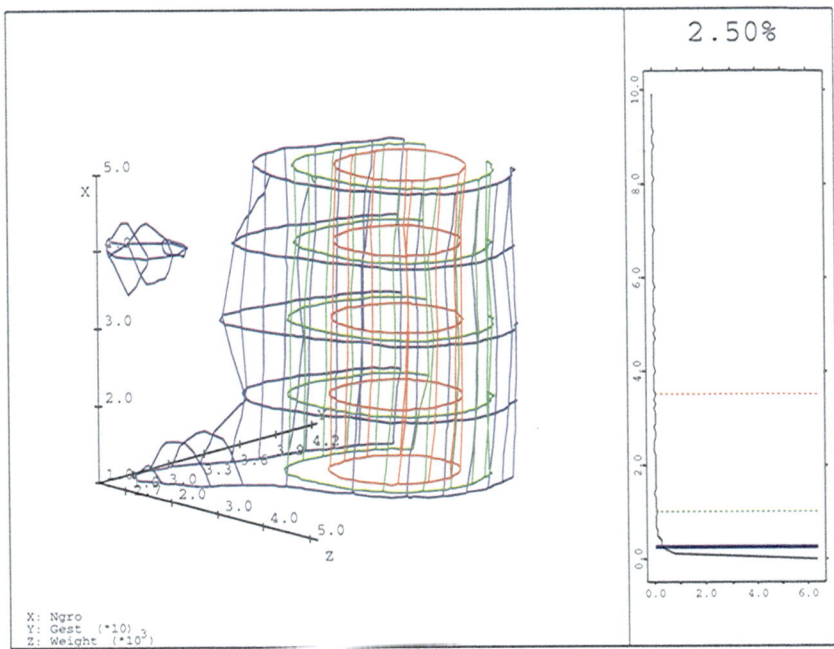

Figure 4.8

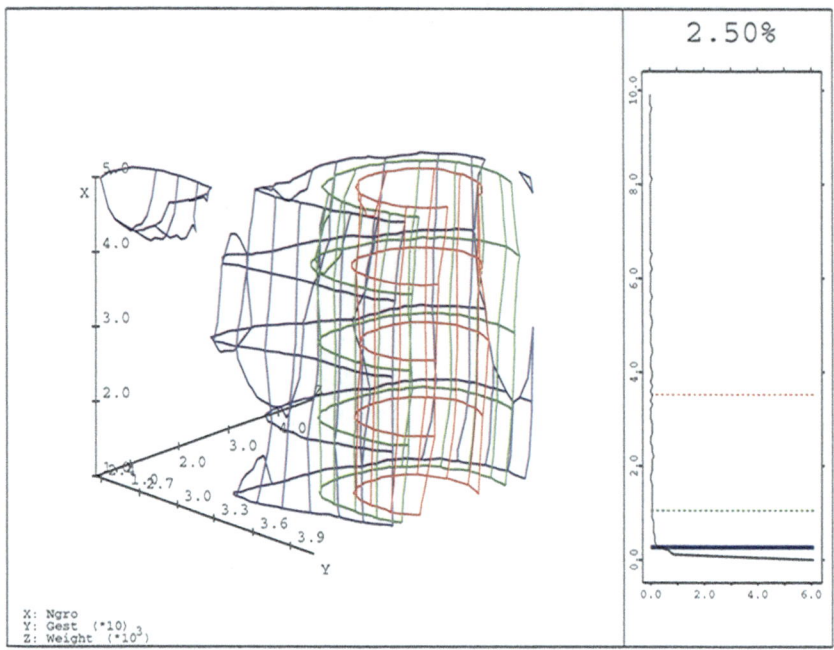

X: Ngro
Y: Gest (*10)
Z: Weight (*10³)

Figure 4.9

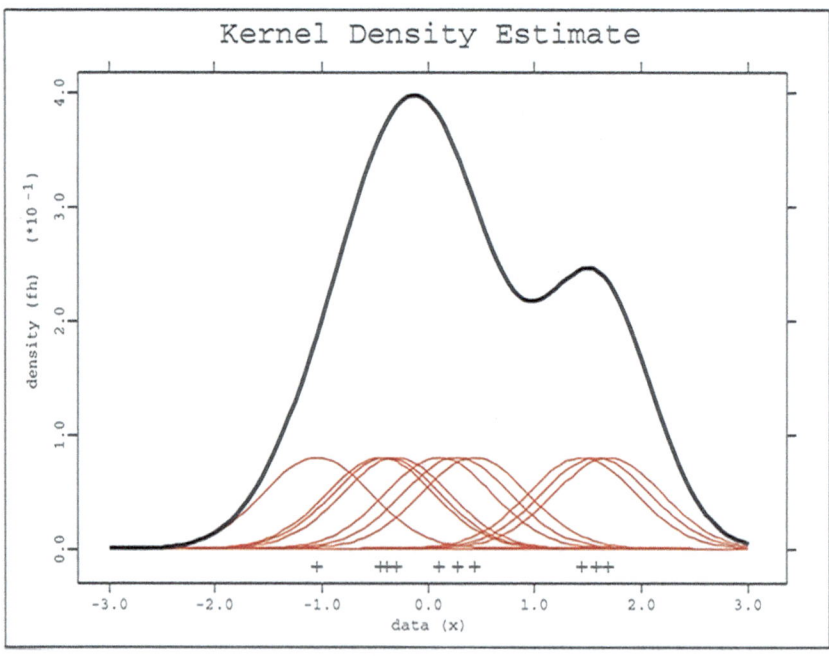

Figure 5.1

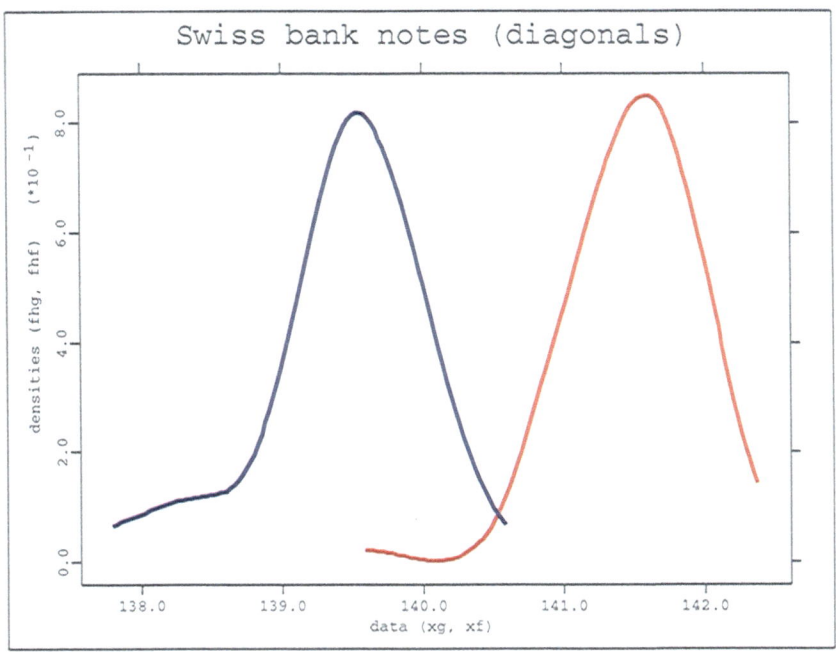

Figure 5.4

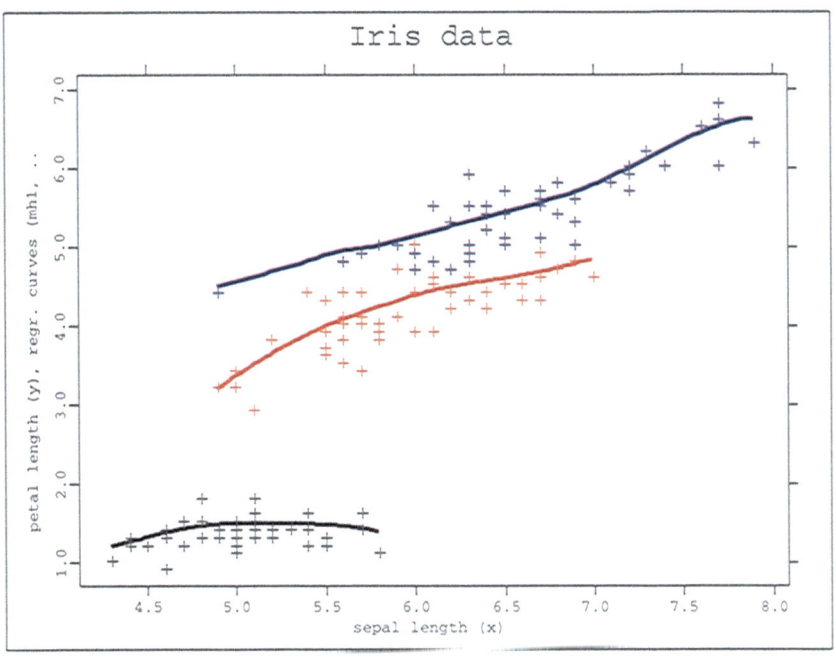

Figure 5.10

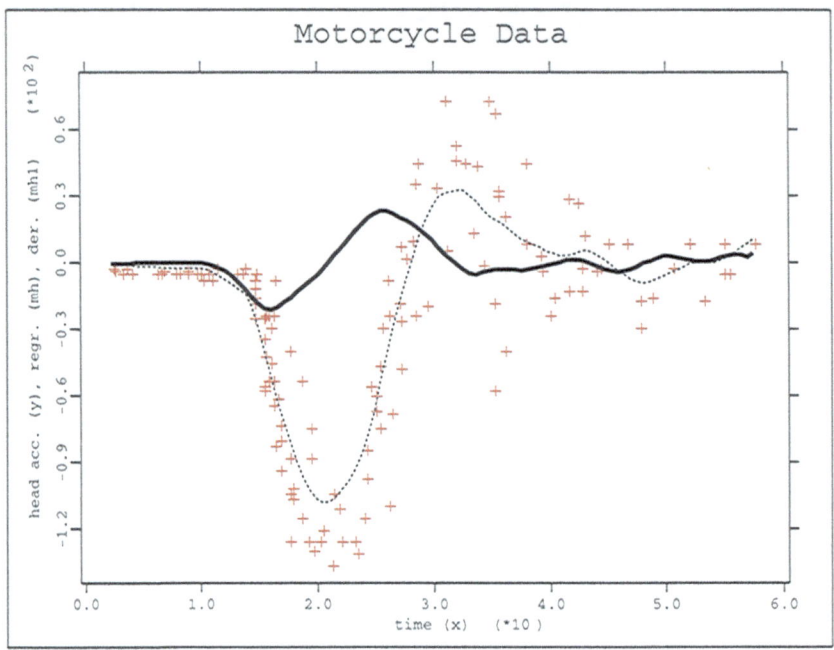

Figure 5.11

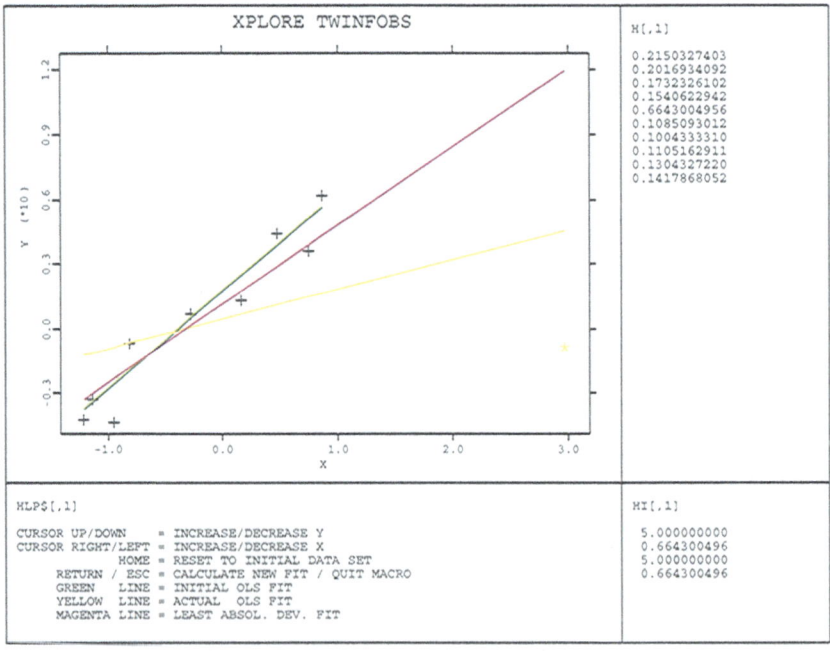

Figure 7.4

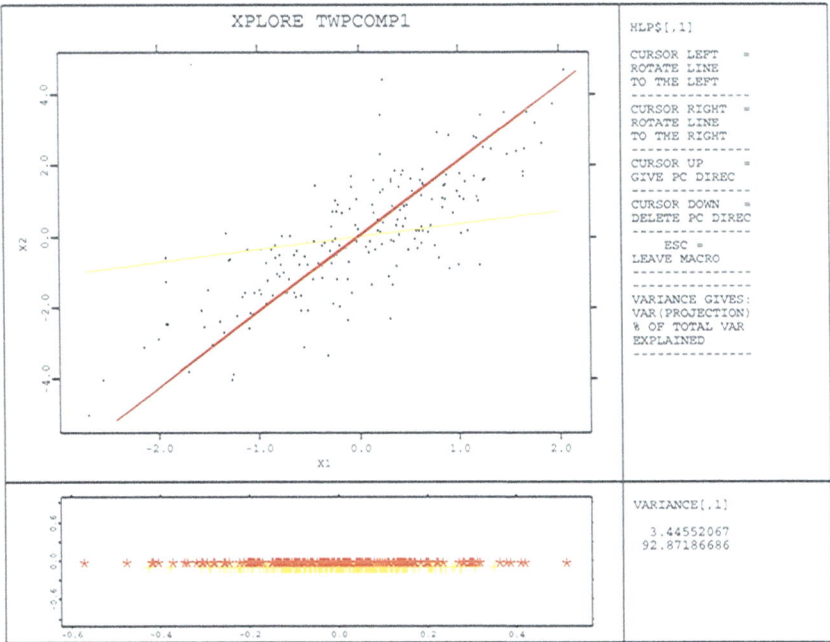

Figure 7.6

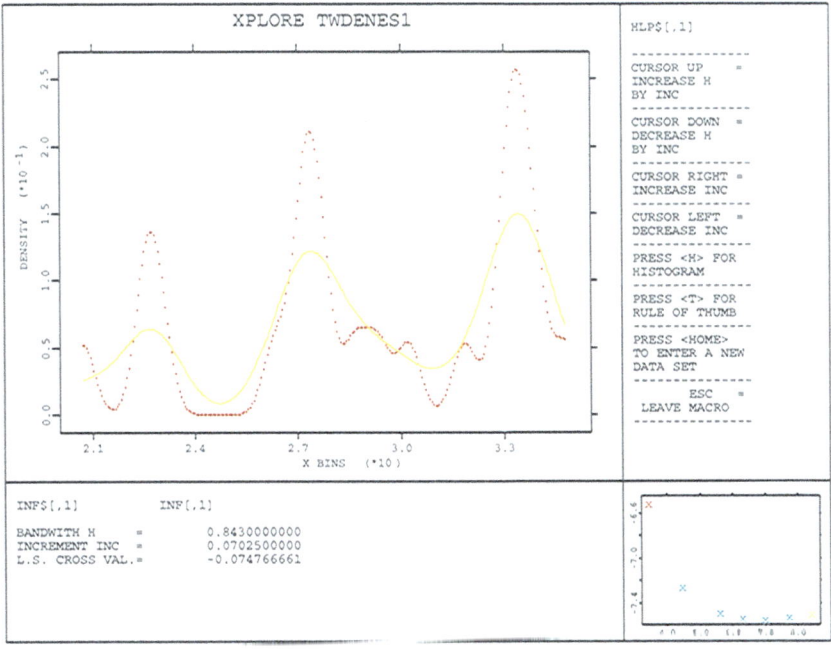

Figure 7.7

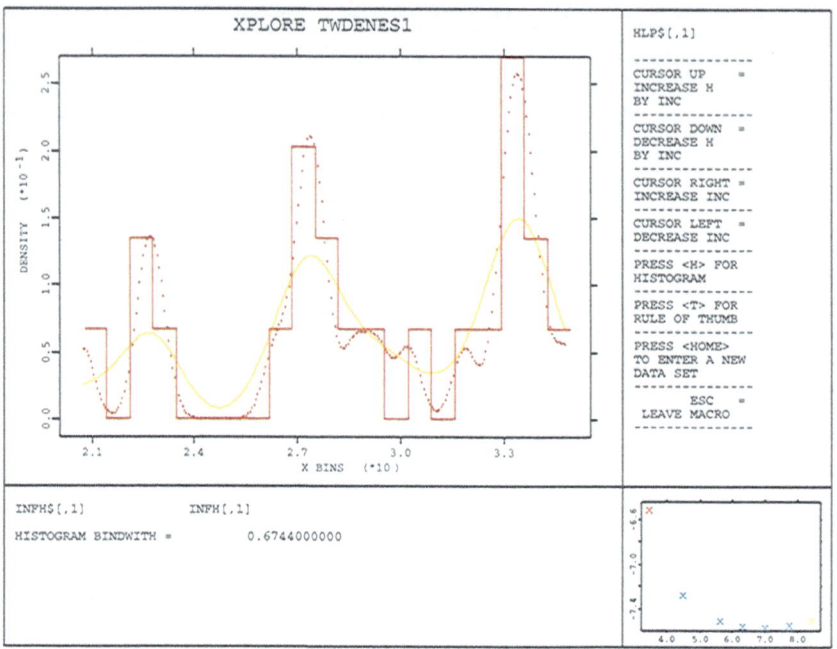

Figure 7.8

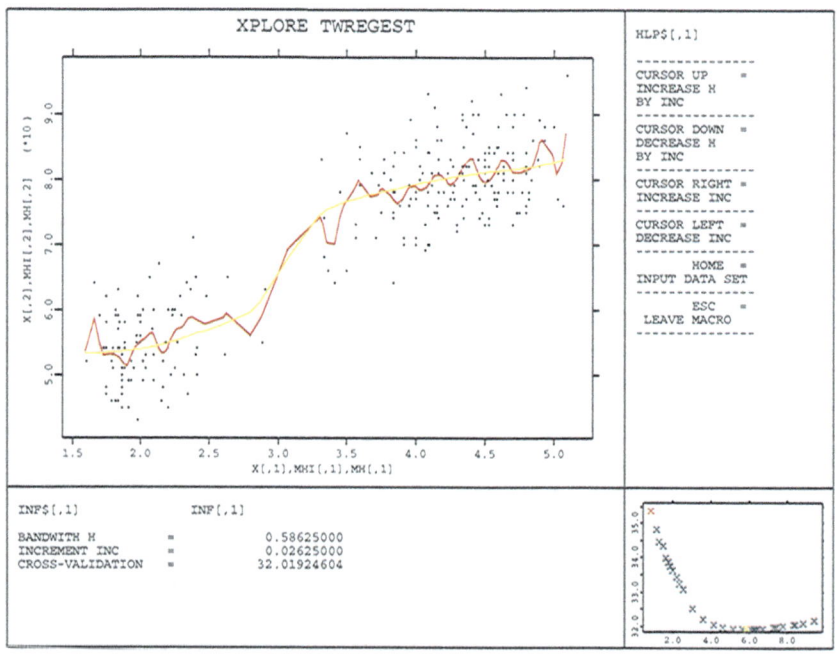

Figure 7.11

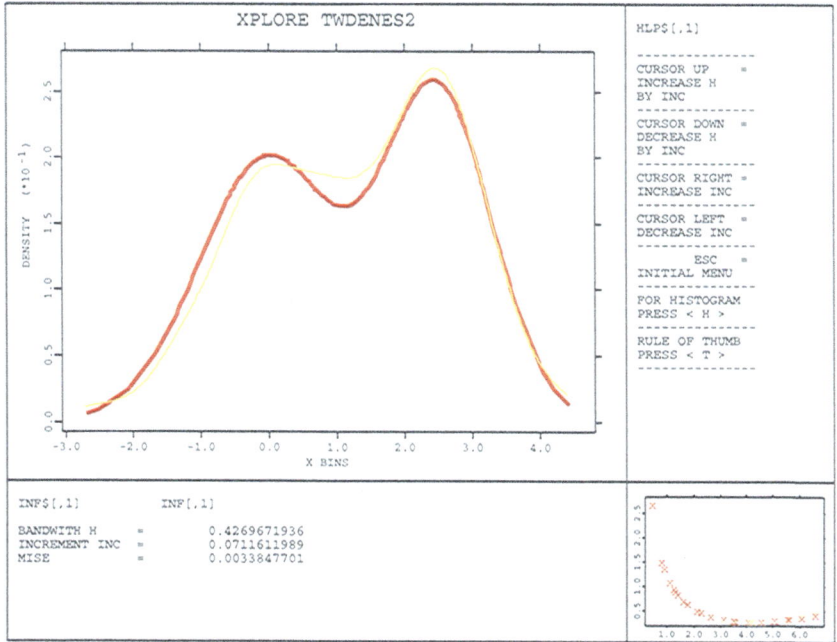

Figure 7.15

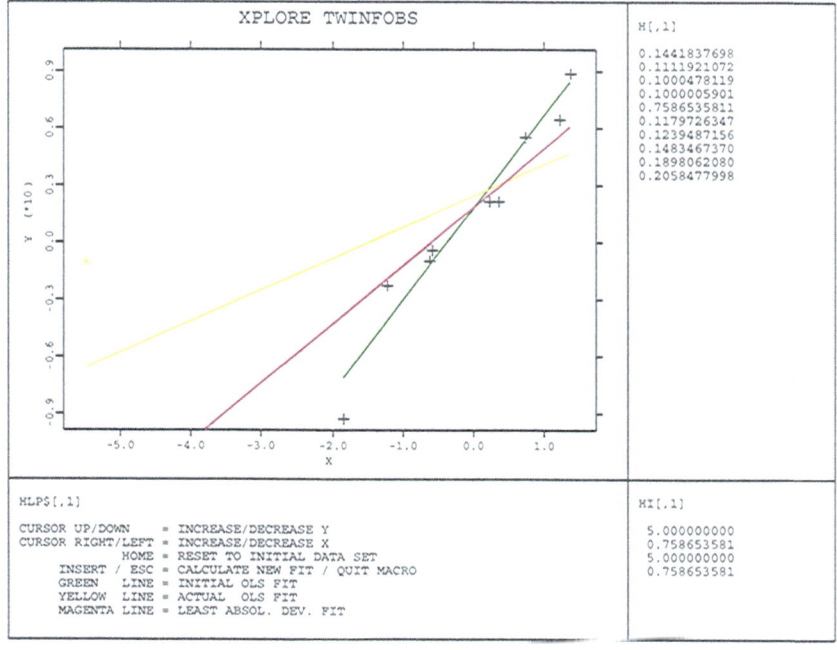

Figure 7.16

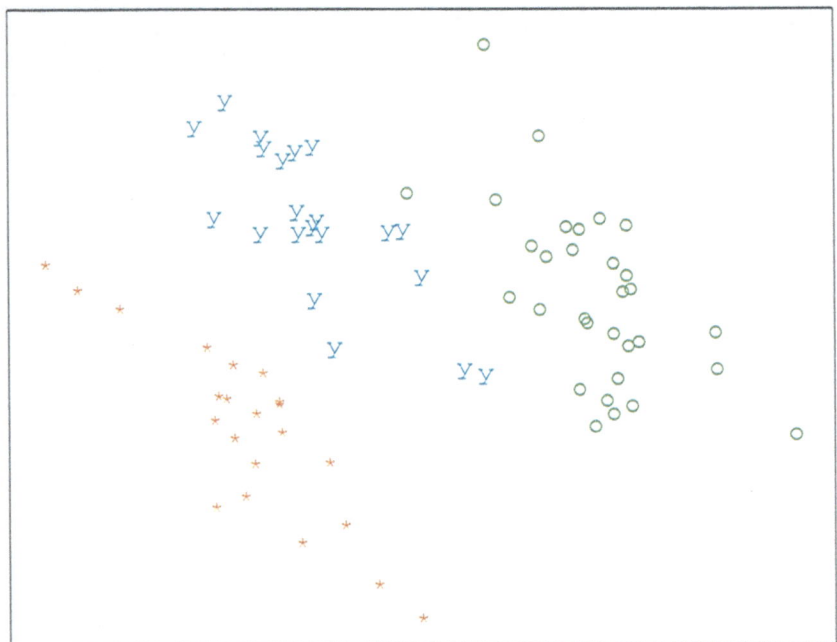

Figure 8.2

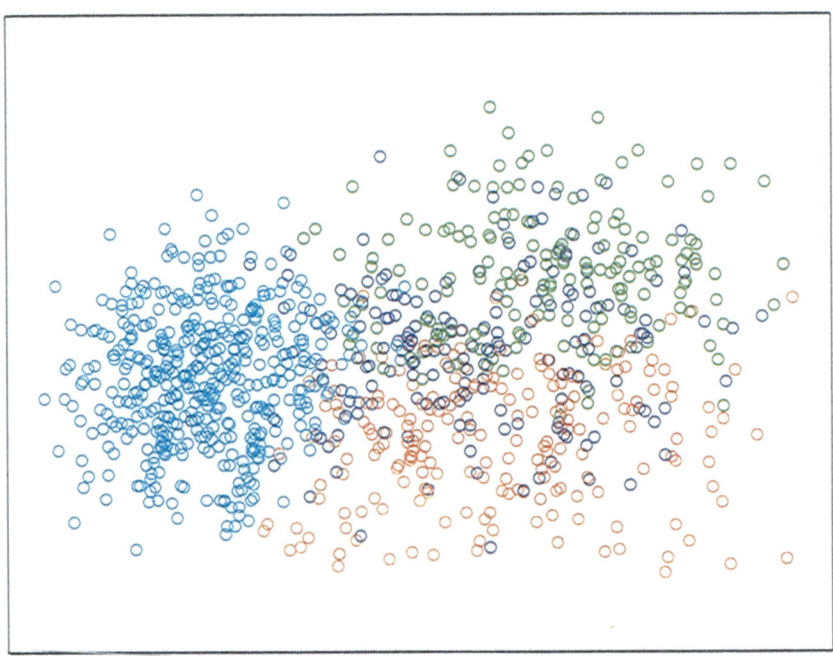

Figure 8.6

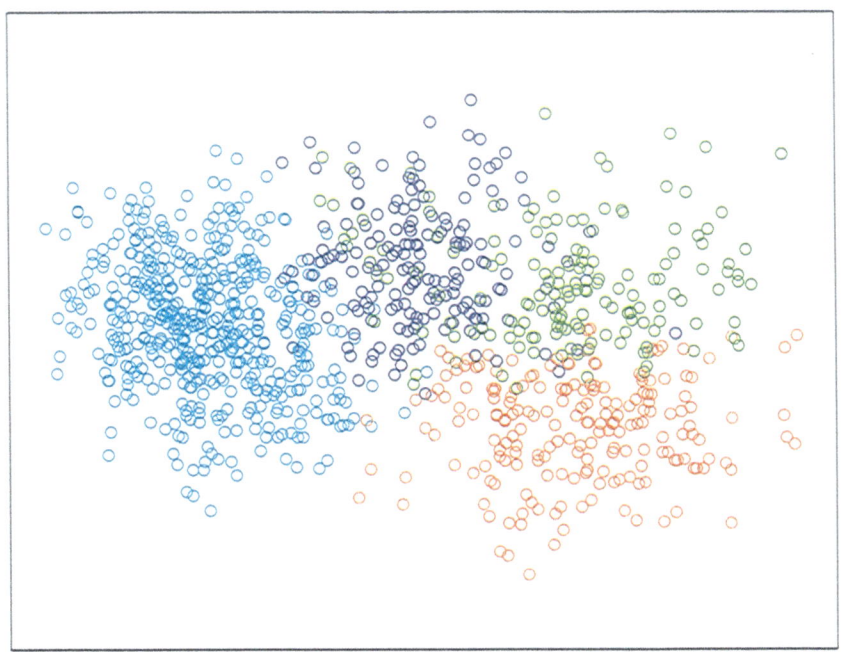

Figure 8.8

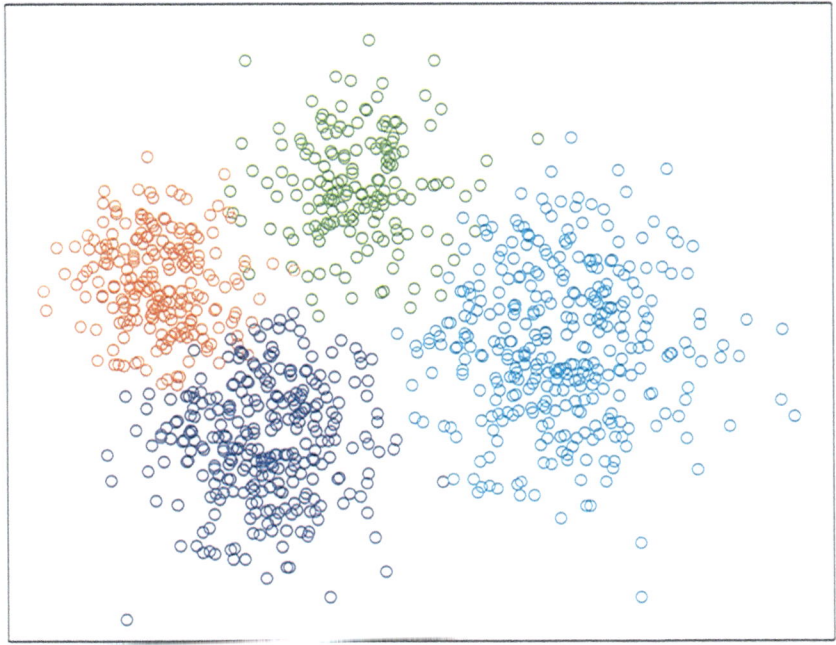

Figure 8.9

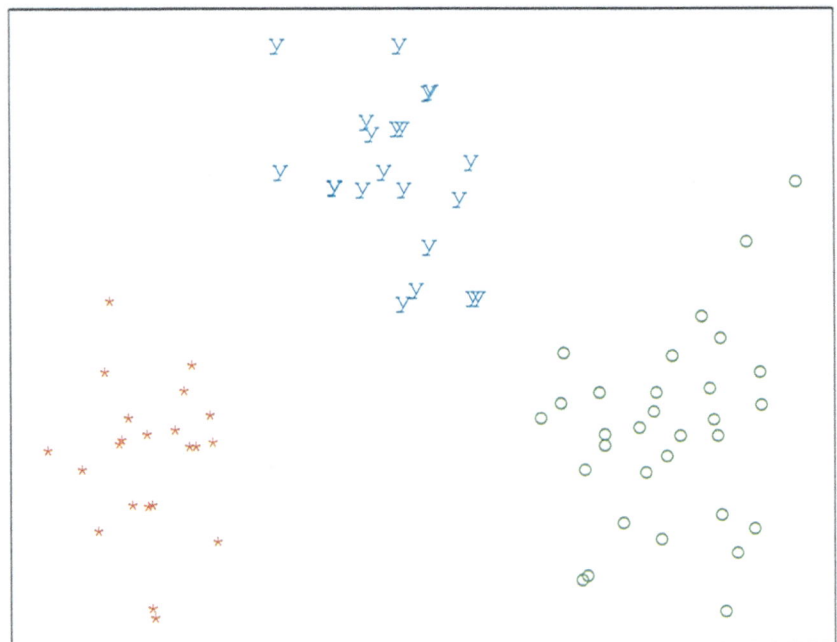

Figure 8.10

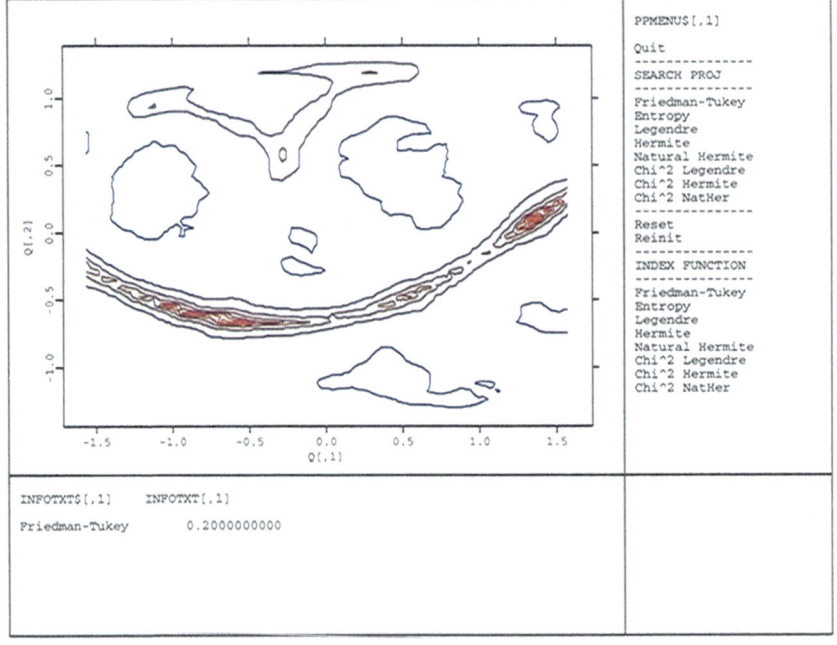

Figure 9.4

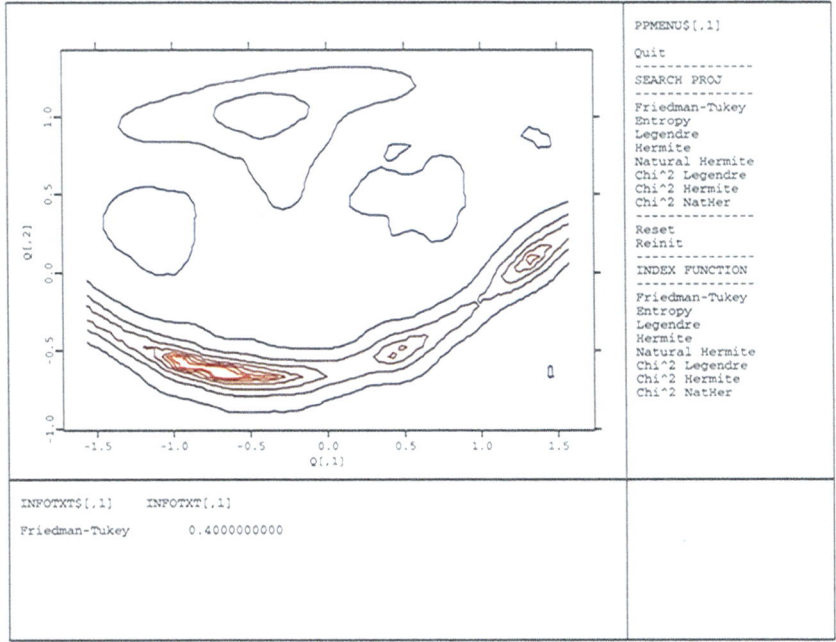

Figure 9.5

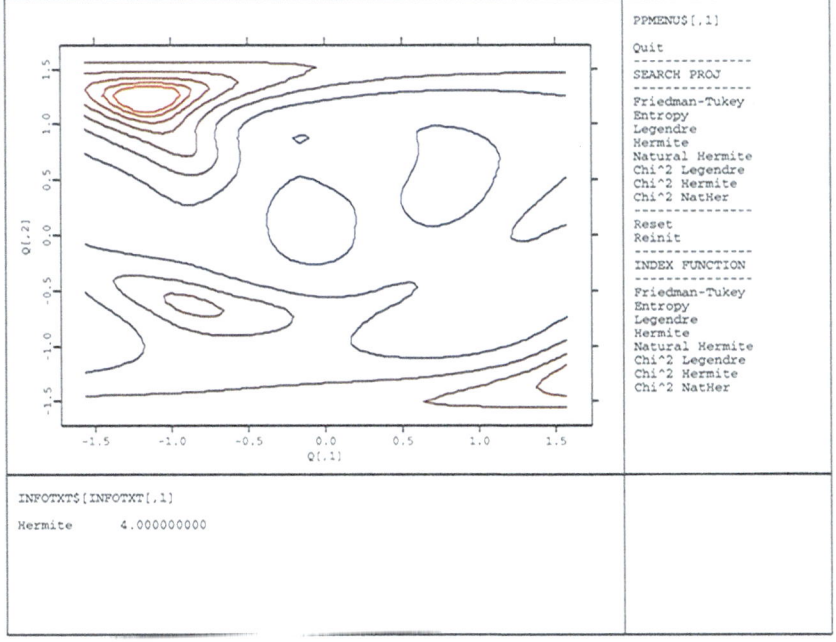

Figure 9.7

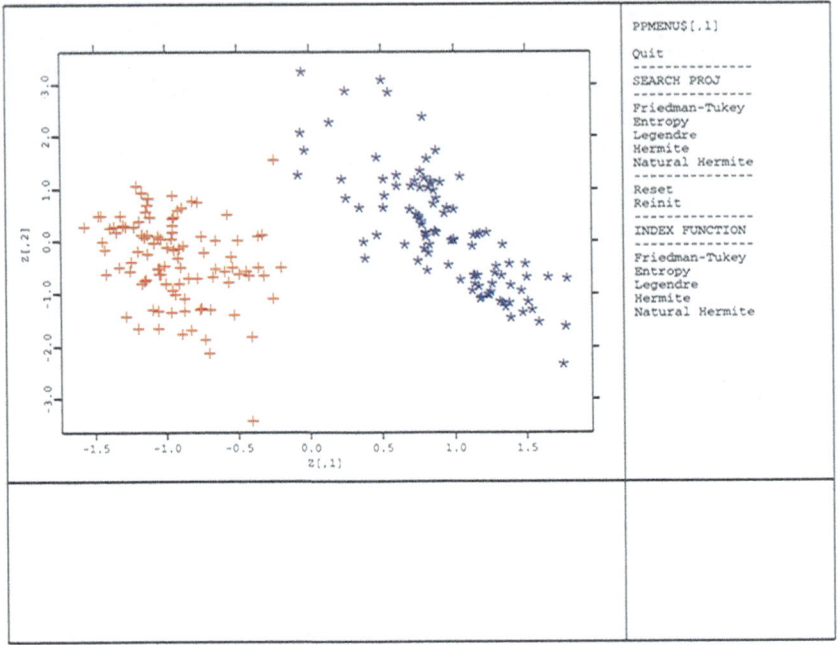

Figure 9.9

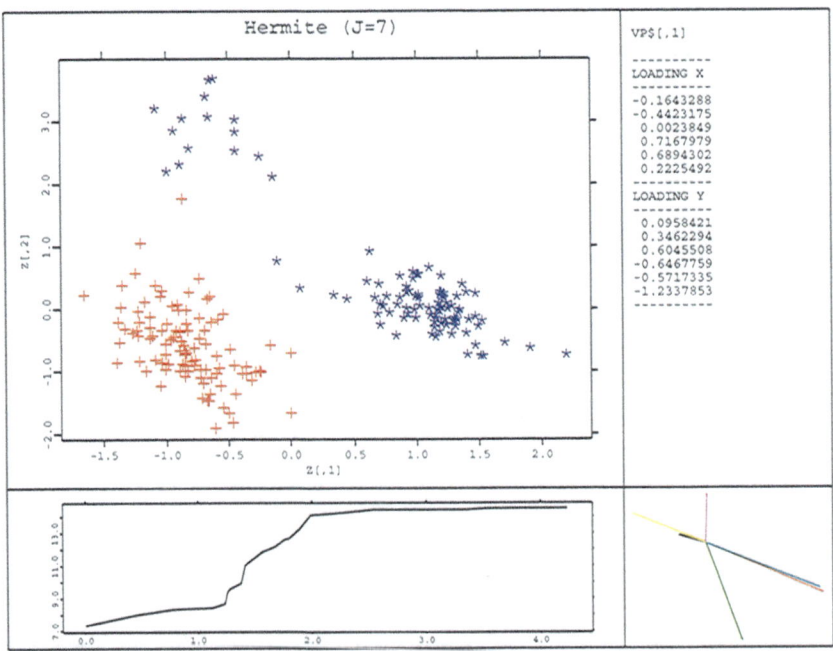

Figure 9.10

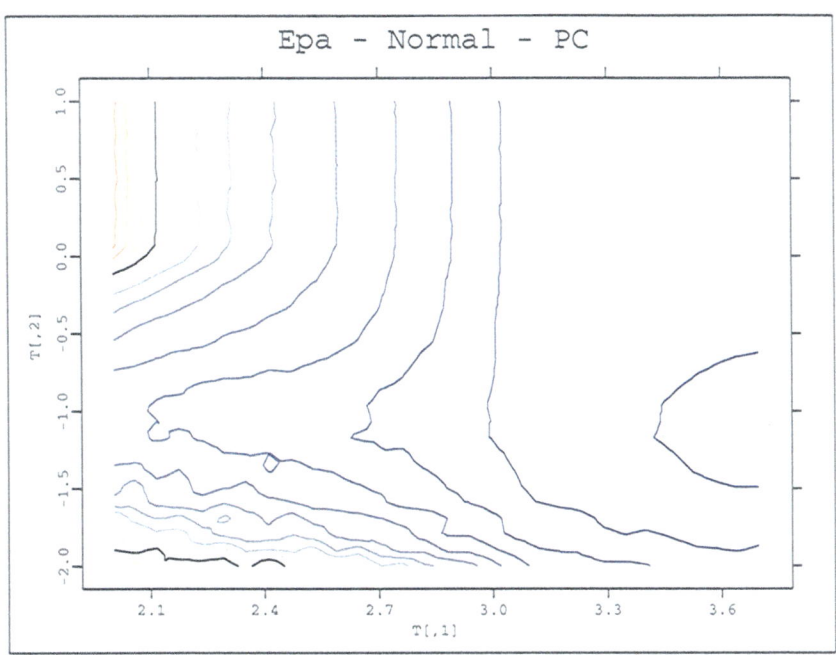

Figure 9.14 (upper left)

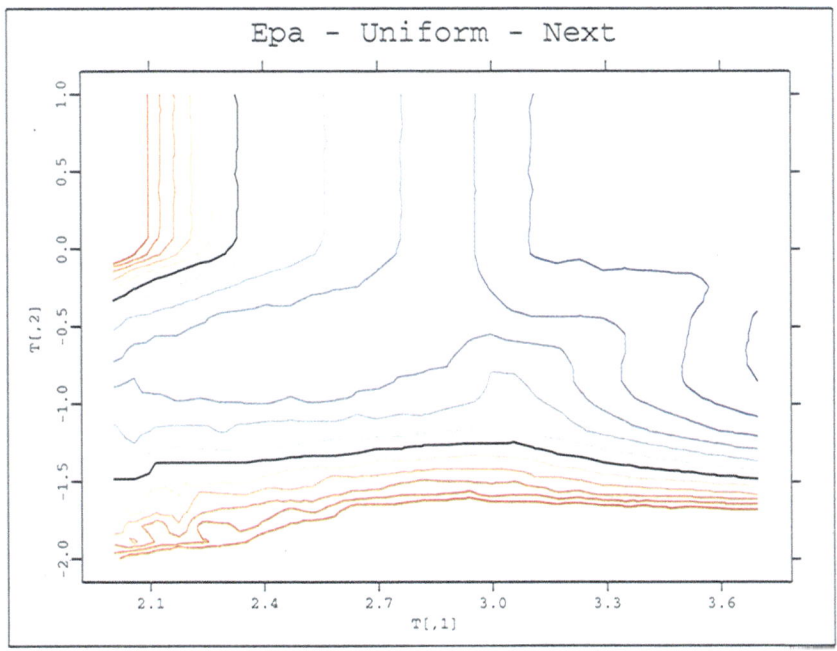

Figure 9.14 (upper right)

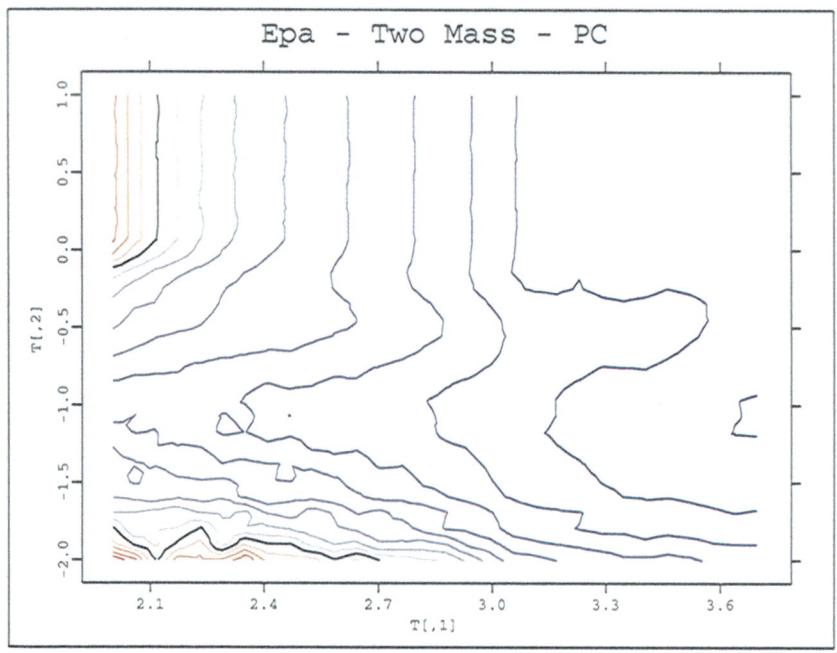

Figure 9.14 (lower left)

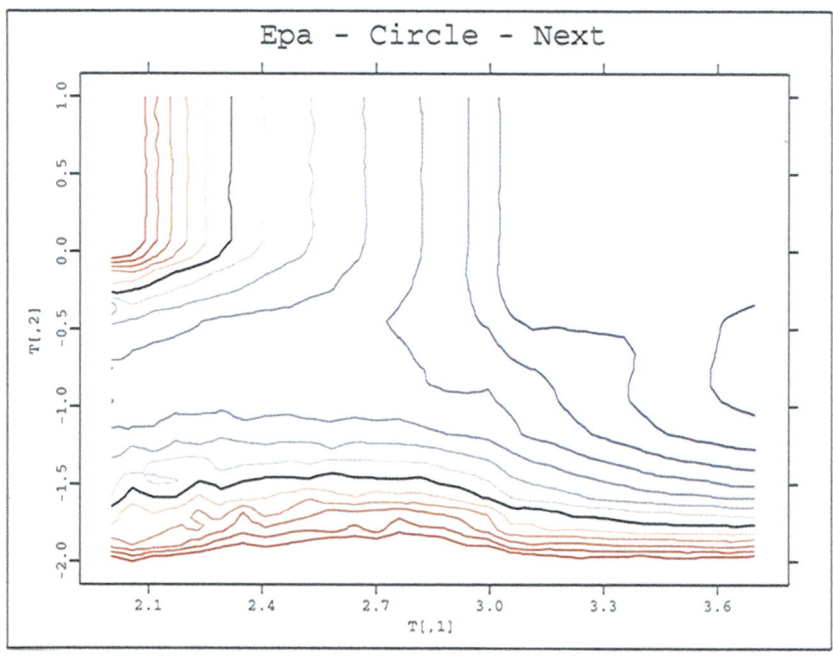

Figure 9.14 (lower right)

Part I

A Beginner's Course

1
Un Amuse-Gueule

Wolfgang Härdle[1]

An amuse-gueule or a canapé is a small appetizer that comes before a menu with several courses. The amuse-gueule is light and should prepare your taste for the later meal. Here, the courses on the menu are different XploRe applications. Some of the applications require one to be a connoisseur with experienced taste in computer-aided statistical modeling. The amuse-gueule given here is therefore designed to develop a good taste.

A French retail sale organization is studying the payment behavior of its clients. Goods can be ordered by catalogue and the customers pay after they receive the ordered goods. Two influential variables were selected:

$$X_1 = \log \text{(total amount of credit)},$$
$$X_2 = \text{age of the client}.$$

The relation to be studied is between these X variables and

$$Y = 1 = \text{good client},$$
$$Y = 0 = \text{bad client}.$$

The classification into good and bad clients was done according to how many reminders were necessary to obtain the full payment. Data were collected from $n = 1814$ clients and stored in a data set named `credit.dat`.

In fact, more than two variables were measured to explain the repayment structure. However, there is general agreement, though, that these two variables are among the most informative ones. Age, for example, is correlated with salary and often with higher social status so that it carries information regarding the other variables as well. The standard technique for this type of data is to select a binary choice model from the family of *Generalized Linear Models* (see McCullagh and Nelder, 1989). The idea is to model the probability that $Y = 1$ as a function of $X = (1, X_1, X_2)^T$. More specifically, as a function of a risk index $X^T\beta$, $\beta = (\beta_0, \beta_1, \beta_2)^T \in \mathbb{R}^3$,

$$P(Y = 1 | X = x) = G(x^T\beta). \tag{1.1}$$

In this formula G denotes an (inverse) link function, for example, in a logit model we have $G(u) = 1/\{1 + \exp(-u)\}$.

---
[1]Institut für Statistik und Ökonometrie, Humboldt-Universität zu Berlin, D-10178 Berlin, Germany.

BETA, S.E.[,1]	BETA, S.E.[,2]	BETA, S.E.[,3]	BVAR[,1]
2.370342871	0.8555478030	2.770555734	0.7319620433
-0.186976837	0.0906837130	2.061856872	-0.074349466
0.031444957	0.0054871856	5.730616502	-0.001487737

STAT[,1]	STAT$[,1]		
1809.000000	DEGREES OF FREEDOM		
1809.000000	N-P		
1188.110123	DEVIANCE		
1830.899059	PEARSON'S CHI^2		
0.656777	SCALE		

FIGURE 1.1. The result of the GLM approach.

In XploRe, we obtain this model (1.1) using the DOGLM function. To do so, we first have to load the generalized linear model library GLM (see also Chapter 10). Then we read the data set, extract the response variable, and construct our design matrix. The code to do this is given below.

```
library(glm)
dat = read(credit)
dat = paf(dat dat[,4].<>200)
y = dat[,1]
x = matrix(rows(dat))~log(dat[,3])~dat[,4]
```

The command

```
(itres beta se t bvar stat) = DOGLM(x y)
```

will lead us through a menu where we should select Binomial as conditional distribution and Logit for the link function $G(u) = 1/\{1 + \exp(-u)\}$. The result is given in Figure 1.1. Note that, as expected, the coefficient of X_1 is negative, whereas the coefficient of X_2 is positive.

A second thought, however, makes it clear that this technique cannot be the best one. A linear relationship between the loan amount with negative coefficient $\hat{\beta}_1$ means that in a given age group, the risk or the probability of not paying back the loan increases linearly with the amount of the loan. Hence, as the loan increases, the risk gets higher and higher for the bank or the retail organization. Empirical evidence shows that banks do not go bankrupt due to this effect. Hence, it does not exist!

In fact, the basic element of our model (1.1) was the linear index $x^T\beta$. Suppose now that the x-dependent part of this index is no longer a weighted sum $\beta_1 x_1 + \beta_2 x_2$ of predictor variables but rather a sum of general unspecified nonparametric functions $g_1(x_1) + g_2(x_2)$. This index allows more flexible transformations on the X variables and thus can model a possible nonlinearity for this situation. The model

$$P(Y = 1|X = x) = G\left(\alpha + \sum_{j=1}^{2} g_j(x_j)\right) \tag{1.2}$$

is a generalization of (1.1) and contains (1.1) with $g_j(x_j) = \beta_j x_j$. The class of models (1.2) are called *Generalized Additive Models* (see Hastie and Tibshirani, 1990).

In order to study this class in more detail let us try the following situation with $n = 200$ observations:

$$X \sim N_2(0, I_2), \qquad g_1(x_1) = x_1$$
$$G = \text{logistic}, \qquad g_2(x_2) = x_2^2 - 1.$$

We will come back later to the credit data set. In XploRe, we create data according to this model by

```
; create 200 Normal random variables in 2 dimensions
x = normal(200 2)
g = x[,1]~(x[,2].*x[,2]-1) ; computes the g function
eta = sumr(g)             ; rowwise sum of g
px = 1./(1 + exp(-eta))   ; the link applied to the sum(g)
y = uniform(200 .< px)    ; the response variables Y
```

The situation for this GAM is depicted in Figure 1.2. In the upper left picture, we see the link function operating on the generalized index $g_1(x_1) + g_2(x_2)$. In the lower right we see the linear g_1, and in the upper right the parabolic g_2. Note that $E_{X_2} g_2(X_2) = 0$; this is, in fact, a condition to identify the g_j in (1.2) above. In the lower right we see a three-dimensional dynamic spin plot with the nonlinear index $\eta = g_1(x_1) + g_2(x_2)$ plotted against x_1 and x_2. We rotated this scatterplot to see the extreme nonlinearity of this surface.

We now estimate the generated data by calling the DOGAM function from the GAM library of XploRe (see also Chapter 11). The sequence of statements is as follows:

```
library(gam)                           ; calls the GAM library
(itres beta alpha fx) = DOGAM(x y)     ; performs the GAM fit
                                       ; similar to the GLM menu
```

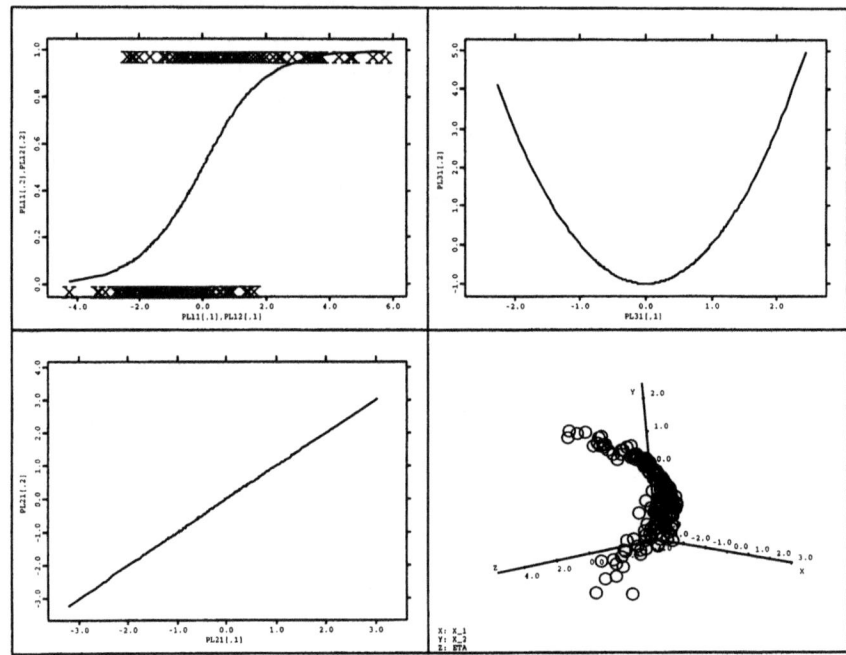

FIGURE 1.2. A simulated Generalized Additive Model.

The estimated g functions are contained in the matrix fx. We can create
a plot as in Figure 1.3 with the following statements:

```
x1  = x[,1]              ; the two response variables
x2  = x[,2]
g1  = fx[,1]             ; and their (estimated) influence
g2  = fx[,2]
eta = sumr(fx)+alpha     ; the generalized index
px  = 1./(1+exp(-eta))   ; and the link function
  ; create a multi window display to show everything....
createdisplay(fig33, 2 2, s2d s2d s2d d3d)
  ; ... the link and y against the (estimated) index
pl11 = eta~y~mask(rows(eta) 1 o)
pl12 = sort(eta~px 1)
  ; ... the variables and their (estimated) influence
pl21 = sort(x1~g1 1)
pl31 = sort(x2~g2 1)
  ; ... the variables and the (estimated) index
pl4  = x1~x2~eta
  ; Show everything and quit the display to...
writecon(27)
show(pl11 pl12 s2d1, pl21 s2d2, pl31 s2d3, pl4 d3d1)
  ; ... update some pictures to show lines
```

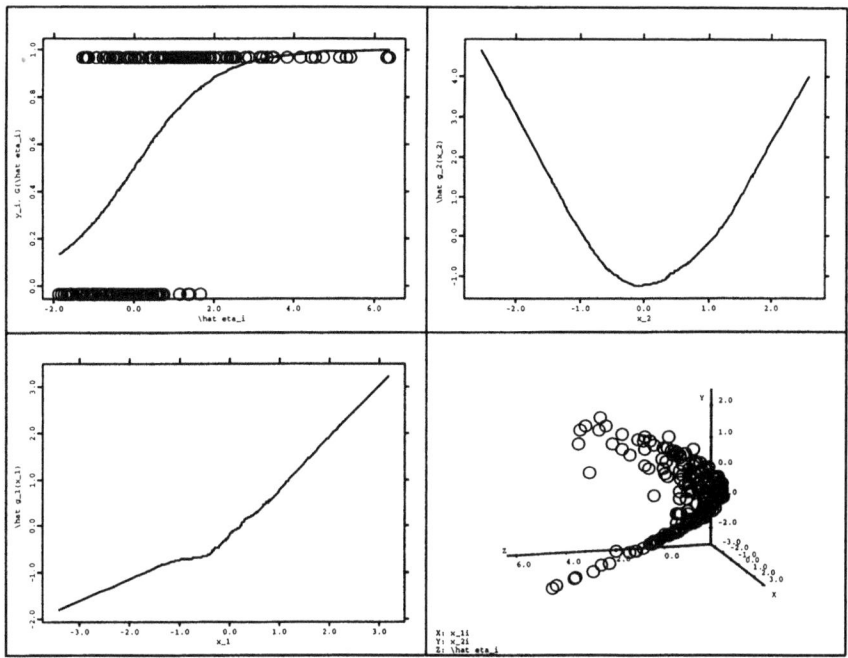

FIGURE 1.3. The estimated link and g functions in the GAM example.

```
dum = update(pl12 2 s2d1 solid line)
dum = update(pl21 1 s2d2 solid line)
dum = update(pl31 1 s2d3 solid line)
  ; ... and change some axis labels
dum = update(pl4  1 d3d1 xaxis "x_1" yaxis "x_2" zaxis "eta")
display(fig33)
```

Figure 1.3 shows the result of these manipulations. The functions g_j were found by a combination of the Newton–Raphson and backfitting algorithms (see Härdle and Turlach, 1992). Inside the backfitting algorithm we used the supersmoother of Friedman (1984) (for a description see Härdle, 1990, p. 181; Bond and Tabazadeh, 1994). The command createdisplay defines a display on the screen with different types of windows. In our case, we want three static two-dimensional windows (S2D) and one dynamic three-dimensional rotation window (D3D). The show command puts the data matrices into the desired windows. The update command changes the linestyle and adds labels to the axis. The display(fig33) puts everything onto the screen.

Let us now apply the GAM algorithm of XploRe to the credit data. We use the following program code:

```
library(gam)
```

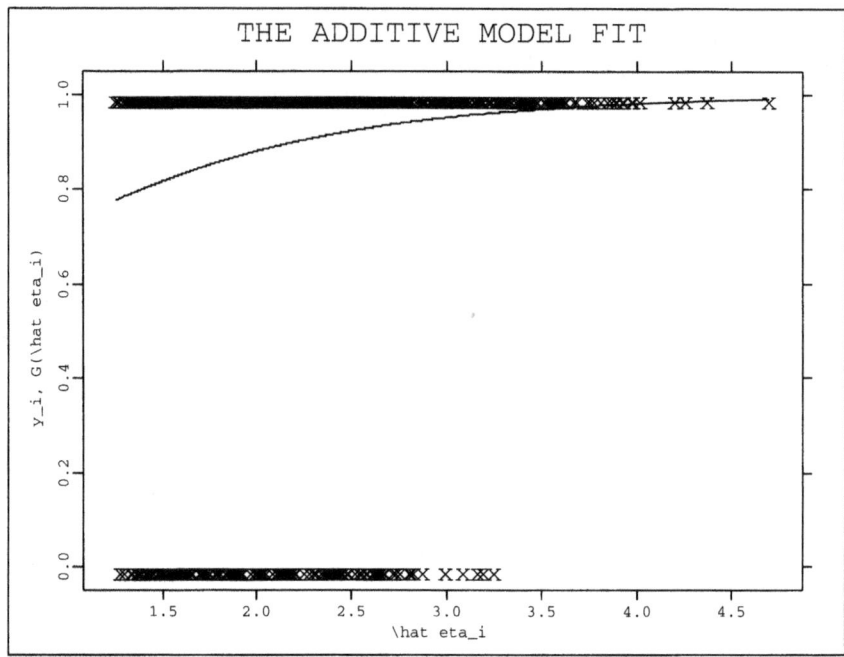

FIGURE 1.4. The resulting link function together with data $(\hat{\eta}_i, Y_i)$.

```
dat = read(credit)
dat = paf(dat dat[,4].<>200)
y = dat[,1]
x = log(dat[,3])~dat[,4]
(itres beta alpha fx) = DOGAM(x y)
```

The resulting link function is plotted against the risk index in Figure 1.4. The function $\hat{g}_1$ (also estimated via the supersmoother) is shown in Figure 1.5. The influence of age is almost linear, as Figure 1.6 shows.

We see that our argument about the linearity of g_1 was justified. It turns out that the influence of the loan amount is nonlinear in such a way that the risk is almost stable after a certain amount of loan. More precisely, the risk increases (negative slope of g_1!) up to $x_1 = 8.5$. Then the curve g_1 stays almost constant up to $x_1 = 12$ and thus does not increase the risk of not paying back the loan (or the goods). The age effect is almost linear and shows a positive effect on age (due to its positive slope). More aged clients are less risky. One could argue that there is some sort of plateau after $x_2 = 60$. To investigate this question further would require inference about the g_j functions. This is somewhat beyond the scope of this amuse-gueule.

One sees that even models of a rather complicated form can arise in simple life situations like buying a TV or a bicycle via catalogue or telephone.

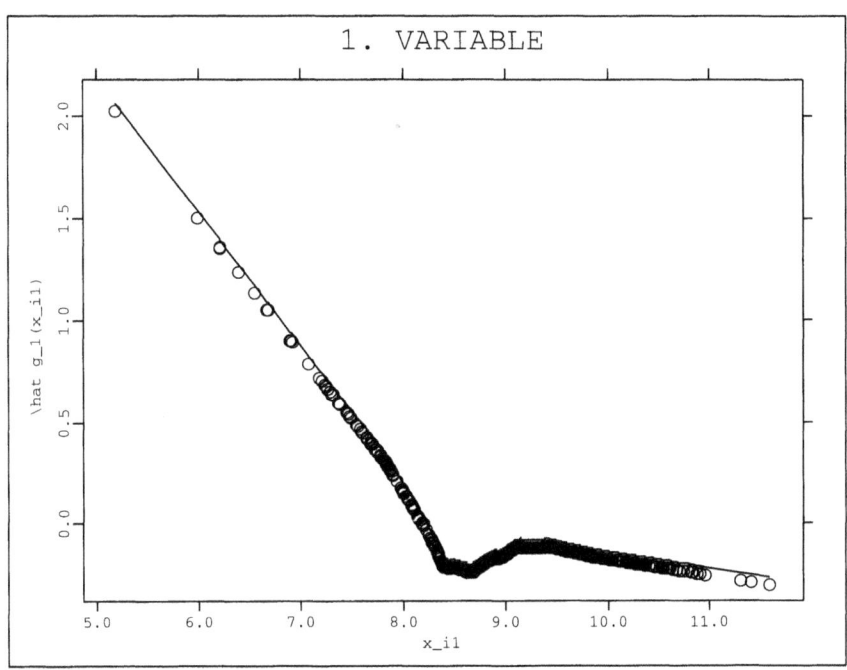

FIGURE 1.5. The function $\hat{g}_1(x_1)$.

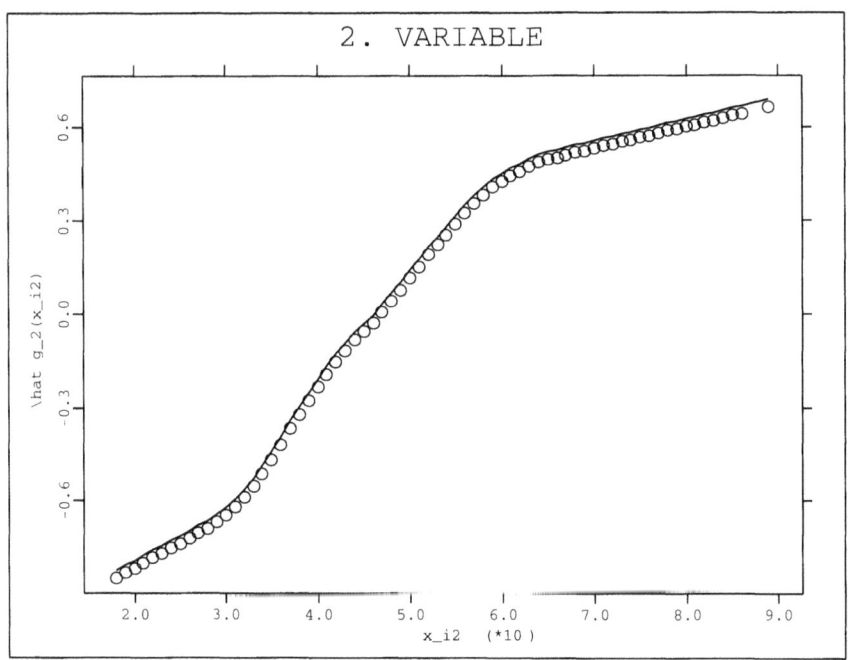

FIGURE 1.6. The function $\hat{g}_2(x_2)$.

What one needs at hand is a handy statistical environment that can assist one in making decisions. We think that XploRe can supply these needs and we hope that the reader has now developed the taste referred to in the beginning.

REFERENCES

Bond, J. and Tabazadeh, A. (1994). Super Smoothing with Xlisp-Stat. Available as PostScript file `supersmu.ps` via anonymous ftp from `ftp.stat.ucla.edu` in the directory `pub/lisp/xlisp/xlisp-stat/code/homegrown/smoothers/supersmoother`.

Friedman, J.H. (1984). A variable span smoother, *Technical Report No. 5*, Department of Statistics, Stanford University, Stanford, California.

Härdle, W. (1990). *Applied Nonparametric Regression*, Econometric Society Monographs No. 19, Cambridge University Press, New York.

Härdle, W. and Turlach, B.A. (1992). Nonparametric approaches to generalized linear models, *in* L. Fahrmeier, B. Francis, R. Gilchrist and G. Tutz (eds), *Advances in GLIM and Statistical Modelling*, Vol. 78 of *Lecture Notes in Statistics*, Springer-Verlag, New York, pp. 213–225.

Hastie, T.J. and Tibshirani, R.J. (1990). *Generalized Additive Models*, Vol. 43 of *Monographs on Statistics and Applied Probability*, Chapman and Hall, London.

McCullagh, P. and Nelder, J.A. (1989). *Generalized Linear Models*, Vol. 37 of *Monographs on Statistics and Applied Probability*, 2nd edn, Chapman and Hall, London.

2
An XploRe Tutorial

Wolfgang Härdle[1]

2.1   Getting Started

XploRe is started by entering the command `xplore` (usually from the directory `C:\XPLORE3`). A graphic appears with a copyright screen. Hitting any key displays the standard environment of XploRe, the screen is divided into

- a command line,
- an icon list, and
- an action window.

The command line has a prompt character `>` and behind that the blinking cursor. Here you enter commands in XploRe language from the keyboard. The icon list is a sequence of changing hotkeys (`<F1>` to `<F10>`). The action window is reserved for displaying graphics and text, for editing commands, and for system information.

Let us create a two-dimensional vector $\mu = \binom{1}{1}$. At the command line we enter

```
mu = # ( 1  1)
```

The coordinates of the vector are given inside the parentheses. The hash `#` indicates that we are defining a vector. Note that there is no comma between the two ones; which is why `mu` is stored as a column vector. We can check that this vector is present by pressing `<F2>` (represented as the second icon from above in the icon list). `<F2>` displays the current objects in XploRe. All objects are given in alphabetical order. Besides the object `mu` with two rows and one column, we find the objects `eh`, `pi`, and `nan`. If we move the cursor over one of those and press the `<F10>` *open objects key*, we see:

```
eh  = 2.718....
pi  = 3.14...
nan = not a number (missing value)
```

---

[1]Institut für Statistik und Ökonometrie, Humboldt-Universität zu Berlin, D-10178 Berlin, Germany.

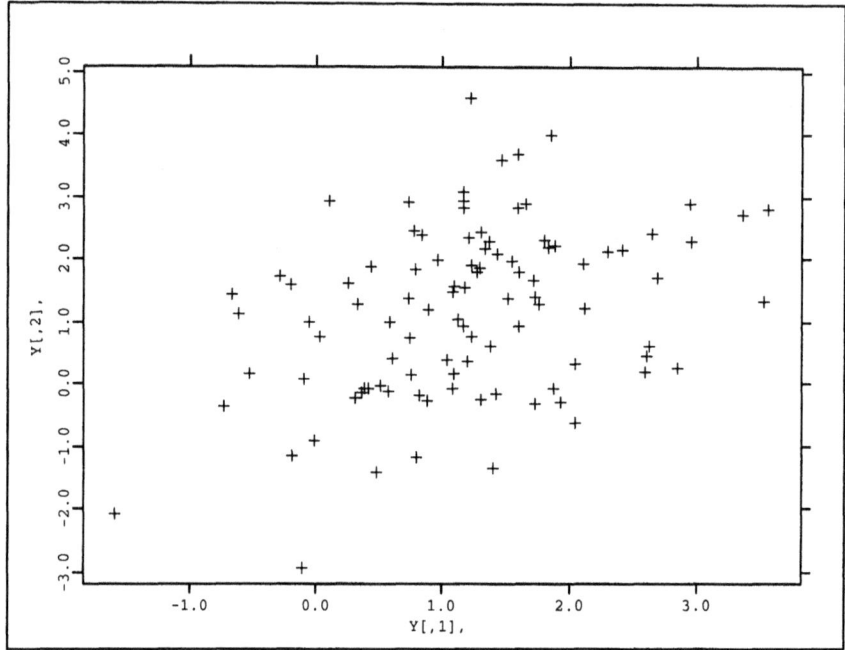

FIGURE 2.1. A normal point cloud.

By pressing <ESC> we return to the command line. On the command line, we could just enter mu and the values of $\mu = \binom{1}{1}$ would appear on the action screen. <ESC> brings us back to the command line.

Let us now create a covariance matrix $\Sigma = \begin{pmatrix} 1 & 0.5 \\ 0.5 & 2 \end{pmatrix}$,

```
sig = # (1  0.5 , 0.5  2)
```

The values of sig can be displayed from the command line by entering sig or via the *show objects key* <F2> and the *open objects key* <F10>.

The determinant of Σ can be computed by det(sig). The displayed value 1.75 is positive. The eigenvalues and the eigenvectors are computed by

```
(eval evec) = eigsm (sig)
```

XploRe has now created the objects eval, a 2×1 column vector with the eigenvalues of Σ, and evec, a 2×2 matrix containing the corresponding eigenvectors in the two columns. A check on the values of eval shows that all eigenvalues are positive. The matrix Σ is indeed a positive definite covariance matrix.

Suppose now that we want to create a two-dimensional normal random variable with distribution $N_2(\mu, \Sigma)$. In XploRe, a pseudo-random number generator is supplied that creates standard normal variables in p dimensions, for example,

```
x = normal (100  2)
```

creates a 100×2 matrix x that contains 100 observations of a two-dimensional normal random variable $X \sim N_2(0, I), I = \begin{pmatrix} 1 & 0 \\ 0 & 1 \end{pmatrix}$.

In order to create normal variables with mean μ and covariance Σ, we need to apply the inverse of the Mahalanobis transformation,

$$Y = \Sigma^{1/2} X + \mu.$$

We compute the square root of Σ by

```
sig2 = evec * diag (sqrt(eval)) * evec'
```

Here, diag creates a diagonal matrix from a column vector and sqrt takes element-wise square roots of a matrix or a vector. The transpose of evec is computed using trn(evec) or evec'. An equivalent and more efficient way of computing the square root of Σ will be given later, for example, in Section 12.4.

The random variable Y is now created by

```
y = x * sig2 + mu'
```

We can now plot the data in a two-dimensional display

```
show (y s2d)
```

The data matrix y is shown graphically in the *display* static2d and there in a *window* of type s2d. The coordinates of y are shown as fine dots. By pressing the *change data part key* <F10> in this display, we may change the icon list once more. By pressing the *data style key* <F3> in this icon list, we see that the dots changed to plus signs +. <ESC> brings us back to the show icon list. The resulting plot is given in Figure 2.1. If you mistyped a key or you don't know in which mode you are running XploRe, there is always the *help key* <F1>. <ESC> is used to leave the (context sensitive) help system.

We shall learn more about the help system in Section 2.4. We leave the show command by <ESC> and are now on the command line. Let us recall the last command using <F3> or by <↑>. (Both keys recall the last 100 commands.) Move the cursor over the word s2d and change it to b1d,

```
show (y b1d)
```

The resulting plot is given in Figure 2.2.

We use the show command to look at the data y through boxplots. We see two parallel boxplots, one for each column of y.

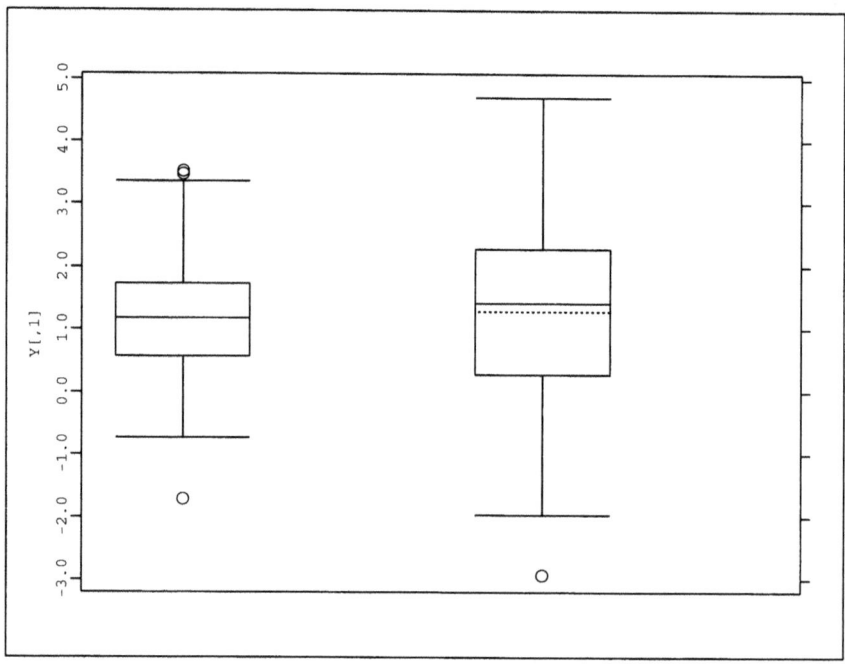

FIGURE 2.2. Two boxplots.

2.2   Two-Dimensional Plots

We have already used the **show** command in the previous section to plot a two-dimensional normal point cloud. The static two-dimensional window that we opened with **show(y s2d)** has more features then just plotting two-dimensional matrices. After the show command, the cursor is in the action screen. We leave the show modus that also changed the icon list via <ESC>.

The icon list still contains the <F1> *help key* but now with help information specific to the **show** command. A new icon is the one corresponding to <F10>; it allows us to change colors, linestyles, etc., even for many overlaid point clouds.

Let us study the effect of the Mahalanobis transformation on the standard normal point cloud by plotting both x and y in our two-dimensional window.

We type **show (x y s2d)** and we see the transformed data y as yellow points in a more elliptic form in Figure 2.3. The yellow points do not stick out very clearly, so we might want to change the color. We press <F10> and see that the cursor sits in a window where we have to mark which of the two displayed data sets we want to change. Since we wanted to change the y matrix (given as the second argument in the **show** command), we move the cursor onto the word **data2** and press <ENTER>.

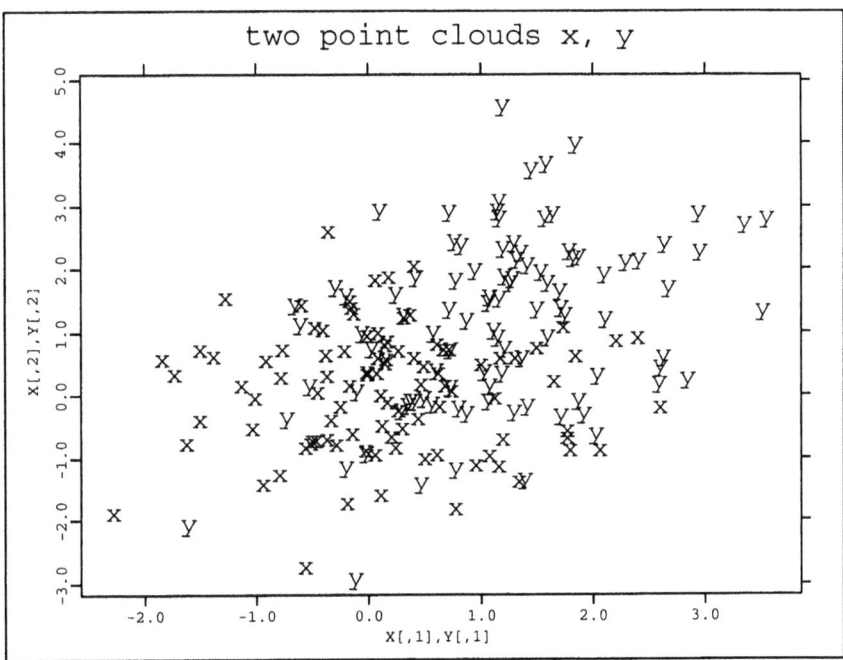

FIGURE 2.3. Two normally distributed point clouds.

The icon list shows the current color, linestyle, etc. set for y. The *color icon* <F2> is yellow for y. By pressing it several times, we see how we can change the color. The button <F3> changes the symbol; we may want to have plus signs, circles, or other symbols. Let us define the symbol "x" for the x matrix and "y" for the y matrix. The button <F4> joins all points in this matrix in their order of entry. <F5> defines the linestyle (dashed, broken, solid, ...). The button <F6> is a toggle for the thickness of the line. <F7> allows us to link the current set with another data matrix (see Chapter 10.4 for an example).

Let us now fit a least squares regression line through the y variables. We first sort the y matrix with respect to the first column by

```
y = sort(y)
```

(the syntax of the sort command can be checked by moving the cursor over the word **sort** and then pressing <F10>, the *open key*). Next we create the design matrix for the least squares line regression problem by adding a column vector of ones to the first column of y

```
yd = matrix(rows(y))~y[,1]
```

Then we call the generalized least squares regression command **gls** with arguments yd and y[,2],

```
b = gls(yd y[,2])
```

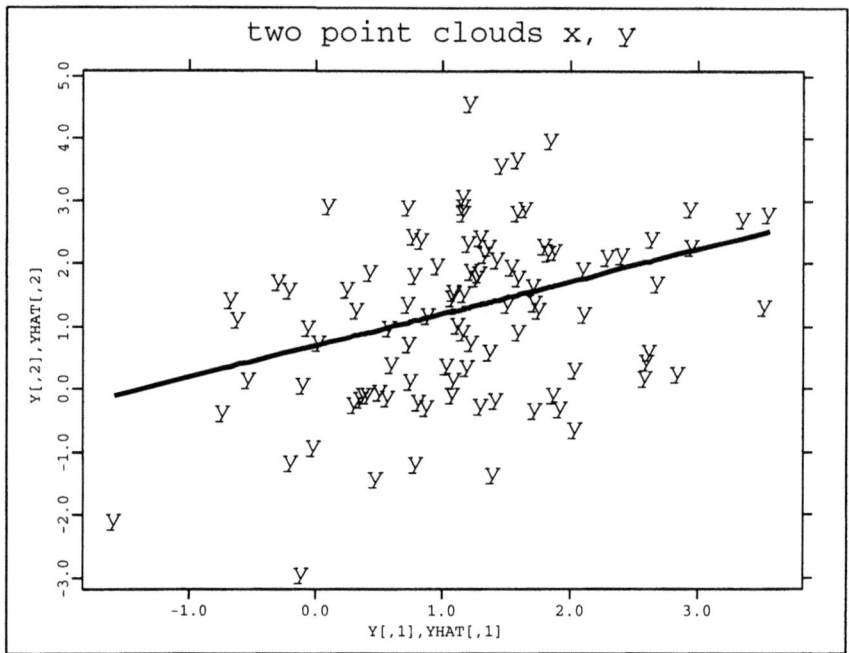

FIGURE 2.4. The **y** matrix and the estimated regression line.

The 2×1 vector **b** now contains the estimated parameters $\hat{\beta}_1, \hat{\beta}_2$ of the linear regression problem

$$y_2 = \beta_1 + \beta_2 y_1 + \varepsilon.$$

Let us add the estimated regression line

$$\hat{y}_2 = \hat{\beta}_1 + \hat{\beta}_2 y_1$$

to a scatterplot of the $y = (y_1|y_2)$ matrix. Define

```
yhat = yd[,2]~(yd * b)
```

and type

```
show(y yhat s2d)
```

The resulting plot is given in Figure 2.4 (the actual plot that you obtain following these instructions could be slightly different since the random number generator changes each time XploRe is called).

The residuals are defined by **res=y[,2]-yhat[,2]**. This is a 100×1 matrix containing the difference between y_2 and the regression line. We can actually measure numerically the value of the residual by using the above-mentioned linking. Let us use the **show (y yhat s2d)** command and link

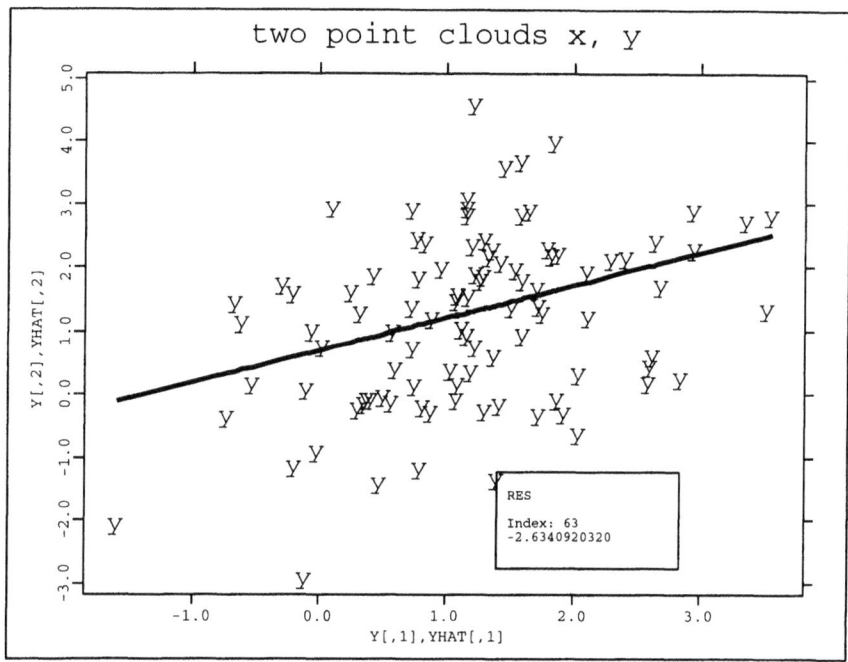

FIGURE 2.5. The point cloud, the linear regression line, and a valuebox with linked residuals.

the y matrix with the **res** matrix by pressing **<F10>** (*change data part key*) then **<F7>** (*link key*), moving the cursor onto **res** and accepting it by **<ENTER>**. We leave the **<F10>** *change data part key* using **<ESC>** and can now study y with linked reference to **res**.

Use the **<F3>** *show value key* to move a box around the screen. As the cursor moves, the box follows and shows the coordinate of the **res** matrix for points close to the cursor. In order to see which point is actually displayed, hit **<ENTER>**; the cursor moves to a corner of the box and it is in this corner that the point lies.

Figure 2.5 shows a situation where we are investigating a large residual. It is the 63rd observation (Index: 63) and has a value of -2.63.

We may now add labels on the axis and introduce a title for this picture. This means that we edit the *window part* of this display by **<F4>**. A new icon list appears with **<F2>** a toggle for the title of the window, and **<F3>** a toggle for a plot with or without axis. The axis ticks and borders are edited by **<F5>**; a box appears that lets you select the desired values. Return to normal **show** mode by **<F10>**.

Brushing is a technique that lets you define and mask certain interesting points interactively. The brushing mode is entered by pressing **<F5>** in normal **show** mode. Again, the icon list changes and a *brush* (a rectangle)

18   W. Härdle

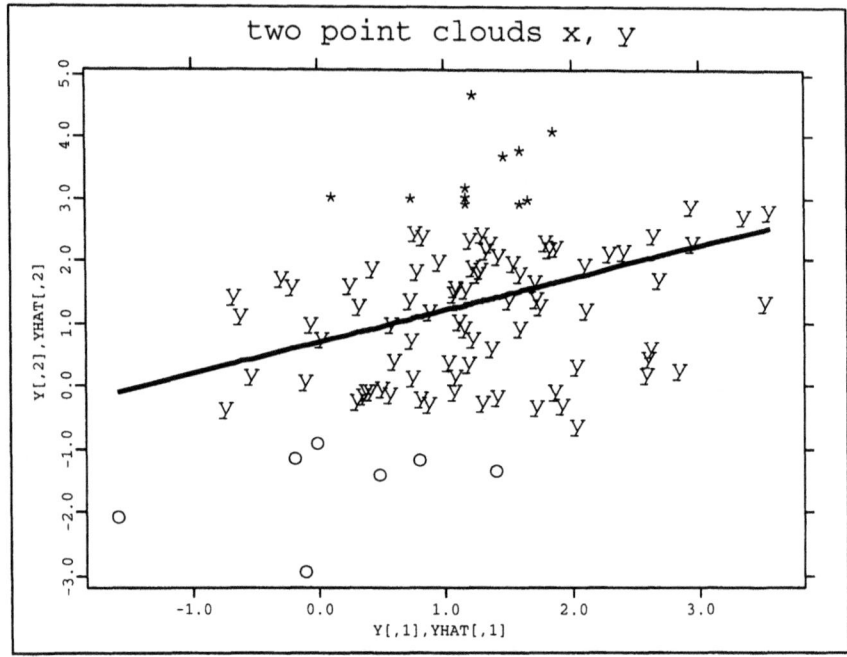

FIGURE 2.6. Result of brushing.

appears on the screen. The idea is that we move this brush over the plot
and mark points that are in this rectangle. The points will then receive a
certain *mask code* that can be studied using other functions.

Let us first resize the brush by <F4>. The cursor keys alter the size of the
brush. Now move the brush with the mouse or the cursor keys and activate
it by <F5>. The color and the symbol of the points inside the brush change.
We may also attach texts to points by this brushing technique. For this
we would use the key <F7>, which we do not describe here in detail (see
Chapter 3).

Figure 2.6 shows the result of brushing where we changed some points
to circles and others to stars.

This brush may define an interesting region of data points, and we would
like to rescale that region to the whole screen. This is done with the <F2>
key in the brush mode. It maps all observations inside the brush onto the
whole screen. The inverse operation is <F3>, in which all observations are
mapped into the brush.

After having brushed the data y, the mask vector can be studied using
the *open key* in the **object**, or by just typing y. The mask vector is denoted
by y%.

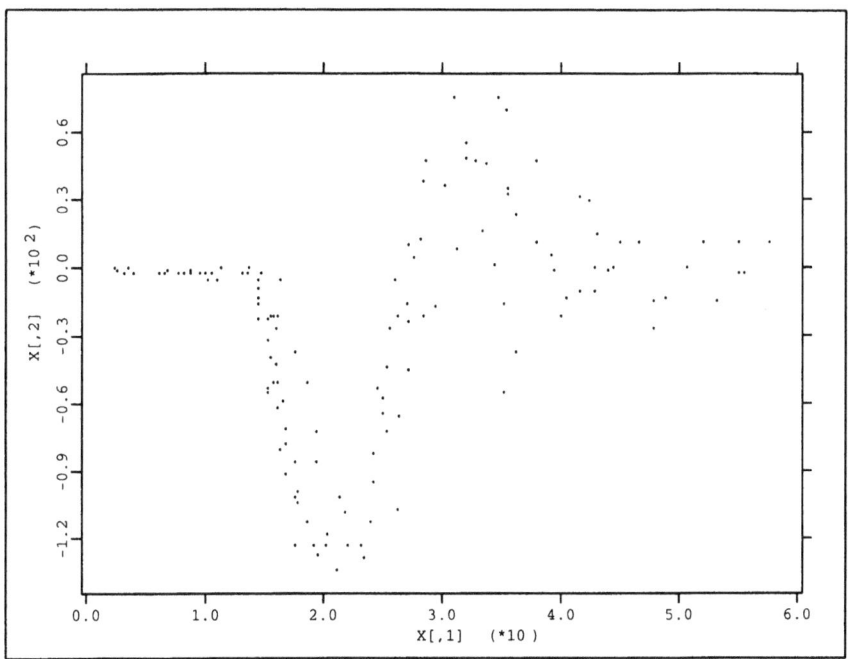

FIGURE 2.7. Result of macro `mymac1.xpl`.

2.3   Creating a Macro

Up to now we have typed all commands from the command line, but this is getting tiring. We can program our statistical tasks interactively by defining macros. A macro is either a main program or a little tool of command sequences that we have already stored on the disk. The simplest macro to start with is probably the `startup.xpl` macro. The *startup macro* is executed every time XploRe is started. All macros have the extension `.xpl`.

Let us define this macro by typing

```
edit("startup")
```

XploRe goes into `edit` mode and we can now select the commands which we want to be executed at each startup of XploRe. It is probably reasonable to have the `<ESC>` key on the left mouse button and the `<ENTER>` key on the right mouse button. So we type

```
mbutton(27 0 13)
```

The argument in the middle is for the middle mouse button. The value 0 means that we do not want to use it or that we only have a two-button mouse. The numbers 27 and 13 correspond to the internal XploRe codes of `<ESC>` and `<ENTER>`.

```
XploRe Help

Command: SKNN
--------

* Function: SKNN computes the k-nearest neighbour smooth regression
            from scatter plot data. As inputs you have to specify the
            explanatory variable x, the dependent variable y and the
            smoothing parameter k.

            z = SKNN (x y k)

->X         x   n x 1 matrix
            y   n x 1 matrix
            k   n x 1 matrix or scalar

X->         z   n x 1 matrix

(!)         SKNN works only on sorted data x-y. Before applying SKNN
            the data must be previously sorted by x. This is done au-
            tomatically by the macro "knn.xpl".

            The smoothing parameter k has to be positive and odd.

* Example:  x =READ(GEYSER)
            x = sort(x)
            y = x[,2]
            x = x[,1]
            z = SKNN (x y 11)
            SHOW (x-y x-z s2d)

*Reference: Haerdle, W. (1990). "Applied Nonparametric Regression,"
            Cambridge University Press.
```

FIGURE 2.8. The syntax of **sknn**.

It also makes sense to record all commands given on the command line in a logfile. So we type in the next line,

```
logfile("xplore")
```

A logfile **xplore.log** will be created when we start XploRe the next time. The contents of the *logfile* can be studied via the **<F6>** key. The active window should now look like this:

```
mbutton(27 0 13)
logfile(xplore)
```

Finally, press **<ESC>** and type "w" to write the file **startup.xpl**. We have just created our first macro.

Now let us program a macro that reads a dataset with two columns and plots them onto the screen. We are on the commandline and type

```
edit("mymac1")
```

This will create a file **mymac1.xpl** in the XploRe directory. We want this as an executable macro so we enter as the first line of the macro,

```
proc() = main()
```

and as the last line,

```
endp
```

```
c:\xplore3\LIB\KNN.XPL

; *****************************************************************
; * KNN macro *****************************************************
; *****************************************************************
; * Function: KNN estimates a regression function from two dimensional *
; *            scatterplot data via the k-nearest neighbor (kNN)        *
; *            method                                                   *
; *                                                                     *
; * Call:      y = KNN (x k)                                            *
; *                                                                     *
; * ->X        x  n x 2  matrix                                         *
; *            k           integer                                      *
; * X->        y  n x 2  matrix                                         *
; *                                                                     *
; * (!)        the cols of X must be equal to 2.                        *
; *            the smoothing parameter k must be an integer > 0.        *
; *                                                                     *
; * Example:   LIBRARY(smoother)                                        *
; *            x = READ(geyser)                                         *
; *            y = KNN(x 11)                                            *
; *            gives the kNN regression estimate for the geyser data    *
; *            using the parameter k=11.                                *
; *            the n x 2 matrix y contains in the first col the sorted  *
; *            first col of x, in the second col the regression         *
; *            estimate.                                                *
; *                                                                     *
; * Comments: symmetrized KNN method, see W. Haerdle                    *
; *            "Applied Nonparametric Regression", Cambridge University *
; *            Press                                                    *
; *                                                                     *
; * See also: SKNN   (command)                                          *
; *                                                                     *
; *****************************************************************
; ** Sigbert Klinke, 901012 **************************************
; *****************************************************************
proc (y)=knn(x k)
  error(cols(x)<>2 "KNN: COLS <> 2")
  z=sort(x)
  a=z[,1]
  b=z[,2]
  y=z[,1]~sknn(a b k)
endp
; *****************************************************************
```

FIGURE 2.9. The macro knn.

In between these two statements we write,

```
x = read (motcyc) ; reads the data
show (x  s2d)      ; plots the data
```

The semicolon is the start of a comment. The **read** command reads the file **motcyc.dat** from the data directory **\data** in ASCII format. There are XploRe internal formats which will be described in Section 2.6. The complete macro should now look like this:

```
proc() = main()
  x = read(motcyc)  ; reads the data
  show(x s2d)       ; plots the data
endp
```

If we now press the **<F4>** *run macro key*, the macro is executed. During execution, we see the currently executed commands flashing by on the command line. We can turn this off by the **trace** command. The **<F10>** *open key* will show you the syntax. The result of the macro is shown in Figure 2.7.

The motorcycle data set **motcyc.dat** is described in Silverman (1985). The statistical question there is how the acceleration curve depends on time. The accelerations y are obviously somewhat noisy, so a smoothing algorithm would give us a better impression of the acceleration curve. XploRe

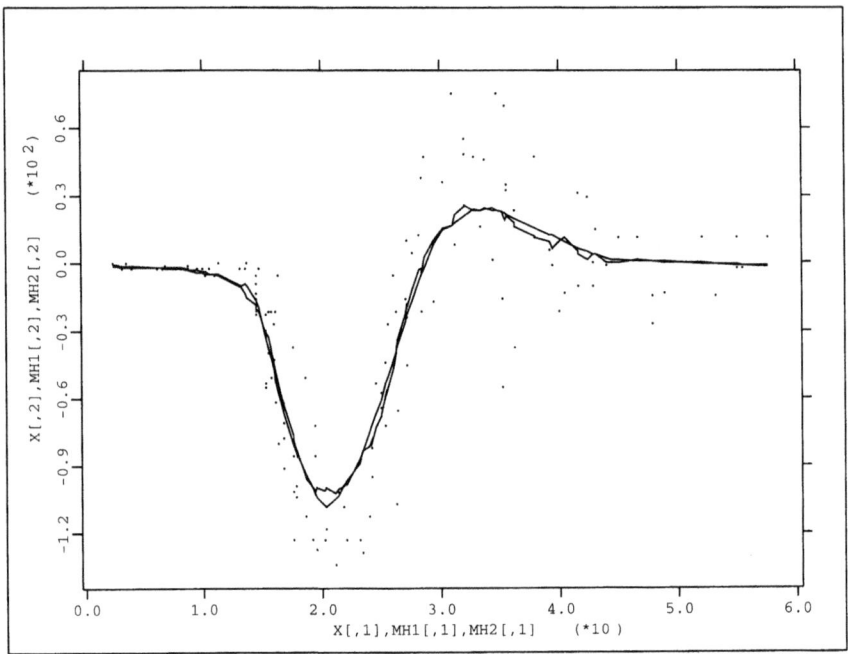

FIGURE 2.10. The motorcycle data set with the **knn** and **lowess** smooth.

has a variety of smoothing algorithms; let us use the symmetrized k-nearest-neighbor algorithm as described, for example, in Härdle (1990, Chapter 3).

The command is called **sknn**, but how do we use it? Type the word **sknn**. Move the cursor over the word **sknn** and press the *open key* <F10>. It also works in *edit* mode. A help screen as shown in Figure 2.8 appears.

We learn from the help file of this command that we need to sort the data and separate the columns of **x** to feed it into **sknn**. But there is an easier way.

In this help file (Figure 2.8) it is indicated that there is another macro **knn.xpl**. We move the cursor over the word **knn** in this help file and press the <F10> *open key*. The macro **knn.xpl** appears as in Figure 2.9.

We see that this macro does all the sorting, etc. for us; the only thing is that we have to call the **library("smoother")**. In fact, this macro is stored in a file **\lib\knn.xpl** and the file **smoother.lib** tells us that this macro is there. So let us add the following statements to our program **mymac1.xpl**:

```
library("smoother")
mh = knn(x 21)
```

The <F4> *run macro key* lets us run the macro. We smooth here over 21 nearest neighbors. Let us plot the curve and the data by

```
show(x mh s2d)
```

```
XploRe Help

Hint to the reader
------------------
Use <Page down> and <Page Up> to move through the file !!!!!!!!!

Commandline
-----------
Note: - To obtain more information for XploRe operators, elements or
        functions move the arrow to the desired keyword and press <F10>.
      - If a help file is shown in the screen the F1-key has no use.

Available Function-Keys:
------------------------
<F1>       : Show this helpscreen.
<F2>       : Show list of available objects.
<F3>       : Get last commandline back.
<F4>       : Show list of data-files.
<F5>       : Show list of library-files.
<F6>       : Show logfile.
<F7>       : Show system info.
<F10>      : Context-sensitive help
<Alt><h>   : Show this helpscreen.
<Alt><r>   : Stops a program during parsing.
<Alt><x>   : Leave XploRe.
<Alt><p>   : Print the actual screen (see PRINT)
<Ctrl><y>  : Delete commandline.
<Ctrl><l>  : Delete left part of commandline.
<Ctrl><r>  : Delete right part of commandline.

Buffer moving:
--------------
<Ctrl><c>  : Copy commandline to buffer.
<Ctrl><v>  : Copy buffer to commandline.
(!) This functions use the same buffer as the XploRe-editor. So it is
    possible to copy a editor-line in the commandline and the command-
    line in the editor.

Available XploRe numbers:
-------------------------
EH              The number e
NAN             Not A Number
PI              The number pi
```

FIGURE 2.11. The initial help screen.

and connect the curve (on the second data part using the <F10> *change data part* inside *show* mode). The curve seems to be somewhat rough on the right side of the picture due to a higher variance pattern. We can play with the smoothing parameter (the number of neighbors), but we will either undersmooth the high variance pattern on the right half of the plot or oversmooth the peak in the left half.

An adaptive smoothing method is called for. One of the most commonly used smoothing methods is an algorithm called LOWESS developed by Cleveland (1979). The syntax in XploRe is as follows:

```
lowess(x[,1] x[,2])
```

We can check via the <F10> *open key* that there is an optional third smoothing parameter f. The algorithm works only on $n \times 2$ data matrices that are sorted with respect to their first column. The output is the curve itself so we must add the first column of x in order to plot it. The macro (with smoothing parameter $f = 0.3$) should look like this:

```
proc() = main()
  x = read(motcyc)        ; reads the data
  library(smoother)
  x = sort(x)
  mh1 = x[,1]~lowess(x[,1] x[,2] 0.3)
```

```
XploRe Help

ASEQ            Additive sequence
ASIN            Arcsine in radian
ASIND           Arcsine in degreee
ASINH           Arcus sinus hyperbolicus
ATAN            Arctangus in radian
ATAN2           Angle of a point (x, y) in radian
ATAND           Arctangus in degree
ATANH           Arcus tangens hyperbolicus
BACKFIT         Backfit regression
BETAF           Beta function
BFGS            Finds a minimum
BINCOF          Binomial coefficients
BINDATA         Binning data
CAPTURE         Set printfile on or off
CARTCOOR        Converts polar in cartesian coordinates
CCTOV           Converts a maskvector to a floatvector
CDFB            cumulative density of Beta-distribution
CDFC            cumulative density of Chi-distribution
CDFF            cumulative density of F-distribution
CDFN            cumulative density of Standard-normal-distribution
CDFT            cumulative density of T-distribution
CEIL            Round up
COLORS          Number of available colors
COLS            Number of columns
CONTOUR2        Contouring of 3-dimensional datasets
CONTOUR3        Contouring of 4-dimensional datasets
CONV            Convolution
COS             Cosine in radian
COSD            Cosine in degrees
COSH            Cosine hyperbolicus
COV             Covariance matrix
CREATEDISPLAY   Generates new displays
CUMPROD         Columnwise cumulative multiplication
CUMSUM          Columnwise cumulative addition
DATE            String with date
DET             Determinant
DIAG            Generates diagonal matrix from a vector
DIFF            Computes the difference of two vector elements
DISPLAY         Pop up an existing display
DISTANCE        Distance between datapoints
DOS             Temporary exit to dos
ECLOCK          Seconds since SCLOCK
```

FIGURE 2.12. A help screen with XploRe commands.

```
mh2 = knn(x 21)
show(x mh1 mh2 s2d)      ; plots the data and the two curves
endp
```

This macro will produce the graph shown in Figure 2.10.

We may want to write our own little lowess macro that does the sorting for us and the splitting into columns. So we leave the editor by <ESC> and type edit("lows").

This time we use the editor just to create a macro that we will use later, so we do not say proc() = main(). Rather, we give input and output parameters as follows:

```
proc(y) = lows(x f)
  error(cols(x) <> 2 "lows: cols <> 2")
  z = sort(x)
  a = z[,1]
  b = z[,2]
  y = z[,1]~lowess(a b f)
endp
```

Having typed this, we press <ESC> and confirm storage by <W>. The macro can now be called in, for example, mymac1.xpl by

```
XploRe Help
Command: DIAG
--------

* Function: DIAG creates a diagonal matrix from a vector.

              z = DIAG (x)

->X           x   p x 1 matrix

X->           z   p x p diagonal matrix

* Example:    x = 1 : 10
              z = DIAG (x)
              Gives a 10 x 10 diagonal matrix

* See also: UNIT, MATRIX
```

FIGURE 2.13. The syntax of **diag**.

```
proc() = main()
  func("lows")
  library(smoother)
  x = read(motcyc)
  mh1 = lows(x 0.3)
  mh2 = knn(x 21)
  show(x mh1 mh2 s2d)
endp
```

The macro **lows.xpl** has been loaded by the command **func**. We will learn later in Section 3.13 how to create our own libraries.

2.4   The Interactive Help System

The help system in XploRe has been designed in order to make a manual obsolete. It not only shows you XploRe commands and macros, it also lets you see your own macros. So it is helpful for the help system if you put comments into your macros, for example, as in the **knn.xpl** that we saw in Figure 2.9.

```
XploRe Help

Command: CCTOV
--------

* Function: CCTOV converts a maskvector in a floatvector. The resulting
            value m represents the color and the outfit of a datapoint
            calculated according to:

                m = 16 * outfit + color

            with the following values for outfit and color,

            value      outfit        color
            -------------------------------
               0       Unused        Black
               1       Point         Blue
               2       Plus          Green
               3       X             Cyan
               4       Y             Red
               5       O             Magenta
               6       Star          Brown
               7       Face          Light Gray
               8       Unused        Dark Gray
               9       Unused        Light Blue
              10       Unused        Light Green
              11       Unused        Light Cyan
              12       Unused        Light Red
              13       Unused        Light Magenta
              14       Unused        Yellow
              15       Unused        White

            m = CCTOV(x)
->X         x   n x p matrix  (already displayed)
X->         m   n x q matrix  with all the mask vectors from the last
                              display of x

* Example:  x = normal(100 2)
            show(x s2d)
            <Esc>
            m = CCTOV (x)
            Gives the matrix  (100 x 1)  #(31 ... 31)
```

FIGURE 2.14. The help screen for CCTOV.

The general *help key* is <F1>. It gives general help in the *command line*, the *show* mode, and the *edit* mode. If we press <F1> with the cursor on the command line, we see the screen shown in Figure 2.11.

We move through this help file by <Pg Dn> or <Pg Up>, or the cursor keys <↑> or <↓>, or the mouse. Figure 2.12 shows the help file after some further <Pg Dn> movements.

If we want to learn more about the syntax of a command, we press, for example, on diag the <F10> *open key* on top of the word diag. The result of this is shown in Figure 2.13.

We leave this help screen by <ESC>. Let us move to cctov, a command necessary to convert color codes (obtained, for example, by brushing, see Section 2.2) into numerical vectors. The screen that appears is shown in Figure 2.14.

We actually moved one line down by <↓> to see that this help screen refers under "see also" to another command, mask. If we press <F10> on this command, the help screen on masking in Figure 2.15 appears.

We see that we can use this help system in a flexible way. If we had written a macro and this macro was in the directory \lib, the help system would load it onto the screen. You leave the help system by <ESC>. You may want to try it with the macro eigensm.xpl. Type eigensm on the command line and press <F10> on top of it. You will see that eigensm makes a spectral decomposition with ordered eigenvalues.

```
XploRe Help

Command: MASK
--------

* Function: MASK creates a mask matrix. This mask matrix can
            be appended to a data matrix.

            z = MASK (n p t s)

->X         n scalar - number of rows of mask matrix
            p scalar - number of columns of mask matrix
            t string with the type of the n datapoints.
            c string with the colour of the n datapoints.

            Possible parameters for t are :
            POINT, PLUS, X, Y, O, STAR, FACE

            Possible parameters for c are :
            BLACK, BLUE, GREEN, CYAN, RED, MAGENTA, BROWN
            LGRAY, GRAY, LBLUE, LGREEN, LCYAN, LRED, LMAGENTA,
            YELLOW, WHITE

X->         z  mask vector

(!)         Default values are POINT and WHITE.
            It is the same to specify first  t  or  c.

* Example:  z1 = MASK (100 1 YELLOW FACE)
            z2 = MASK (50  1 PLUS RED)
            x = NORMAL (150 2)
            x = x ~ (z1 | z2)
            SHOW (x s2d)
            Shows the first 100 points as yellow faces and the other
            50 points as red plus

* See also: UPDATE, SHOW, CCTOV, VTOCC, PALETTE
```

FIGURE 2.15. The help screen for **MASK**.

2.5   Three-Dimensional Plots

An important issue in computer-assisted interactive statistical data analysis is the dynamic three-dimensional spinning plot. XploRe has a three-dimensional dynamic graphics realized through the **show** command with window type **D3D**. A three-dimensional data matrix x is plotted via **show(x d3d)**. Let us try this with the bank data set of Flury and Riedwyl (1988). We read the bank data set x = **read(bank2)** and then we extract the first, second, and sixth columns by x = x[,1:2]~x[,6]; we then use the three-dimensional show command **show(x d3d)**. We go into *spinning mode* by **<F6>** and we then rotate the point cloud by the cursor keys. Figure 2.16 shows the result of spinning this $n \times 3$ matrix. We have labeled the axis using **<F4>** and have changed the plotting style using **<F10>**, the *change data part key*.

There seem to be two different groups of bank notes, namely forged and true ones. We brushed the point cloud (as described in Section 2.2) in the projection shown in Figure 2.16 and changed the plotting style from circles to x inside the brush. The result is shown in Figure 2.17.

Later, we can check the observations that we brushed by putting them onto the screen. In Figure 2.18 we show the result of this interactive computer-guided discriminant analysis.

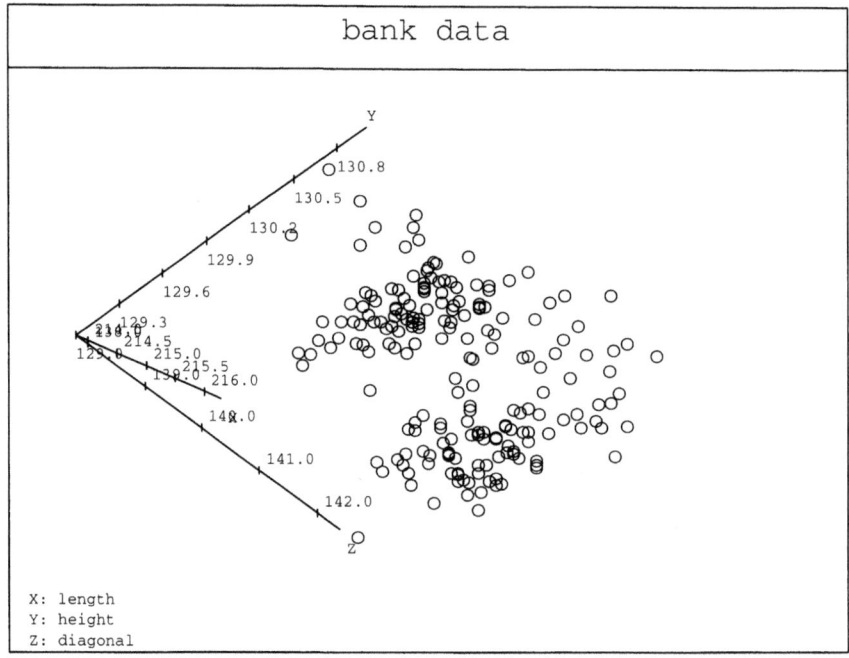

FIGURE 2.16. The rotated bank data set.

The *mask code* given in the last column allows us to tell which observations we classified to be in which point cloud. "95" corresponds to a white circle and "44" to red plus; the mask codes are given in the appendix (in the description of the command **vtocc**).

The XploRe three-dimensional dynamic graphic is not there only to spin point clouds. You can create three-dimensional contours as well. Let us stay with the bank data set and make a two-dimensional density estimate of the variables x[,2] and x[,6] (the columns here refer to the original data matrix with six columns). We refer to library(smoother) and call

```
fh = denest2(x[,2]~x[,6] 0.2)
```

The output is a two-dimensional density estimate on a regular grid on the range of the observations. The command show(fh d3d) lets us join lines by pressing <F10> and <F4>; the result is given in Figure 2.19.

XploRe can also show three-dimensional density contours, a technique developed by Scott (1992). If we go back to the data matrix x[,1:2]~x[,6] with the three columns already shown in Figures 2.16 and 2.17 and apply the density estimate in three dimensions, to be described in Chapter 4, we obtain Figure 2.20. This estimate will be considered in more detail in Chapter 15.

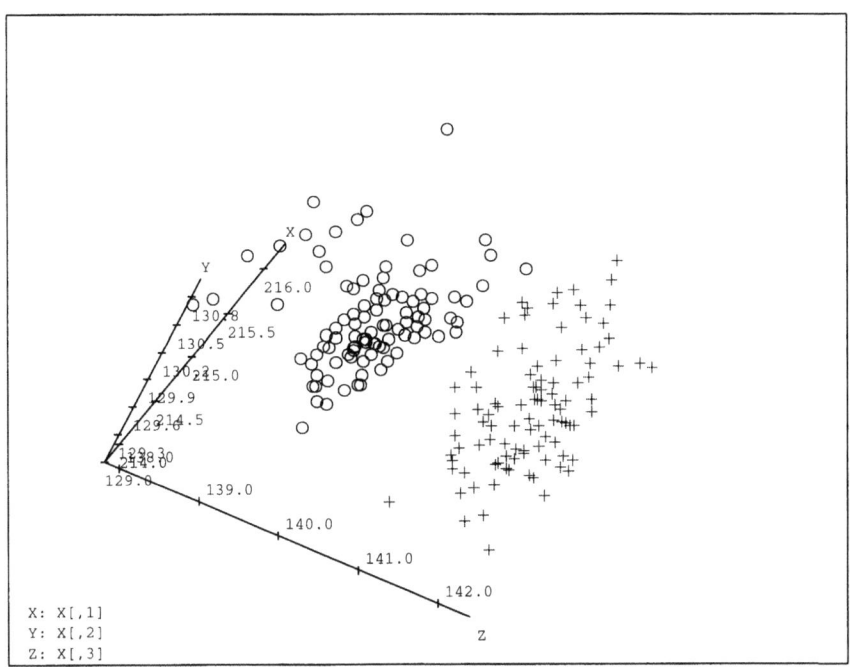

FIGURE 2.17. The bank data set with one cluster brushed (see color insert).

2.6 Reading and Writing Data

XploRe reads data sets in two formats. The first is the regular ASCII format, and the second is the XploRe internal *X-format*. XploRe will recognize whether a data set is to be read in XploRe internal format. When reading data (it may contain columns with text), XploRe classifies values separated by blanks as belonging to different columns.

Data may be read by skipping certain columns, for example,

```
x = read(bank2 "d-dddd-d")
```

will read only the second through fourth and the sixth columns since the first and the fifth columns are to be skipped with the "d-" notation. The parameter inside the quotation marks is a format string. Details on the read command can be obtained in the help system.

Data can be read from the command line by the command **readcon**. The command **readval** opens an interactive box and lets you enter float values. The command **readstr** does the same but expects a string as input parameter.

X[,1]	X[,2]	X[,3]	X%[,1]
214.6000000	129.9000000	141.8000000	95
215.1000000	129.7000000	140.6000000	95
214.9000000	129.8000000	141.0000000	95
215.2000000	129.7000000	141.9000000	95
215.2000000	130.1000000	141.3000000	95
215.4000000	130.7000000	141.2000000	44
215.1000000	129.9000000	141.5000000	95
215.2000000	129.9000000	141.6000000	95
215.0000000	129.6000000	142.1000000	95
214.9000000	130.3000000	141.5000000	95
215.0000000	129.9000000	142.0000000	95
214.7000000	129.7000000	141.6000000	95
215.4000000	130.0000000	141.4000000	95
214.9000000	129.4000000	141.5000000	95
214.5000000	129.5000000	141.5000000	95
214.7000000	129.6000000	142.0000000	95
215.6000000	129.9000000	141.7000000	95
215.0000000	130.4000000	141.1000000	44
214.4000000	129.7000000	141.2000000	95
215.1000000	130.0000000	141.5000000	95
214.7000000	130.0000000	141.2000000	95
214.4000000	130.1000000	139.8000000	44
214.9000000	130.5000000	139.5000000	44
214.9000000	130.3000000	140.2000000	44
215.0000000	130.4000000	140.3000000	44
214.7000000	130.2000000	139.7000000	44
215.0000000	130.2000000	139.9000000	44
215.3000000	130.3000000	140.2000000	44
214.8000000	130.1000000	139.9000000	44
215.0000000	130.2000000	139.4000000	44
215.2000000	130.6000000	140.3000000	44
215.2000000	130.4000000	139.2000000	44
215.1000000	130.5000000	140.1000000	44
215.4000000	130.7000000	140.6000000	44
214.9000000	130.4000000	139.9000000	44
215.1000000	130.3000000	139.7000000	44
215.5000000	130.4000000	139.2000000	44
214.7000000	130.6000000	139.8000000	44
214.7000000	130.4000000	139.9000000	44
214.8000000	130.5000000	140.0000000	44
214.4000000	130.2000000	139.2000000	44
214.8000000	130.3000000	139.6000000	44

FIGURE 2.18. The three-dimensional data matrix and the mask vector.

Writing data to a file is done using `write(x filename)`. The matrix `x` is written into a file given in `filename`. The parameter `filename` is optional; if it is not given, XploRe writes into a file `nofile.dat`. If the data are to be written in *X-format*, the additional parameter `"x"` must be given. The sequence of commands

```
x = read(bank2)
write(x "bank2x" x)
```

writes the data matrix into the file `bank2x.dat`. The `read` command works at the speed of light when you apply it to X-formatted data sets. Large data sets should therefore be stored in this XploRe internal format.

REFERENCES

Cleveland, W.S. (1979). Robust locally-weighted regression and smoothing scatterplots, *Journal of the American Statistical Association* **74**(368): 829–836.

Flury, B. and Riedwyl, H. (1988). *Multivariate Statistics, A Practical Approach*, Cambridge University Press, New York.

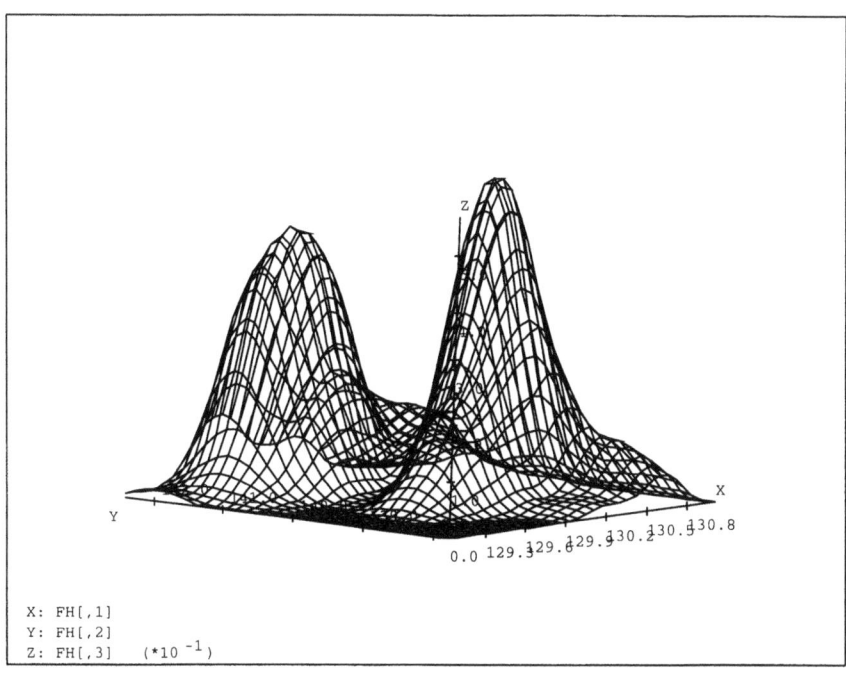

X: FH[,1]
Y: FH[,2]
Z: FH[,3] $(*10^{-1})$

FIGURE 2.19. A two-dimensional density estimate of the bank data.

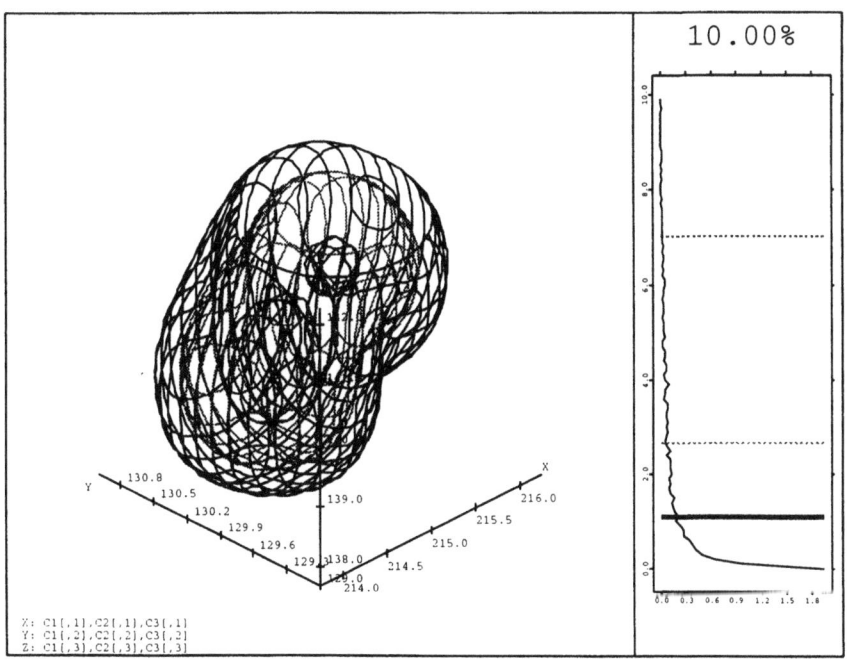

FIGURE 2.20. A three-dimensional density contour (see color insert).

Härdle, W. (1990). *Applied Nonparametric Regression*, Econometric Society Monographs No. 19, Cambridge University Press, New York.

Scott, D.W. (1992). *Multivariate Density Estimation: Theory, Practice, and Visualization*, John Wiley & Sons, New York, Chichester.

Silverman, B.W. (1985). Some aspects of the spline smoothing approach to nonparametric curve fitting (with discussion), *Journal of the Royal Statistical Society, Series B* **47**(1): 1–52.

3
The Integrated Working Environment

Claudia Gajewski[1]

3.1 Introduction

The aim of this chapter is to take a deeper look into the working environment of XploRe. This working environment consists of several tools as, for example, an editor, a help system, a command line interpreter, and an interactive graphic. We will show how these tools work together.

Section 3.13 contains a complete list of the user libraries, which consist of routines for advanced statistical methods. The application of the several methods is studied in detail in the ensuing chapters of Part II.

We will not give a description of the commands, operators, and language elements of XploRe; this is done in the Appendix, the XploRe user manual, or the built-in help facility (function keys `<F1>` and `<F10>`; see Section 3.4). Here we concentrate on describing the integrated working environment.

3.2 The Editor

The XploRe editor is intended to be used to edit data matrices as well as text files, for example, programs. There are several hot keys that make it possible to quasi-simultaneously edit a program, run it, and get information about other programs or data. Table 3.1 to Table 3.4 show all the available editor commands and corresponding keys, which you can also display during an XploRe session via the help key `<F1>`.

Every XploRe screen has a command line at the bottom and on the right margin a list of icons representing the function keys `<F1>` to `<F10>`. Suppose that the cursor in the command line is blinking, which indicates that you are in the so-called *command* mode and can now type a command (see Section 3.4 for the meaning of the function keys in this mode).

Start the editor with the command `edit("foo")` where `foo` denotes either an existing matrix or a possibly nonexisting file. Then `foo` is either loaded or created and the cursor moves to the edit window. Note that the

---

[1]Sonderforschungsbereich 373, Humboldt-Universität zu Berlin, D-10178 Berlin, Germany.

Basic commands	
Char left	<←>
Char right	<→>
Line up	<↑>
Line down	<↓>
Start of line	<HOME>
End of line	<END>
Greater distances	
Top of page	<Pg Up>
Bottom of page	<Pg Dn>
Start of matrix or file	<CTRL Pg Up>
End of matrix or file	<CTRL Pg Dn>

TABLE 3.1. Moving around within the XploRe editor.

Toggle insert mode	<INS>
Delete char left from cursor	<BACKSPACE>
Delete char under cursor	<DEL>
Cut line into buffer	<CTRL Y>
Delete from cursor to end of line	<CTRL R>
Delete from start of line to cursor	<CTRL L>
Copy line into buffer	<CTRL C>
Paste line from buffer	<CTRL V>

TABLE 3.2. Insertion and deletion commands within the XploRe editor.

function keys change their meaning (see Table 3.4 for the function keys that are now available).

When you have completed your work you can either leave the editor by pressing <ESC>, after which you are prompted either to save the changes (press <w>) or to discard them (press <y>), or, in case of a program, save and run the program immediately with function key <F4>. The first action brings you back to the command line. The second action puts you back into the editor. For further explanation, see Section 3.3.

As an example, we edit a data matrix. We use a six-dimensional performance data set from U.S. companies, which is stored in the file uscomp2.dat in six columns of real numbers. The different kinds and structures of data that XploRe can deal with are mentioned in Section 2.6 of the tutorial. The following two command lines first read the data to a matrix x and then edit this matrix:

```
x = read("uscomp2")
edit(x)
```

Go one vector to the left	<CTRL ←>
Go one vector to the right	<CTRL →>
Go to the first vector	<CTRL HOME>
Go to the last vector	<CTRL END>

TABLE 3.3. Vector commands in the XploRe editor.

Get information about function keys	<F1>
Get context-sensitive help	<F10>
Run an XploRe macro	<F4>
Zoom editor window	<F2>
Save current file	<CTRL W>
Change current filename	<CTRL O>
Move to next tab position	<TAB>
Set tabsize	<SHIFT TAB>
Leave editor	<ESC>

TABLE 3.4. Miscellaneous commands in the XploRe editor.

Note that XploRe assumes by default the suffix .dat for data files in the **read** or **write** command. Now we can modify the data of x. After all work is done, save the modified matrix by confirming to write it back after pressing <ESC>.

As a second example, let us write a short program.

```
edit("new.xpl")
```

As mentioned above, XploRe looks in the working directory to see if there is a file with the name given to the **edit** command, in our example, **new.xpl**. If it exists, it is loaded into the editor; if not, it is created. We now write the following lines:

```
proc() = main()
  x = read("uscomp2")
  z = x[,4]./x[,3]
  z
endp
```

and, after pressing <ESC>, confirm to save the work. This brings us back to the command line. Of course, you can also run the program directly from within the editor using the hot key <F4>. In this case, the quotient z taken element by element will be calculated, and the result will be shown on the screen. By pressing <ESC>, you will return to the editor.

3.2.1 Getting Help Within the Editor

The XploRe help system is intended to be fully on-line; thus, we can access it while editing anything. For illustration, we load our example again:

```
edit("new.xpl")
```

First, suppose that we need some information about the **read** function. Move the cursor to the word **read** and press <F10>. A new window with a description of **read** appears. After reading this, we can leave it with <ESC>. We are now back in the edit window and want to look at the data stored in **x**. Therefore, we move the cursor to one of the three occurrences of **x** and press <F10> again.

A new window appears, displaying the matrix **x**, which we can move through, for instance, column by column with <CTRL ←> and <CTRL →>. Table 3.3 shows a complete list of all possible movements.

If you write a program and can remember only the name of an XploRe command but not how to use it, a little trick helps: write the name of this partially known command on an extra line of your program, move the cursor there, and press <F10>. After you have closed the help window, delete the name again.

3.3 How to Run and Debug a Program

In XploRe you run a main program, for example, **foo.xpl**, by loading it into the editor and pressing <F4>. Now three things may happen.

- If you are lucky, the program runs successfully, and if it has finished its work, you get back to the editor window where the source code is displayed. Of course, if your program produces any output, it is displayed on an extra window, which has to be closed via <ESC>.

- In the second, less pleasant case, the program contains a bug and execution stops after the occurrence of an error. An error message is displayed and pressing any key will bring you back to the command line.

- Maybe you realize during execution that the program does not work sensibly. In this case, you are able to stop the execution yourself by the *user break* hot key <ALT R> (see Table 3.5). Pressing any key will bring you back to the command line.

Suppose a program has been created and started in the way explained above. While it runs, the command that is currently being executed appears in the command line. This can be avoided by stopping the **TRACE** mode. If an error occurs, the execution stops and the following sources of information give a hint as to what the reason for the error might be.

Hot key	Action
<ALT R>	stops a program during execution
<ALT X>	leaves XploRe. This is equivalent to QUIT.
<ALT H>	shows a helpscreen. This is equivalent to <F1> and to the command HELP.
<ALT P>	prints the active display. It does the same as PRINT.
<CTRL Y>	copies current line into buffer and deletes it
<CTRL R>	deletes text right of cursor position
<CTRL L>	deletes text left of cursor position
<CTRL C>	copies current line into buffer
<CTRL V>	copies buffer into current line

TABLE 3.5. Hot keys in *command* mode.

1. The command that was executed last is shown in the command line, and it may very well have something to do with the error; it may even be the reason for the error.

2. A window appears with a short error message. You press any key and then the cursor is in the command line.

3. Using a hot key <F2> (see the table about the action of the function keys in *command* mode in Section 3.4) a table can be examined that contains all active objects in XploRe, including the matrices of the macro the moment after the occurrence of the error. In the table you find information about type, size, and depth of these objects. That means you know the dimension of matrices and the program depth in which they have been defined. Matrices with depth 1 are global matrices in XploRe. Matrices with depth greater than 1 have been defined in the macros.

   The contents of the XploRe objects listed in the table can be examined using the *help key* <F10>. In practice, it means that move the cursor to the object or matrix and press <F10>. A window appears, displaying the contents of the object or the data that are saved in the matrix. Note that you cannot modify anything in this situation.

Let us look at an example. We write a short program that tries to multiply a 2×2 by a 3×3 matrix, which is obviously not very sensible.

```
proc() = main ()
  x = matrix(2 2 3) ; yields a matrix containing threes
  y = matrix(3 3 5) ; yields a matrix containing fives
  z = x*y
  z
endp
```

```
OBJECTNAME          TYPE        SIZE    DEPTH

BOXPLOT1D           DISPLAY        1
CORR                FUNCTION
DYNAMIC3D           DISPLAY        1
EH                  MATRIX       1,1       1
EIGENSM             FUNCTION
FACES               DISPLAY        1
FIVENUM             FUNCTION
HELPDISPLAY         DISPLAY        1
MAIN                FUNCTION
NAN                 MATRIX       1,1       1
PCA                 FUNCTION
PI                  MATRIX       1,1       1
STATIC2D            DISPLAY        1
TEXT                DISPLAY        1
X                   MATRIX       2,2       2
Y                   MATRIX       3,3       2
```

FIGURE 3.1. The active objects in XploRe (example).

After starting this program (see the beginning of this section), the commands appear on the command line one after the other until the execution stops. The current command is z=x*y and an error message appears which says,

```
Error(#123) in *
Left matrix dimension <> right matrix length!
Press any key
```

Here, of course, we already know the reason for the error, but for completeness we take a look at x and y in the table of active XploRe objects. For that purpose, press <F2>, and we get Figure 3.1, which shows the list of active objects. In this case, the SIZE column points out that x and y do not fit together. Both have DEPTH 2 because they are defined in the macro. In the final step, we look at the data saved in x and y by moving the cursor there and pressing <F10>, and we note again the incompatible dimensions (see Section 3.4 about context-sensitive help).

3.4   The Context-Sensitive Help System

In Section 2.4 of the tutorial, the help system has been studied in a practical setting, and in Section 3.2, we have already mentioned some aspects of it when editing an object or running a program. Here we will look at the help system in a more systematic way.

When working with XploRe, you are always in one of three modes:

- *edit* mode, when you edit an object or program;

- *graphic* mode, when a graphic is displayed (for details, see Sections 3.5 to 3.12); and

- *command* mode, when the cursor blinks in the command line and you can enter a command.

Here we want to give an overview of the help facilities in these modes, which are all provided by a selection of the function keys <F1>...<F10>, depending on the mode. Table 3.6 shows the help facilities and their corresponding keys for the different modes. The *command* mode, however, needs further explanation:

- <F1> gives you the general help screen for the *command* mode, including a short description of all tools, commands, operations, and language elements of XploRe. We call it the *help key*.

- <F2> yields the table of active XploRe objects, which we explained in Section 3.3 (Figure 3.1 shows an example).

- <F3> brings the last entered command back onto the command line so that you can enter it again. Repeatedly pressing <F3> allows you to cycle also through older commands.

- <F4> will give you an overview about the data files *.dat, which you can directly load from \XPLORE3\DATA.

- <F5> lists all the macros and libraries contained in the directory \XPLORE3\LIB that you can work with.

- <F6> opens the log file, which documents the whole XploRe session.

- <F7> shows a screen with technical data of the XploRe system; for instance, it describes the connected printer, the coprocessor, the mouse, and XploRe paths.

- <F10> activates the context-sensitive help. Therefore you have to move the cursor to the object of interest. It is the so-called *open key*.

The context-sensitive help works as follows. Suppose the cursor is at the word foo. First, XploRe looks for a help file named foo.hlp in the directory \XPLORE3\HELP. If it exists, its contents are shown. If not, XploRe assumes that foo is a macro and tries to load the source code of foo.xpl in \XPLORE3\LIB.

	edit mode
`<F1>`	*help key*—show a help screen
`<F10>`	*open key*—context-sensitive help
	graphic mode
`<F1>`	*help key*—show a help screen
	command mode
`<F1>`	*help key*—show a help screen
`<F2>`	show list of active objects
`<F3>`	recall last commands
`<F4>`	show list of data files
`<F5>`	show list of program files
`<F6>`	show log file
`<F7>`	show system info
`<F10>`	*open key*—context-sensitive help

TABLE 3.6. Help keys within the different modes.

So it is easy to create a help file for a command, for example, `foo`, by writing a text file with a description of `foo` and then putting it, with the name `foo.hlp`, into the directory `\XPLORE3\HELP`. Pressing `<F10>` while the cursor is on `foo` displays the contents of that text file.

Context-sensitive help for a user macro is provided by writing the description as a comment on top of the source code, which is displayed by the help system in that case (see Section 3.13 for more details about macros and libraries).

Figure 3.2 shows as an example for this procedure the source code of the built-in macro `runmed`.

3.5   Graphic Tools in XploRe

Graphics are generated in XploRe with the `show` command. The command `show` is a multi-purpose command and can display data in the following forms:

1. boxplot, `b1d`;

2. static two-dimensional picture, `s2d`;

3. dynamic three-dimensional picture, `d3d`;

4. text, `text`; and

5. flury faces, `fcs`.

```
c: XPLORE3 LIB\RUNMED.XPL

; ********************************************************************
; * RUNMED macro ****************************************************
; ********************************************************************
; * Function: RUNMED estimates a regression function from two        *
; *                  dimensional scatterplot data via the running    *
; *                  median method                                   *
; *                                                                  *
; * Call:      y = RUNMED (x k)                                      *
; *                                                                  *
; * ->X        x  n x 2  matrix                                      *
; *            k          scalar                                     *
; * X->        y  n x 2  matrix                                      *
; *                                                                  *
; * (!)        the cols of x must be equal to 2.                     *
; *                                                                  *
; * Example:   LIBRARY(smoother)                                     *
; *            x = READ(geyser)                                      *
; *            y = RUNMED(x 13)                                      *
; *            gives the RUNMED regression estimate for the geyser data *
; *            the n x 2 matrix y contains in the first col the sorted *
; *            first col of x, in the second col the regression estimate *
; *                                                                  *
; * Comments: uses the optimal running median smoother, see W. Haerdle *
; *            "Applied Nonparametric Regression", Cambridge University *
; *            Press                                                 *
; *                                                                  *
; * See also: RMED    (comnd)                                        *
; *                                                                  *
; ********************************************************************
; ** Wolfgang Haerdle, 910426 **************************************
; ** Berwin Turlach,    901012 ************************************
; ********************************************************************
proc (y)=runmed(x k)
  error(cols(x)<>2 "RUNMED: COLS <> 2")
  z=sort(x)
  a=z[,1]
  b=z[,2]
  y=a~rmed(b k)
endp
; ********************************************************************
```

FIGURE 3.2. The source code of XploRe macro `runmed.xpl`.

We have added in short form the XploRe names for these forms. They will be explained in a minute. Additionally, a combination of these forms is also possible (see Section 3.6). Note that the structure of the data must conform to the form displayed (see Sections 3.9 to 3.12 about the different graphic formats for a more detailed explanation).

The simplest syntax of the **show** command is

`show(x1 ... type)}`

where **x1** ... denotes the name of a single matrix or a list of names of matrices, separated by space, and **type** denotes the type of desired graphic. This command opens a new window and displays all the data stored in **x1** .... The list in the beginning of this section shows the graphic formats together with their **type**. Creating a graphic is one thing, but getting rid of it is another: After a graphic window is opened, you can only work within this window; therefore you have to quit in order to edit again or do something else. This is easily done with <ESC>.

3.6   Multiple Window Displays

3.6.1   How to Create, Activate, and Delete a Display

In order to create a display with only one window, it is sufficient to use the show command as explained above. Very little additional effort is needed to display data in several windows simultaneously. First, you have to create an object called display, which contains all the different windows you need, and declare the type of these windows. The complete syntax for creating a display is

```
createdisplay(disp, p q, type_1 ... type_k)
```

The parameters have the following meaning:

disp   the display's name

p, q   determine the kind of partition of disp into small windows:

1. symmetric partition (that is, disp consists of windows of the same size):

$$p = n, \quad q = m$$

2. asymmetric partition (that is, disp consists of windows of different size). The following pairs of parameters realize different asymmetric partitions:

$$p = (-n), \quad q = (-m);$$
$$p = n, \quad q = (-m); \text{ and}$$
$$p = (-n), \quad q = m;$$

where n and m are integers, which denote the numbers of rows respectively columns disp is divided into.

type_i   denote the types of the windows disp consists of (i = 1,...,k and k=n*m). Here the window on top of the leftmost column has type_1, the one on bottom of this column type_n, and so on column by column until the window on bottom of the right-most column, which has type_k.

If there is only one type of window to be used, you can shorten this to:

```
createdisplay(disp, p q, type)
```

where disp, p, and q have the same meaning, and type denotes the type of graphic you need.

If you don't need disp anymore, you can delete it with

```
free(disp)
```

After a createdisplay(disp,...) command, disp becomes the active display. Only one display can be active at a time.

Once you have created a display `disp`, you can always reactivate it by the command

```
display(disp)
```

This should only be done if the data that were shown in the window still exist, that is, if they were not temporary.

3.6.2 Multi-`show` and an Example

In general, all multiple `show` commands must match the type of the active display. Note that for a single window, the corresponding single display is created automatically.

Suppose you have created a display with the `createdisplay(disp,...)` command above and there are k_l windows of type l, where l ranges from 1 to 5 and denotes the types of graphics available according to the list at the beginning of this section. Thus, `k` is the sum of the k_l. The corresponding `show` command looks like this:

```
show(x1 t1, x2 t2, ..., xk tk)
```

where `x1,...,xk` are the matrices to be displayed, and `ti` is the type of the ith window. `ti` is determined as follows. Suppose the type of the ith window in the `createdisplay(disp, ...)` command is `type`, for example, `s2d`, and it is the jth window of this type in `disp`, for example, the third. Then `ti` would be `type_j`; thus, in our example, `s2d3`. This type of notation of window types will perhaps be more understandable later.

Consider the following example. We again use `uscomp2.dat`, which is a 79×6 matrix containing a six-dimensional data set from 79 U.S. companies. The data set contains the following variables: assets, sales, market value, profits, cash flow, and employees. This data set is listed and investigated in Härdle and Simar (1994, Section 9.2).

The following XploRe code first computes some transformations of the six variables and then displays some of them in a symmetric 2×2 window with `sd2` windows, and others in an asymmetric 2×2 window with a `d3d`, a `b1d`, and two `text` windows:

```
proc()=main ()
  x=read("uscomp2")
  ; ********definition of the variables to plot********
  profitm=x[,4]./x[,3]    ;profit      / market value
  profita=x[,4]./x[,1]    ;profit      / assets
  salesm = x[,2]./x[,3]   ;sales       / market value
  salesa=x[,2]./x[,1]     ;sales       / assets
  casha = x[,5]./x[,1]    ;cash flow / assets
  zz1=salesa~profitm
  zz2=salesm~profitm
  zz3=salesa~profita
```

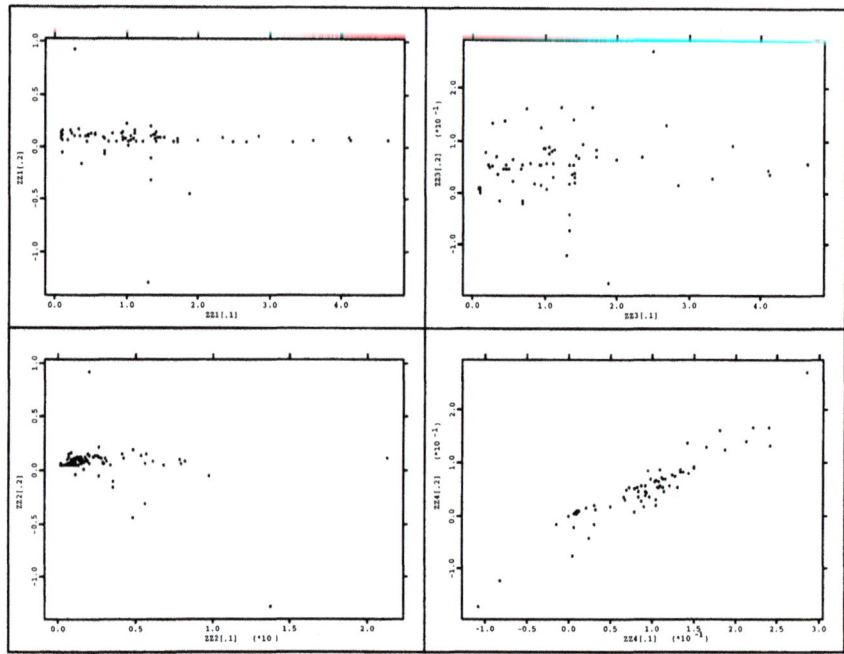

FIGURE 3.3. The s2d-pictures of variables zz1, zz2, zz3, zz4.

```
zz4=casha~profita
zz6=casha~profita~salesa
; ********end of definition**************************
createdisplay(usplot1,2 2, s2d )
show(zz1 s2d1, zz2 s2d2, zz3 s2d3, zz4 s2d4)
createdisplay(usplot2,(-2) (-2), d3d text b1d text)
show(zz6 d3d1, zz6 text1, profita b1d1, profita text2)
endp
```

The two resulting multiple displays are shown in Figures 3.3 and 3.4.

3.7   Manipulating Windows

You can always change the layout of a graphic window, manipulate the data displayed there, and adjust a set of parameters in order to fit the display of data to your needs as much as possible.

In general, after you have started to display any data, a graphic window consisting of one or more windows is opened and a cursor can be moved around within one window using the arrow keys. In the multiwindow case, there is more than one window, and you have in addition the function keys <F8> and <F9> to move from one window to the next respective previous

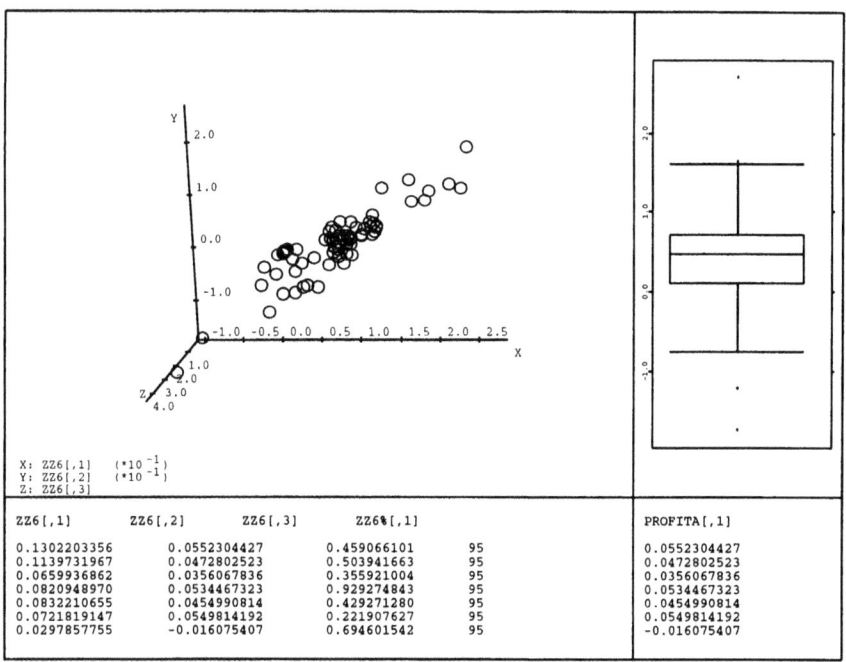

FIGURE 3.4. The **d3d**-picture of the variable zz6 and the boxplot of profita with the corresponding data in the text-windows below.

one, where the windows are ordered first, column by column from left to right, and within a column from top to bottom. Of course, in a single window case, <F8> and <F9> do not have any meaning.

Depending on what kind of window you are in, the meaning of the hot keys changes. We will describe this and the way you can modify layout and display parameters in Sections 3.9 to 3.12, which deal with the special features of the window types.

3.8 How to Print a Graphic

By now we know how to produce a graphic, display it on the screen, and how to quit it. Now we will explain first, how to print it, and secondly, how to save it in a graphic file.

1. The command **print** sends the graphic that has been just created to a connected printer. The command **print** always refers to the current display; it is equivalent to the hot key <ALT P> (for questions concerning the administration of your printer refer to the XploRe user manual or the **setprinter** command).

2. The command **print** is also able to save a graphic to a file. You have to do the following:

   - Specify the file where the output should be saved:

     ```
     capture ("filename")
     ```

   - Tell XploRe that it has to print:

     ```
     print
     ```

   - Tell XploRe that nothing more should be printed to the file:

     ```
     capture ("off")
     ```

   An alternative is `capture("on")`; then you are prompted for the filename.

Let us make this clearer with an example. As in the previous section, we use the data file **uscomp2.dat**. The following program plots transformations of variables of the data set in a graphic and saves the plot in a PostScript[2] file **test.ps**:

```
proc() = main ()
  capture("test")
  x   = read("uscomp2")
  x2 = x[,4]./x[,1]    ; quotient : profit / assets
  x3 = x[,2]./x[,1]    ; quotient : sales  / assets
  zz5 = x2~x3
  show(zz5 s2d)
  print
  capture("off")
endp
```

The resulting graphic is given in Figure 3.5. Note that for better visibility the plotstyle was changed.

3.9   The Static 2-D Window

In this section, we will describe how two-dimensional (2-D) plots of one or more 2-D data sets are created and how their shape can be manipulated. In XploRe, it is possible to plot more than one 2-D data set in one picture, so you are able to compare different data sets or the data with the estimated function (see, for instance, Figure 2.4 in Section 2.2 of the tutorial).

As mentioned in Section 3.5, we create a static 2-D window with the command **show**:

```
show(x... s2d)
```

---

[2]PostScript is a trademark of Adobe Systems, Inc.

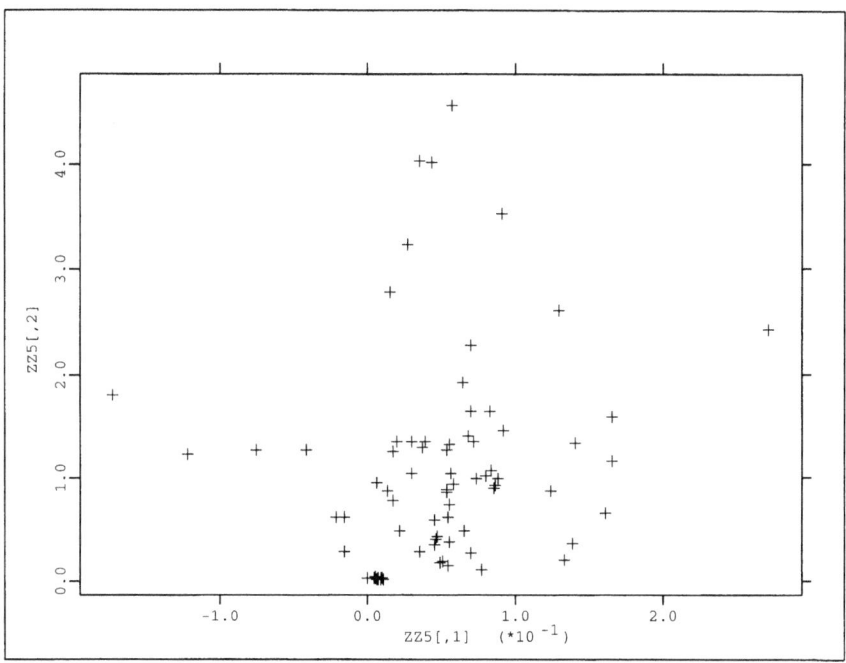

FIGURE 3.5. The picture saved in test.ps.

where x... denotes the name of a single matrix or a list of names of
matrices, separated by space. The matrices are assumed to contain one or
more pairs of data vectors, where the number of rows equals the number
of observations in each data vector, and the number of columns equals the
number of data vectors. Every pair of data vectors yields a point-cloud,
which is plotted in the s2d-display. Data of different matrices mentioned
in the list x... are plotted in different colors.

Let x be a matrix containing 4 data vectors of 30 observations, that is,
x is a 30×4-matrix. The command

```
show(x s2d)
```

yields a plot of two point-clouds in the same color; the first point-cloud is
built by the first and the second column of x, and the second by the third
and the forth one.

The two point-clouds are plotted in different colors, if you put the corresponding data into different matrices; v1 containing the first two columns
of x, v2 containing the second two columns of x, and both matrices included
in the parameter list:

```
v1 = x[,1:2]
v2 = x[,3:4]
show(v1 v2 s2d)
```

A static 2-D window appears, and the cursor is placed in it. Of course if you have chosen a multi-show, there will be some more windows, and you have to move the cursor to the appropriate s2d window using <F8> and <F9>.

In the upper right of the screen there is an icon list, where the active function keys are shown. They are intended to manipulate display or window parameters. Some of them act directly, others change the meaning of the function keys themself (like a sort of submenu system). We will now discuss them briefly.

- <F1> brings up a short help description of the function keys currently available.

- <F3> allows you to read the indices of data points by selecting them with the cursor.

- <F4> opens a function key menu which allows you to change several window parameters, for example, window title, form of axes, and so on.

- <F5> opens a function key menu which allows you to mark or brush special regions or to study them with a focusing lens by zooming in and out.

- <F8> moves the cursor to the next window (only in multiwindow mode).

- <F9> moves the cursor to the previous window (only in multiwindow mode).

- <F10> opens a function key menu, which allows you to change several display parameters like style or color of points, line style, and so on.

Although the icon list shows the function keys <F6> and <F7> as well, they are only meaningful in d3d mode.

It is often the case that the appearance of a picture can be improved by a change of line style of colors. In addition, it can be useful to brush the data or to identify outliers. In Section 2.2 of the tutorial, you find some examples that describe the wide range of uses of the function keys in s2d modus.

The system of function keys in s2d modus is listed in Table 3.7 for a better overview.

If you are inexperienced with XploRe, it may be more convenient for you to control the displayed data and the display style with the commands update and mask. The former enables you to add, delete, or replace a data vector in a s2d window, while the latter modifies color and point style. For this technique we refer to the XploRe manual and the Appendix.

<F1>	Show help screen.	
<F2>	Zoom (not implemented).	
<F3>	Show link data.	
<F4>	Change projection mode.	
	<F1>	Show help screen.
	<F2>	Headline on/off.
	<F3>	Axis on/off.
	<F4>	Edit headline and axis text.
	<F5>	Edit axis borders (Max., Min., Steps).
	<F6>	Edit rotation point (only **d3d** mode).
	<F10>	Return to ground level.
<F5>	Select brush operations.	
	<F1>	Show help screen.
	<F2>	Zoom in brush area.
	<F3>	Zoom out brush area.
	<F4>	Change size of brush.
	<F5>	Mark the brush area.
	<F6>	Select color for marking.
	<F7>	Select point style for marking.
	<F10>	Return to ground level.
<F6>	Rotate data (only in **d3d** mode).	
<F7>	Reset display (only in **d3d** mode).	
<F8>	Next window forward (only for multiple window).	
<F9>	Next window backward (only for multiple window).	
<F10>	Changing data or window parameter.	
	<F1>	Show help screen.
	<F2>	Change colors.
	<F3>	Change data point style.
	<F4>	Change connections.
	<F5>	Change line style.
	<F6>	Change line thickness.
	<F7>	Select data for linking.
	<Enter>	Select data part to change its attributes.
	<ESC>	Return to show icons.

TABLE 3.7. Function keys for the **s2d** window.

3.10   The Dynamic 3-D Window

In this section we will describe how three-dimensional (3-D) plots of 3-D data sets are created and manipulated. As explained in the last section about the **s2d** window it is also possible in the **d3d** window to plot more than one 3-D data set, each in either a different color or in the same color.

The dynamic 3-D window is opened by the following command:

```
show(x d3d)
```

where x... denotes the name of a single matrix or a list of names of matrices, separated by space. The matrices are assumed to contain one or more triples of data vectors, where the number of rows equals the number of observations in each data vector, and the number of columns equals the number of data vectors. Every triple of data vectors yields a point-cloud, which is plotted to the **d3d** display. Data of different matrices mentioned in the list x... are plotted in different colors.

Let x be a matrix containing six data vectors of 30 observations, that is, x is a 30×6 matrix. The command

```
show(x d3d)
```

yields a plot of two point-clouds in the same color; the first point-cloud is built by the first, second, and third column of x, and the second by the fourth, fifth, and sixth one.

As explained in Section 3.9 the two point-clouds are plotted in different colors if you put the corresponding data in different matrices; v1 containing the first three columns of x, v2 containing the second three columns of x, and both matrices included in the parameter list:

```
v1 = x[,1:3]
v2 = x[,4:6]
show(v1 v2 d3d)
```

After running the command **show**, the dynamic 3-D window of XploRe is opened and the point-cloud of the data is visible on the screen. All function keys of this level are active (see Section 3.9 for a detailed description of the function keys <F1> to <F5> and <F8> to <F10>). In addition, the function keys <F6> and <F7> are active.

- <F6> opens a function key menu, which allows you to rotate the point-clouds, to change the viewpoint, or to display projections onto 2-D planes.

- <F7> allows you to reset the display with the third axis orthogonal to the screen.

An overview of the function keys, their actions, and connections is given in Table 3.8. This table should make it easier to find the right combination of function keys for a special action in the **d3d** window that you want to carry out. Using these function keys, you can experiment with the data in many ways. The different possibilities are demonstrated in practical examples in Section 2.5 of the tutorial (see, for instance, Figure 2.16 on page 28).

<F1>	Show help screen.
<F2>	Zoom (not implemented).
<F3>	Show link data.
<F4>	Change projection mode.
	<F1> Show help screen.
	<F2> Head line on/off.
	<F3> Axis on/off.
	<F4> Edit head line and axis text.
	<F5> Edit axis borders (Max., Min., Steps).
	<F6> Edit rotation point .
	<F10> Return to groundlevel.
<F5>	Select brush operations.
	<F1> Show help screen.
	<F2> Zoom in brush area.
	<F3> Zoom out brush area.
	<F4> Change size of brush.
	<F5> Mark the brush area.
	<F6> Select color for marking.
	<F7> Select point style for marking.
	<F10> Return to groundlevel.
<F6>	Rotate data.
	<F1> Show help screen.
	<F2> Rotate data points.
	<F3> Move data points.
	<F4> Move view point.
	<F5> Press data point to one axis.
	<F6> Select projection 1.
	<F7> Select projection 2.
	<F8> Select projection 3.
	<F9> Select projection 4.
	<F10> Return to groundlevel.
<F7>	Reset display.
<F8>	Next window forward (only for multiple window).
<F9>	Next window backward (only for multiple window).
<F10>	Changing data or window parameter.
	<F1> Show help screen.
	<F2> Change colors.
	<F3> Change data point style.
	<F4> Change connections.
	<F5> Change line style.
	<F6> Change line thickness.
	<F7> Select data for linking.
	<Enter> Select data part to change its attributes.
	<ESC> Return to show icons.

TABLE 3.8. Function keys for the d3d window.

3.11   The Boxplot Window

The third way to display data in XploRe is to draw boxplots of data vectors. Boxplots are a well-known graphic tool that give an initial idea of the empirical distribution of an observed random variable. Boxplots are constructed in the following way (see, for example, Emerson and Strenio, 1983):

> In the center of every boxplot is a box with two lines (one solid, one dashed). The lower border of the box represents the 25%-quantile of the column vector. The middle solid line is the median of the column vector, and the dashed line the mean value of the vector. The upper border of the box represents the 75%-quantile.

> The horizontal solid line under the box, connected with the box by a solid vertical line, is the smallest value x of the vector complying with the condition:

> $x >$ 25%-quantile -1.5 (75%-quantile $-$ 25%-quantile).

> The horizontal solid line above the box is the largest value x of the vector complying with the condition:

> $x <$ 75%-quantile $+1.5$ (75%-quantile $-$ 25%-quantile).

> Every value of a vector outside of the box is shown as a point. These are the "outliers".

The following command creates a boxplot for every column of a matrix x:

```
show(x b1d)
```

In the **b1d** window, a subset of the function keys as described in the previous sections about **s2d** and **d3d** windows (see the corresponding tables in Sections 3.9 and 3.10).

The shape of boxplots is fixed, so most of the function keys have no meaning for the **b1d** window. However, some of them are very useful; for instance, you can discover "outliers" of the data vector with the function key <F3>.

The next example will give a clear idea of boxplots in XploRe. Two transformations of variables of the U.S. company data set (see the example in Section 3.8) are plotted as boxplots by the following macro:

```
proc() = main ()
  x=read("uscomp2")
  y=x[,4]./x[,3]     ;  profit / market value
  z=x[,2]./x[,1]     ;  sales  / assets
  zz=z~y
  show(zz b1d)
endp
```

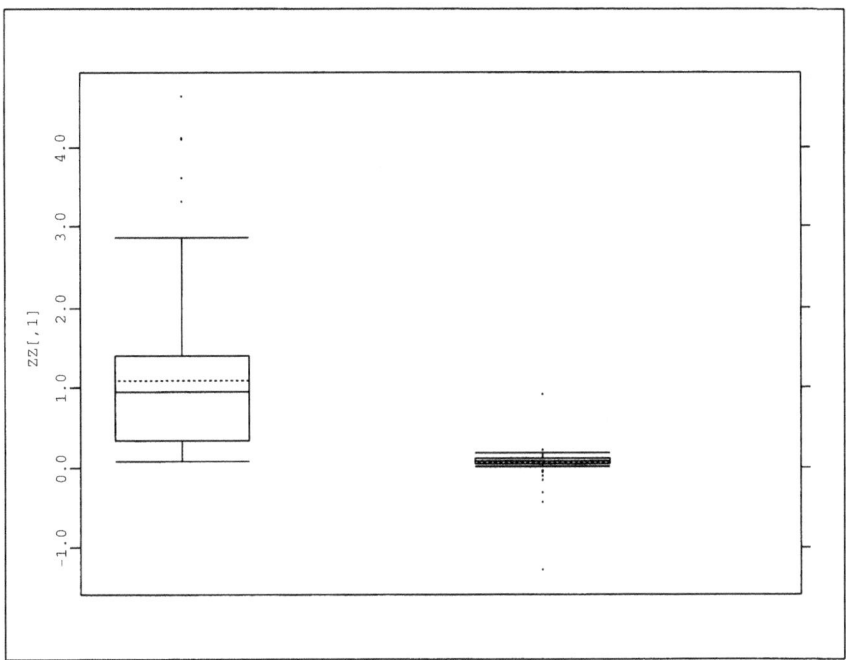

FIGURE 3.6. The boxplots of z (left) and y (right).

The program yields Figure 3.6, showing that the normalized profits are much more concentrated than the sales/asset variables.

3.12   The Flury Faces Window

Face plots are used to detect the relationship between random variables of higher dimensional data sets. Since one intuitively distinguishes human faces very sensitively, the idea of Chernoff (1973) is to represent each variable of a data set by a special part of the face. Thus, the data are analyzed by studying faces corresponding to data-points.

In XploRe the faces are constructed according to an advanced method which is explained in Flury and Riedwyl (1988, Chapter 4). A design of faces is used in which the two halves of the face can be varied separately; so each face is described by 36 parameters, 18 for the left side and 18 for the right side.

Table 3.9 shows the list of the face parameters that can be specified by the variables of the data set. Each face corresponds to one observation of the data set.

Now we explain the XploRe command for the creation of flury faces:

Face parameter	Right	Left
eye size	1	10
pupil size	2	20
position of pupil	3	21
eye slant	4	22
horizontal position of eye	5	23
vertical position of eye	6	24
curvature of eyebrow	7	25
density of eyebrow	8	26
horizontal position of eyebrow	9	27
vertical position of eyebrow	10	28
upper hair line	11	29
lower hair line	12	30
face line	13	31
darkness of hair	14	32
hair slant	15	33
nose line	16	34
size of mouth	17	35
curvature of mouth	18	36

TABLE 3.9. Face parameters.

- Suppose x is an $n \times p$ data matrix, where n is the number of observations, that is, the number of faces, and p is the number of variables.

  All vectors of x have to be normalized to a range of $[0, 1]$. Let z be the normalized x.

- The $k \times 1$ column vector ind assigns for each column vector of x the corresponding parameter of the face.

  For example, if ind[1,]=k, the kth column of x will determine the right eye size (see Table 3.9).

  If ind[j,]=0, then the size of the jth face parameter is set by default to 0.5. The vector ind does not need to have 36 elements. If the number k of rows in ind is less than 36, then the remaining face parameters are set to default value 0.5.

- xwin and ywin are parameters determining the number of faces to be plotted.

  xwin is the number of faces in a horizontal line, and ywin is the number of faces in a vertical line. The product xwin*ywin is the number of faces on the screen.

The following command yields the flury faces:

<F1>	Show this help screen.
<F8>	Next window backward.
<F9>	Next window forward.
<ESC>	Leave window.
<↑>	One line up.
<←>	One line up.
<↓>	One line down.
<→>	One line down.
<Pg Up>	One page up.
<Pg Dn>	One page down.
<HOME>	To first page.
<END>	To last page.

TABLE 3.10. Hot keys for the flury faces window.

```
show( z xwin ywin ind fcs)
```

After this command, the flury faces window is open. Table 3.10 presents the action of the function keys and other hot keys.

Finally, we give an example. The program below yields flury faces for 16 company data vectors of the data set of 79 U.S. companies (see Section 3.8).

```
proc()=main ()
  capture ("on")
  x = read("uscomp2")
  x = x[4:11,]|x[64:67,]|x[69:72,]
  ; x[4:11,] are Energy-Companies
  ; x[64:67,] and x[69:72,] are Retail-Companies
  ;******Transformation of the variables on a range of [0,1]***
  cc=max(x)-min(x)
  erg=x-min(x)'
  z=erg./cc'
  ;******End of transformation of the variables*****************
  ;******Definition of the index-vector************************
  m=matrix(18 1 0)
  m[1,1]=5  ; RIGHT EYE SIZE              - cash flow    = x[,5]
  m[6,1]=3  ; VERTICAL POS. OF RIGHT EYE - market value = x[,3]
  m[7,1]=1  ; CURVATURE OF RIGHT EYEBROW - assets        = x[,1]
  m[13,1]=4 ; RIGHT FACE LINE            - profit        = x[,4]
  m[14,1]=6 ; DARKNESS OF RIGHT HAIR     - employees     = x[,6]
  m[18,1]=2 ; RIGHT CURVATURE OF MOUTH   - sales         = x[,2]
  m=m|m     ; the faces will be symmetric
  ;******End of definition of index-vector*********************
  xwin=4
  ywin=4
  show(z xwin ywin m fcs)
endp
```

FIGURE 3.7. The flury faces for 16 U.S. companies.

As you can see in the program, six face parts represent the six variables of the company data. For instance, the face line corresponds to `profit` and the curvature of mouth corresponds to `sales`, so the fat smiling faces are the companies with high profit and many sales. The resulting graphic is given in Figure 3.7.

3.13   How to Use and Create Libraries

With the XploRe basic package, you are equipped with all sorts of commands. In addition, you may use special macros, mostly related to regression estimation or multivariate statistical analysis.

For practical reasons and for a better survey, the macros are classified in libraries. The following libraries are available in XploRe:

ADDMOD	library for flexible additive modeling
COMPLEX	library for operations with complex numbers and for complex mathematical functions
GAM	library for *generalized additive models*
GLM	library for *generalized linear models*

Macros for Regression Estimation	
KNN	symmetrized kNN smoother
MONREG	isotonic regression
REGEST	kernel WARPing smoother
REGSSM	supersmoother
RUNMED	running median
LOWE	LOWESS
REGAUTO	Automatic kernel smoothing
REGLNBC	Linton and Nielsen's bias correction method
REGMHFH	computes m and f
REGUNCB	computes m and uniform confidence band
SKERREG	direct kernel regression
SKERDENS	direct kernel density
Macros for Wavelet Estimation	
WGEN	generation of father and mother wavelet
WAVEREG	wavelet regression
WAVEDENS	wavelet density

TABLE 3.11. Some macros of the library SMOOTHER.

HIGHDIM	library of graphical methods for data sets of more than two dimensions
SMOOTHER	library for kernel estimation and smoothing methods
TEACHWAR	library for examples teaching basic statistics with XploRe
XCLUST	library for clustering methods
XPLORE	standard library of XploRe

If you intend to use a macro of a special library, you must load this library before calling the macro by the command:

```
library("name of the XploRe-library")
```

In Table 3.11 some macros are listed that one can work with after the library **smoother** has been loaded. This library provides smoothing methods for estimating regression functions or density functions. Examples of the use of this library are given in Section 2.3 of the tutorial.

User-written XploRe code macros can be added to the program package XploRe. Examples, describing the way in which macros are made in XploRe, are given in Section 2.3 of the tutorial.

The simplest way to include your own macros into the XploRe environment is to save the self-made macro in a file under its name with extension *.xpl and to place it in the subdirectory \XPLORE3\LIB. Later, you can use it in the same way as the original XploRe commands.

An XploRe library and its corresponding macros are saved in the directory \XPLORE3\LIB. For every library there is a corresponding file named after the library with the extension .lib. This file contains the names of all macros that belong to this library.

Let us see how the XploRe library XCLUST has been constructed. This library contains macros for clustering methods.

In directory \XPLORE3\LIB we find the file xclust.lib. The contents of xclust.lib are:

```
; *****************************************************************
; ********* the XploRe  XClust   library  *************************
; *****************************************************************
; ************ date : 941011 **************************************
; *****************************************************************
Adaptive          ; Adaptive K-means cluster analysis
KMCont            ; K-means clustering of contingency tables
WardCont          ; Ward's hierarchical clustering of contingencies
MaxCont           ; Maximum correspondences in contingency tables
PSwap             ; Change of categories of a variable (partition)
Cor2Dist          ; Transform of correlations into distances
Mat2Vec           ; Store a symmetric matrix into a vector
Measure           ; Evaluate measures of correspondence of tables
ReCode            ; Collapse of categories of variables
Conting           ; Contingency table by crossing two variables
LpDist            ; Evaluate Lp-distances
Vec2Mat           ; Store a vector into a symmetric matrix
Divisive          ; Adaptive divisive K-means cluster analysis
```

These thirteen macros belong to the library xclust, and they are all saved in \XPLORE3\LIB.

In the same way, user-written libraries can be created. Suppose there are macros A.xpl, B.xpl,..., G.xpl which are to be put into a library XY. The following must be done:

- Save the macros in directory \XPLORE3\LIB.

- Create a file XY.lib in the same directory.

- Edit the file XY.lib and write down the names of the macros A, B,..., G without extensions.

Now you can load the library XY via the command library("XY") and work with your macros.

REFERENCES

Chernoff, H. (1973). Using faces to represent points in k-dimensional space graphically, *Journal of the American Statistical Association* **68**(342): 361–368.

Emerson, J.D. and Strenio, J. (1983). Boxplots and batch comparison, *in* D.C. Hoaglin, F. Mosteller and J.W. Tukey (eds), *Understanding*

Robust and Explanatory Data Analysis, John Wiley & Sons, New York, chapter 3, pp. 58–96.

Flury, B. and Riedwyl, H. (1988). *Multivariate Statistics, A Practical Approach*, Cambridge University Press, New York.

Härdle, W. and Simar, L. (1994). Applied multivariate statistical analysis. Lecture script.

Part II

XploRe in Use

4
Graphical Aids for Statistical Data Analysis

Sigbert Klinke[1] and Christian Ritter[2]

4.1 Introduction

Graphical methods are best explained by working through examples. In this chapter we shall conduct a preliminary analysis of a data set on neonatal mortality. Data on the birth of 3331 children were registered from 1990 to 1992 at Clinique Saint Luc, Brussels, Belgium (Ritter and Bouckaert, 1993). Of these newborns, 56 died in utero, at birth, or within the first seven days after birth. Table 4.1 shows an excerpt of the data matrix.

Obs.	Death	Weight	Gest	Sex	Mult	ngro	apg3
1435	0	3130	39	1	1	2	9
1436	0	3880	41	1	1	3	9
1437	0	2680	35	2	1	2	9
1438	1	950	27	1	3	3	0
1439	0	960	27	1	3	3	6
1440	1	750	27	1	3	3	0
1441	0	3660	39	2	1	3	9
1442	0	4160	41	1	1	2	9
1443	0	3460	39	2	1	2	9
1444	0	3520	38	1	1	1	9
1445	0	1050	28	2	1	6	7
1446	0	2650	37	2	1	3	9
1447	0	3000	37	2	1	4	9
1448	0	3270	39	2	1	3	9

TABLE 4.1. Excerpt from the raw data file.

---

[1]Institut de Statistique, Université Catholique de Louvain, B-1348 Louvain-la-Neuve, Belgium; and Institut für Statistik und Ökonometrie, Humboldt-Universität zu Berlin, D-10178 Berlin, Germany.

[2]CORE and Institut de Statistique, Université Catholique de Louvain, B-1348 Louvain-la-Neuve, Belgium.

In our analysis we shall use the response death (0 for survival, 1 for dead; *mort*) and the explanatory variables birthweight (*weight*), gestational age (length of pregnancy; *gest*), sex (*sex*), multiplicity (1 for singleton, 2 for twin, etc.; *mult*), and the number of pregnancies (*ngro*).

The following XploRe code reads data from the raw input file `birth.dat` and writes them in XploRe format to the file `birthx.dat`. This is an important first operation, for reading raw data can take up a considerable amount of time whereas restoring a data set from a file in XploRe format is very fast. Retrieving it as an XploRe object requires much less time:

```
x = read ("birth")      ; reading raw data
write (x "birthx" x)    ; storing as XploRe format
```

Once the data are available in XploRe format, they can be readily retrieved. The following macro restores the data from the file `birthx.dat` in XploRe format and constructs named data vectors for later use:

```
proc (x weight gest sex mult ngro mort) = main ()
   x       = read ("birthx")       ; reads data as xplore format
   mort    = x[,1]                 ; create named data vectors
   weight  = x[,2]
   gest    = x[,3]
   sex     = x[,4]
   mult    = x[,5]
   ngro    = x[,6]
endp
```

We have created a `main` program in XploRe. As we know from the tutorial in Part I it is treated differently: we can run it from the editor with `<F4>`. Moreover if we define output-parameters, for example, x, `weight`, and so on, they will not get lost after the execution of the program, but remain as global variables. Thus we can use them in our later programs.

4.2   First Pictures

It is convenient to display the data in the variables *weight* and *gest* since they are the only variables that cover wide ranges of values. The other variables can be used to label points and to stratify the analysis. The following code lines are used to create Figure 4.1:

```
msk = vtocc(90-45.*mort) ; mark mort=1 as magenta + (45)
                         ; and mark mort=0 as green O (90)
pic = gest~weight~msk
show (pic s2d)
```

The command `vtocc` is used to create the necessary mask vector. Internally, a magenta plus is coded with 45 and the green O as 90. The graph

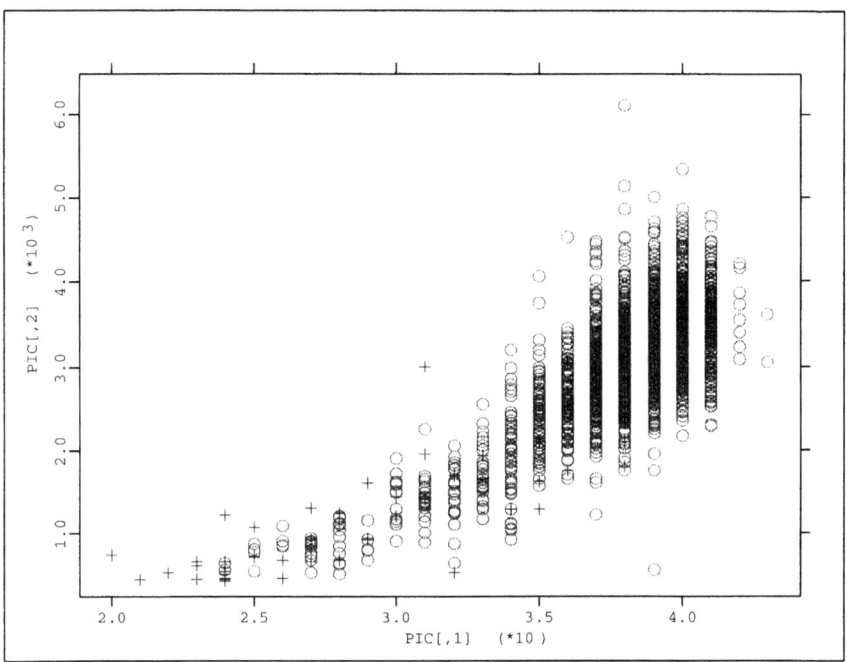

FIGURE 4.1. Birth weight and gestational age for all births. Survivals are noted by green circles, and deaths by magenta crosses (see color insert).

shows the weights and gestational ages of all births. Deaths are labeled by magenta crosses. Since gestational age is given in weeks, there is a substantial amount of overprinting of points, and therefore it is difficult to judge mortality rates. Nevertheless, one can see that mortality is highest for early births.

Figure 4.2 shows the data when uniform noise between −0.5 and 0.5 is added to the gestational age. This breaks the ties, eliminates overprinting, and makes it easier to assess the local mortality rates in different parts of the data range:

```
gestj = gest+uniform(rows(gest))-0.5
pic   = gestj~weight~msk
show (pic s2d)
```

Another problem that hampers our analysis is that the observations are concentrated in the upper right. We can try to improve the plot by using a sunflower plot (Chambers, Cleveland, Kleiner and Tukey, 1983). Here, the number of observations that fall in a specified bin (here, 0.5 weeks and 250 gram) is shown by the different symbols. A point means that few observations fall in this bin, whereas a star represents a high number of observations. The selected symbol in the sunflower plot depends on the bin

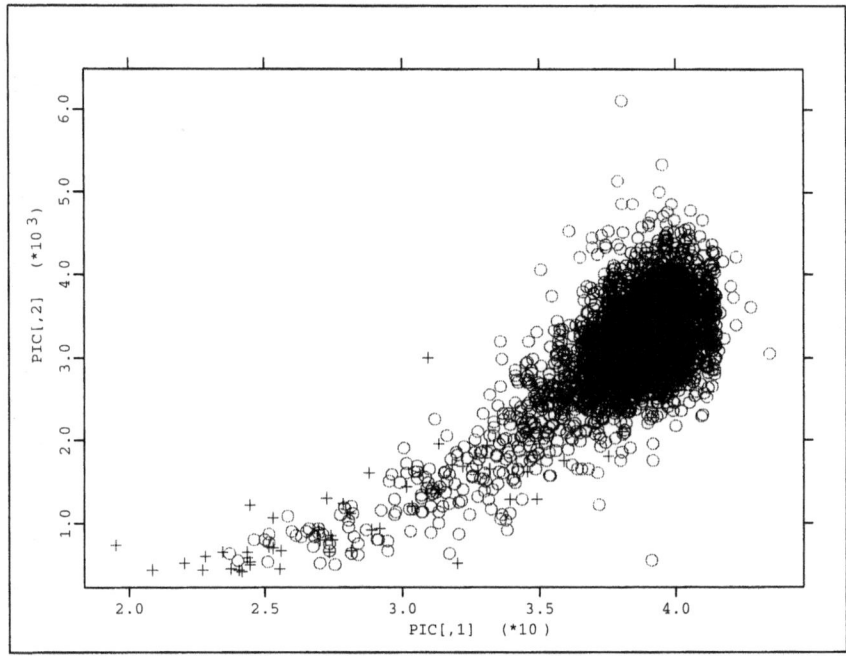

FIGURE 4.2. Birth weight and dithered gestational age for all births. Survivals are noted by green circles, and deaths by magenta crosses.

with the highest number of observations. To do this in XploRe, we load the HIGHDIM-library and call the sunflow-macro. Figure 4.3 shows the result.

```
library ("highdim")
sunflow (gestj~weight #(0.5 250))
```

Note that a title and axis labeling have been added. This was done using the interactive facilities of the plotting environment.

Another method of visualizing the density of points is by displaying 2-D kernel density estimates. To do this, we load the SMOOTHER-library and call the DENEST2-macro. Figure 4.4 shows the result. The keystrokes <F10>, <F4>, and <ESC> are used to get combined lines and <F6>, <F2>, and the cursor keys are used to rotate.

```
library ("smoother")
fh = denest2 (gest~weight #(2 500) #(1 100))
show (fh d3d)
```

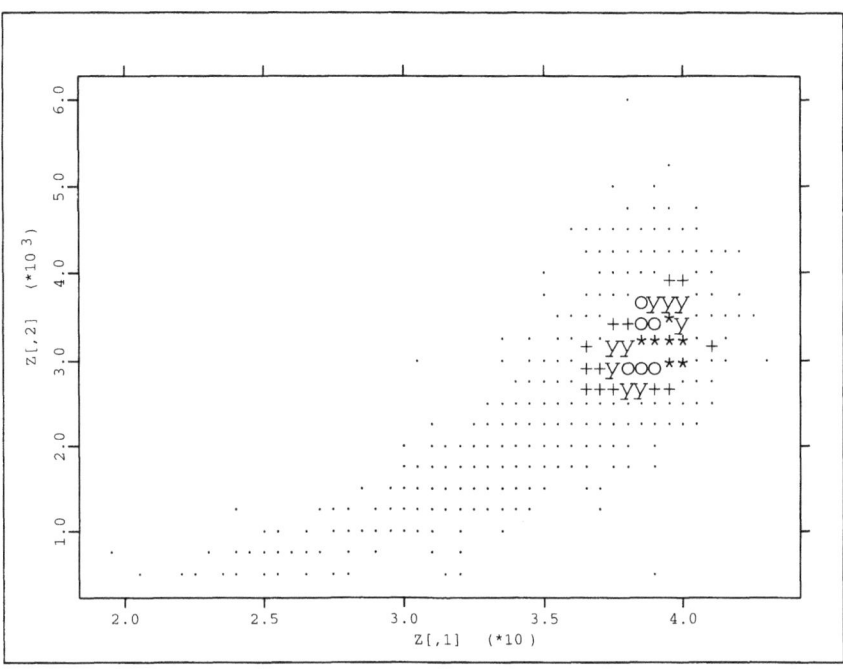

FIGURE 4.3. Sunflower plot of birth weight and dithered gestational age for all births.

4.3   Stratifications

Now we study whether stratification of the data is necessary. At first we separate the observations into singleton and multiple births. In Figure 4.5, created by the macro below, we mark each observation by a number expressing birth multiplicity, that is, 1 for singletons, 2 for twins, etc. We see that most of the higher order births occur after shorter than normal pregnancies, and that these newborns are somewhat lighter than singletons. We conclude that singletons and multiplets should be treated separately in further analyses. In fact, very few births belong to triplets or beyond, and even the number of twins is small. Therefore, we can study singletons in detail and check whether our findings also apply to higher order births.

```
proc()=main()
  data  = gestj~weight
  data1 = paf (data mult.=1)   ; singleton
  data2 = paf (data mult.=2)   ; twins
  data3 = paf (data mult.>2)   ; higher order multiplets
  mask1 = mask(rows(data1) 1 yellow star)   ; create
  mask2 = mask(rows(data2) 1 lgreen x)      ; corresponding
  mask3 = mask(rows(data3) 1 magenta plus)  ; marks
```

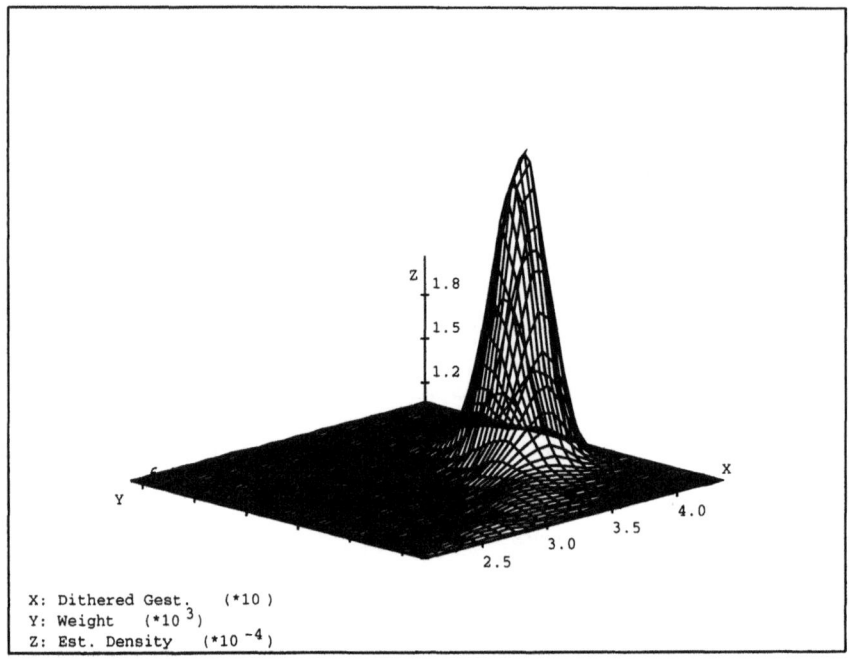

FIGURE 4.4. Kernel density estimation of birth weight and dithered gestational age for all births. Selected bandwidths are 2 weeks and 500 grams.

```
    show((data1|data2|data3)~(mask1|mask2|mask3) s2d)
endp
```

Another natural division of the data is into male and female newborns. However, before splitting the data into these two subsets, and thereby effectively halving our sample size, we should check whether we can see differences that would warrant this split. Figure 4.6 shows male newborns marked by red x's and female newborns marked by blue o's. Here, we use a faster method to color the datapoints than before.

```
mask1 = vtocc(65.*sex-31)
data  = gestj~weight~mask1
show (data s2d)
```

We see weak evidence that male newborns are on average heavier than female newborns for equal gestational ages, but the result is far from conclusive. A closer study of the weight distributions of males and females is needed. For example, empirical bivariate densities of births in the variables *weight* and *gest* can be computed for males and females separately. With these density estimates, we can construct contour plots which, when overlaid, provide a convenient display for this comparison.

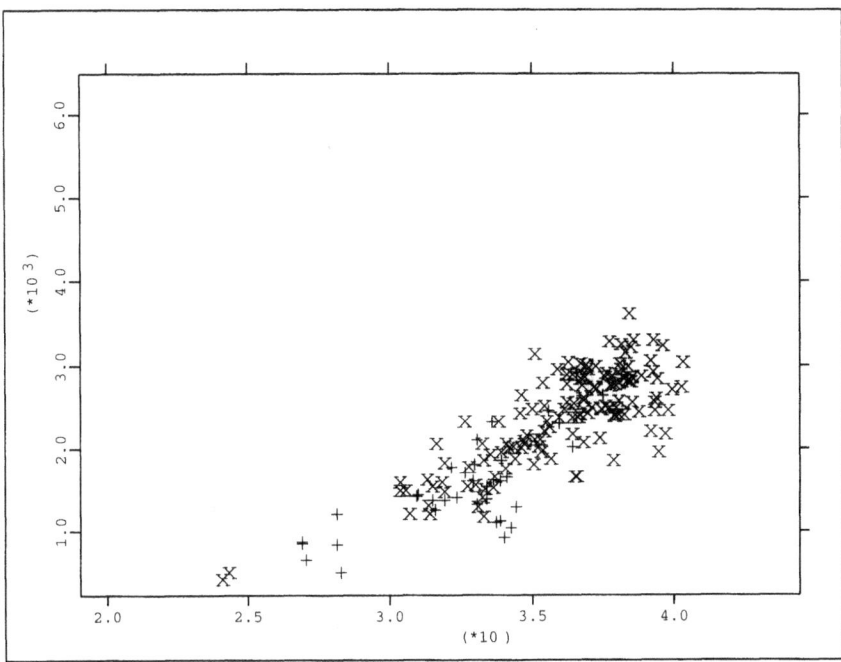

FIGURE 4.5. Births marked by their multiplicity. Singletons are marked by yellow stars, twins by green x's, and higher order births by magenta crosses (see color insert).

The next XploRe macro, which creates Figure 4.7, is longer and more complicated than the previous ones. The complications occur since both colors and line styles are manipulated. Colors are properties of data objects and are therefore attached to the data using masking vectors. On the other hand, line styles are properties of the picture and they have to be modified using the update function. Note how the construction of the display proceeds. First, the densities are estimated. Then the resulting matrix is converted into the three column format (x_i, y_i, z_i) required by the contouring routine. Here, x_i and y_i stand for scaled versions of *weight* and *gest*, while z_i captures the density estimate at the point (x_i, y_i). Using this representation, we compute the desired contours and supplement them by mask vectors, indicating the desired colors. The plot is shown in two stages. First, a pre-plot is constructed using a show command and two update commands. Without the command **writecon(27)**, the user would have to type <ESC> after **show** to continue. Then a sequence of **update**s adjust the line style for each contour. We wish to distinguish between contours for males and females, and we do so by choosing different colors. In addition, we reset all line styles to solid. Note that the two update commands are also used to set the axis labels.

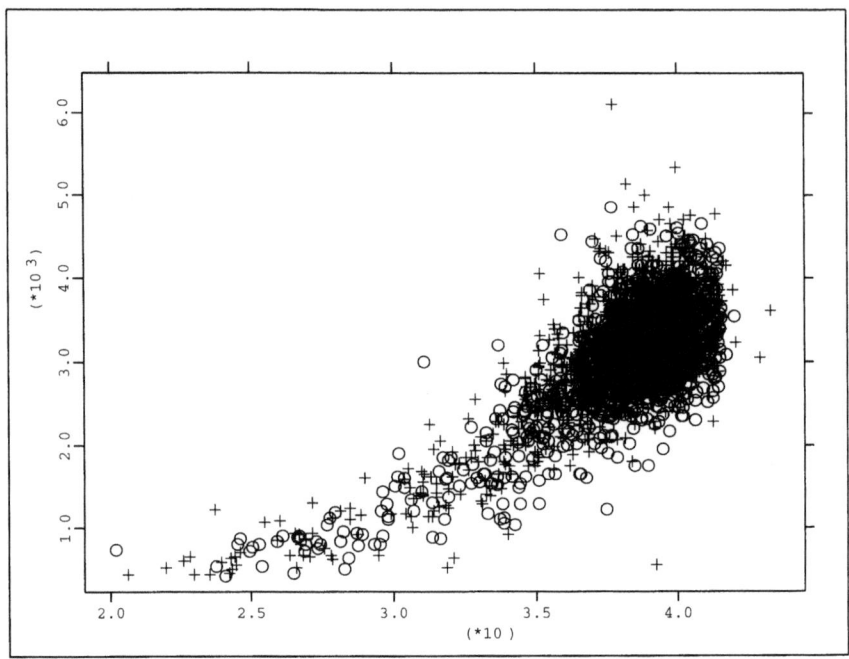

FIGURE 4.6. Male singletons are marked by red +'s, and female singletons are marked by light blue o's.

Figure 4.7 shows clearly that there is a difference of a few hundred grams
between the birthweight distributions of males and females. This finding
encourages us to treat males and females separately when analyzing pat-
terns linking birthweight, gestational age, and mortality.

```
proc()=main()
  library ("smoother")
  data    = gest~weight~sex
  data    = paf (data mult.=1)          ; take only singletons
  males   = paf (data data[,3].=1)      ; take only male
  males   = males[,1:2]                 ; forget now sex
  ym      = denest2(males 3|200 1|100)  ; density estimate for
                                        ; newborn male
  yym     = split(ym (-3))              ; resplit for contour
                                        ; lines
  x1m     = min(males[,1])+(1:cols(ym)/3)-1 ; calculate x-grid
  x2m     = min(males[,2])+100*(1:rows(ym))
  cl      = (1e-5.*#(0.5 1 2 4 8))~mask(5 1 point lblue)
  cm      = contour2(x1m x2m yym[,3] cl)  ; calculate 5 contour
                                          ; lines
  females = paf (data data[,3].=2)      ; do the same with
  females = females[,1:2]               ; females
```

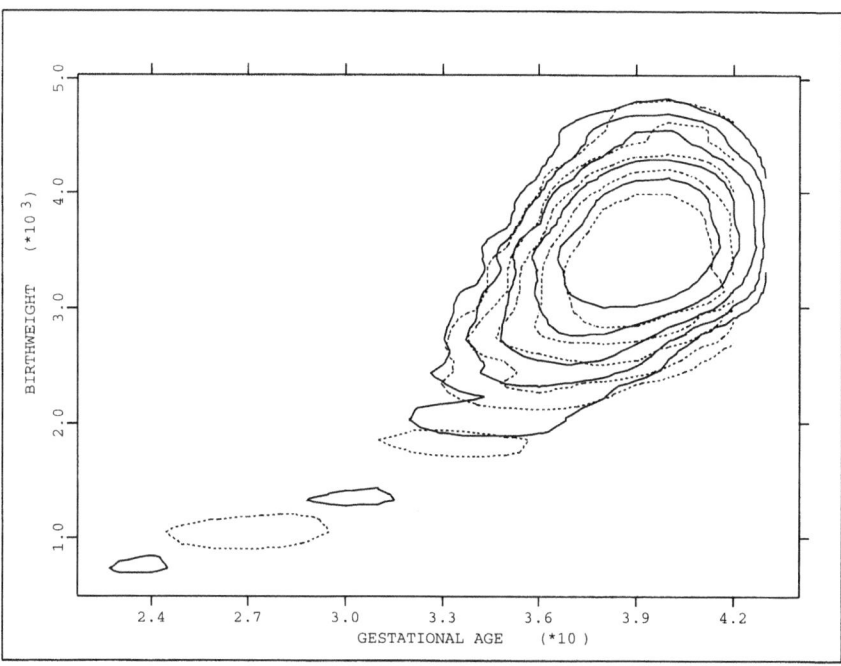

FIGURE 4.7. Density contours for male (blue solid) and female (red dotted) for singletons (see color insert).

```
yf        = denest2(females 3|200 1|100)
yyf       = split(yf (-3))
x1f       = min(females[,1])+(1:cols(yf)/3)-1
x2f       = min(females[,2])+100*(1:rows(yf))
cl        = (1e-5.*#(0.5 1 2 4 8))~mask(5 1 point lred)
cf        = contour2(x1f x2f yyf[,3] cl)
writecon(27)                    ; flash through plot and
show (cm cf s2d)                ; draw lines instead of
                                ; points
dum = update(cm 1 line solid  xaxis "gestational age")
dum = update(cf 2 line dotted yaxis "birthweight")
show (cm cf s2d)                ; show again
endp
```

Now let us include a third variable, the number of pregnancies. Of course, this is a discrete variable and has been tabulated in Table 4.2.

We now construct a 3-D density plot as described, for example, in Scott (1992). For this we must ensure that we have enough observations in each level of the third variable; otherwise, we just try to interpret boundary effects. This leads us to omit all singleton births of mothers who had six or more births.

No. of pregnancy	Male	Female	Total
1	487	573	1060
2	502	444	946
3	301	270	571
4	176	150	326
5	69	70	139
6	34	30	64
7	21	13	34
8	4	7	11
9 or more	7	7	14
Sum	1601	1564	3166

TABLE 4.2. Number of pregnancy of the mother for singletons.

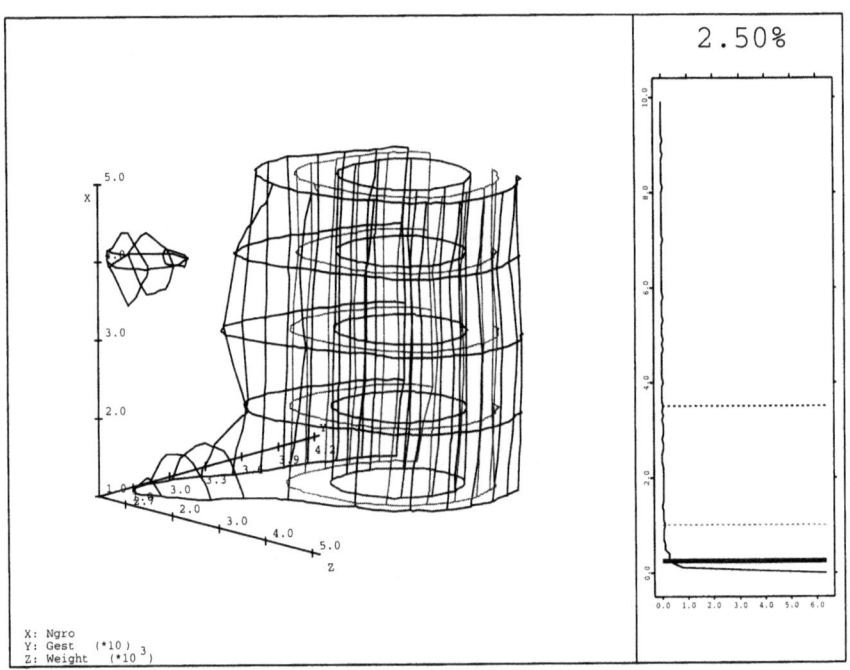

FIGURE 4.8. Density contours for male singletons (see color insert).

Figures 4.8 and 4.9 show 3-D density estimates of births using *gest*, *weight*, and *ngro* for male and female newborns.

The density plots show little or no effect of the variable *ngro* labeled as **X** on the contours. This suggests that the number of pregnancies is not related to gestational age or weight.

Figures 4.8 and 4.9 were created by the following macro:

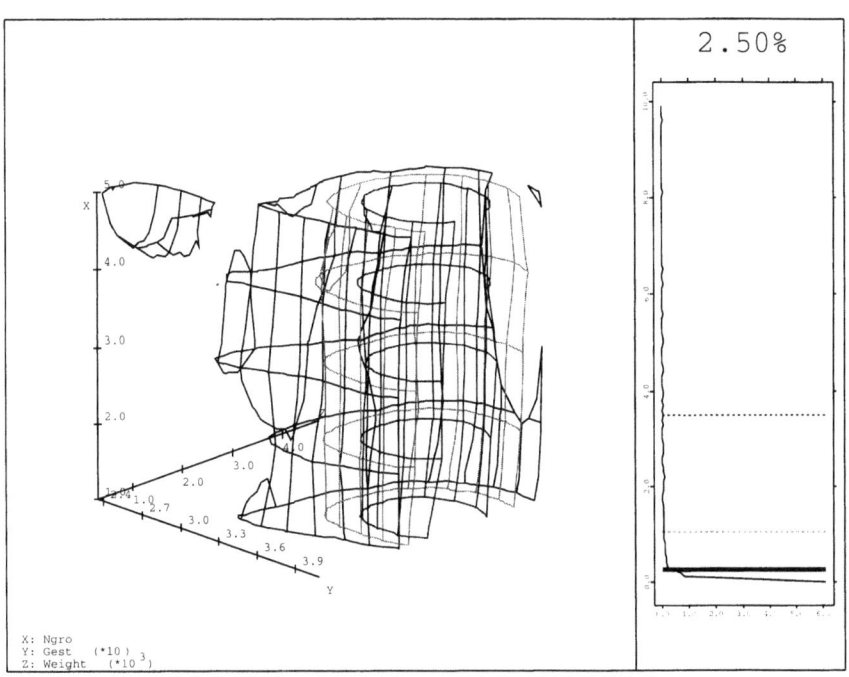

FIGURE 4.9. Density contours for female singletons (see color insert).

```
proc(y)=quac(x)                         ; use a special kernel
  y = qua(x[,2:3]).*(x[,1].=0)
endp
;
proc()=main()
  LIBRARY("smoother", "highdim")
  data = ngro~gest~weight~sex
  data = paf (data mult.=1)             ; take only singletons
  data = paf (data data[,1].<6)         ; take only mother with less
                                        ; than 6 births before
  males = paf (data data[,4].=1)        ; take only male
  males = males[,1:3]                   ; forget now sex
  nm = #(487 502 301 176 69)./100       ; take values of the table
  n  = #(5 15 15)                             ; estimate density with
  d  = (MAX(males)-MIN(males))./(n-1)   ; WARPing algorithm
  nh = #(1 3 3)
  h  = nh.*d
  (xb yb) = BINDATA(males d)
  wx = #(0 0 0)
  wy = SYMWEIGH(wx d./h nh &quac)
  wx = GRID(wx 1 nh)
  (xc yc or) = CONV(xb yb wx wy)
```

```
  s  = (xc.*d')~(yc./(ROWS(x)*PROD(d)))
  s[,4]=s[,4]./index(um xc[,1])     ; make densities comparable
  CONTOUR(s)                        ; make contour lines
;
  females = paf (data data[,4].=2) ; again for females
  females = females[,1:3]
  nf = #(573 444 270 150 70)./100
  n  = #(5 15 15)                          ; estimate density with
  d  = (MAX(males)-MIN(males))./(n-1)  ; WARPing algorithm
  nh = #(1 3 3)
  h  = nh.*d
  (xb yb) = BINDATA(males d)
  wx = #(0 0 0)
  wy = SYMWEIGH(wx d./h nh &quac)
  wx = GRID(wx 1 nh)
  (xc yc or) = CONV(xb yb wx wy)
  s  = (xc.*d')~(yc./(ROWS(x)*PROD(d)))
  s[,4]=s[,4]./index(nf xc[,1])     ; make densities comparable
  CONTOUR(s)                        ; make contour lines
endp
```

The macro begins by defining a special kernel quac for our density estimate. The kernel is a "two-dimensional" kernel, which means it smooths
only in the y- and z-directions. In the x-direction, we have no smoothing,
since *ngro* is a discrete variable. Then the library SMOOTHER, which contains the smoothing procedures, and the library HIGHDIM, which contains
the contour plots, are loaded. In the next step, all nonsingleton births and
all births where the mother has at least six births are excluded. In the
main section, the male newborns are selected first and the kernel density
estimate is computed. Up to now, XploRe does not contain a macro that
creates 3-D density estimates. Therefore, it has to be custom designed.
Here, the DENEST- and DENEST2-macro provide what is necessary. Before
plotting, we make the density estimates comparable by dividing each slice
in x by the corresponding number of observations. Finally, we call the contour plot routine. Interactive facilities of this routine, like changing the
intersection values (via Cursor keys, <Pg Up> and <Pg Dn>), or entering in
the picture (via <ENTER>) and changing the view, have to be used to get
the figures. According to Stuetzle (1987), two windows are used to display
all information. The same is done for the female newborns.

REFERENCES

Chambers, J.M., Cleveland, W.S., Kleiner, B. and Tukey, P.A. (1983). *Graphical Methods for Data Analysis*, Duxbury Press, Boston.

Ritter, C. and Bouckaert, A. (1993). Modeling the morbidity of newborns. Submitted for publication. Available as manuscript from authors.

Scott, D.W. (1992). *Multivariate Density Estimation: Theory, Practice, and Visualization*, John Wiley & Sons, New York, Chichester.

Stuetzle, W. (1987). Plot windows, *Journal of the American Statistical Association* **82**(398): 466–475.

5
Density and Regression Smoothing

Jianqing Fan[1] and Marlene Müller[2]

5.1 Introduction

A useful tool for examining the overall structure of data is *kernel density estimation*. It provides a graphical device for understanding the overall pattern of the data structure. This includes symmetry and the number and locations of modes and valleys. The basic idea is to redistribute the point mass at each datum point by a smoothed density centered at the datum point. An important question is how much the point mass should be smoothed out. This will be discussed in the next section. More detailed discussions on this subject can be found in Chapter 6.

Important graphical tools for understanding the association between a covariate and a response are *nonparametric regression* techniques. These techniques estimate the underlying regression function without any restrictive parametric form. This feature is particularly appealing when one analyzes a data set without much prior knowledge. Thus, nonparametric regression techniques could assist us in choosing a parametric model (if needed) by visually inspecting the estimated curve. They can also be useful as diagnostic tools by comparing the difference between parametric and nonparametric fits. Of course, the nonparametric estimates themselves also describe the association between the independent and dependent variables. The results can be used directly to understand and interpret the relationship between the independent and dependent variables.

Section 5.2 discusses various issues concerning density estimation and its applications. Nonparametric regression based on the Nadaraya–Watson estimator is depicted in Section 5.3. In Sections 5.4 and 5.5, we summarize the nonparametric regression based on the local polynomial fit. Finally, in Section 5.6, we describe how to use variable smoothing to enhance spatial adaptation.

---

[1]Department of Statistics, University of North Carolina, Chapel Hill, NC 27599-3260, USA.

[2]Institut für Statistik und Ökonometrie, Humboldt-Universität zu Berlin, D-10178 Berlin, Germany.

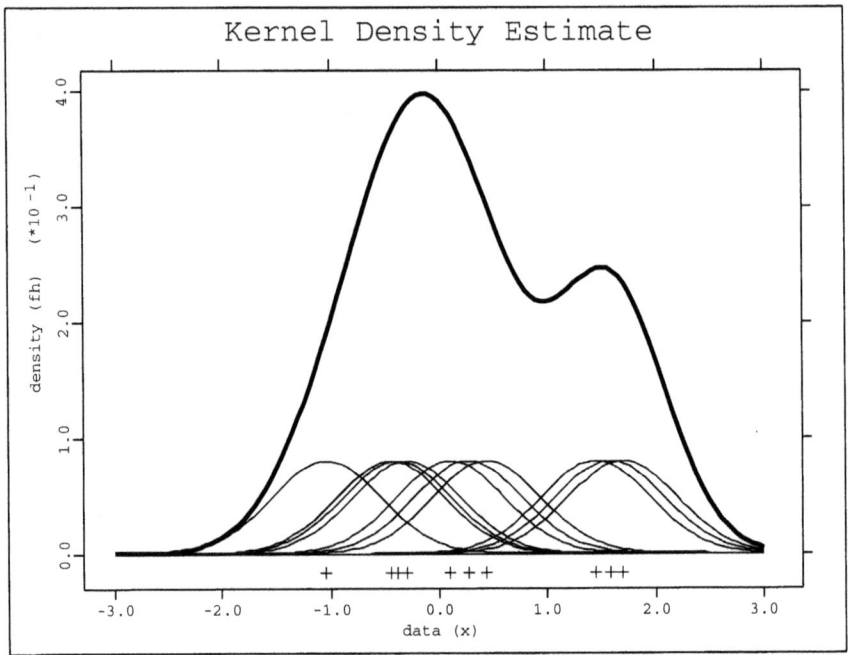

FIGURE 5.1. Kernel density estimate for a hypothetical data set. It smoothly redistributes the point mass at X_i by the function $n^{-1}K_h(x - X_i)$. The small bumps show how point masses are redistributed (see color insert).

5.2   Density Estimation

Consider a set of data $\{X_1, \ldots, X_n\}$ that can be thought of as a random sample from a population with a probability density $f(x)$. Useful structure of the data set can be found by attempting to estimate $f(x)$. The empirical cumulative distribution function $\hat{F}_n(x)$ is obtained by putting $1/n$ mass at each datum point. However, the data structure can hardly be examined by the plot of the function $\hat{F}_n(x)$. A better visualization device is to attempt to plot its density function $\hat{F}'_n(x)$, but this function does not exist. A better idea is to *smoothly* distribute the $1/n$ mass at each datum point to its vicinity (see Figure 5.1).

This is usually done via a symmetric probability density function K and a nonnegative parameter h. The function K is called a *kernel function*, controlling the shape of the mass redistribution. The parameter h is called a *bandwidth* or a *smoothing parameter*, governing the spread of the mass redistribution. Let $K_h(t) = K(t/h)/h$ be a rescaling function of K. The kernel density estimator is defined by

$$\hat{f}_h(x) = n^{-1} \sum_{i=1}^{n} K_h(x - X_i). \qquad (5.1)$$

Useful books describing the applications of density estimation include Silverman (1986) and Scott (1992).

The most commonly used kernel function is the Gaussian density function given by

$$K(t) = \frac{1}{\sqrt{2\pi}} \exp(-t^2/2). \qquad (5.2)$$

Other popular kernel functions include the Epanechnikov kernel

$$K(t) = \frac{3}{4}(1 - t^2)_+ \qquad (5.3)$$

and the Quartic (Biweight) kernel function

$$K(t) = \frac{15}{16}(1 - t^2)_+^2, \qquad (5.4)$$

where + denotes the positive part so that the support of K is $[-1, 1]$. It is well known that the estimate $\hat{f}_h(x)$ is not very sensitive to the choice of K, scaled in a canonical form as discussed by Marron and Nolan (1988). The default kernel function in XploRe is the Quartic kernel. Experienced XploRe users are encouraged to modify the XploRe macros DENAUTO and DENEST to explore the effect of using different kernel functions on their own.

However, the quality of the estimate is sensitive to the bandwidth choice. A bandwidth that is too small results in a rough estimate, producing artificial modes, while a bandwidth that is too large gives an oversmooth estimate, obscuring the finer structure of the data set (see Figure 5.3, for example). Indeed, it can be shown that under some mild conditions

$$\mathrm{E}\hat{f}_h(x) - f(x) = \frac{f''(x)}{2} \mu_2(K)h^2 + o(h^2) \qquad (5.5)$$

and

$$\mathrm{Var}(\hat{f}_h(x)) = \frac{R(K)f(x)}{nh} (1 + o(1)), \qquad (5.6)$$

where $\mu_2(K) = \int t^2 K(t)\, dt$ and $R(K) = \int K^2(t)\, dt$. Thus, as evidenced by (5.5) and (5.6), a large bandwidth h results in a large bias, while a small bandwidth produces an estimate with a large variance. A good choice of bandwidth would balance the bias-variance trade-off. This is conveniently assessed by the *Asymptotic Mean Integrated Squared Error* which is defined as

$$\mathrm{AMISE}(h) = \frac{\mu_2^2(K)h^4}{4} \int \{f''(x)\}^2\, dx + \frac{R(K)}{nh}. \qquad (5.7)$$

Minimizing (5.7) with respect to h gives the ideal (asymptotically optimal) bandwidth:

$$h_I = \left(\frac{R(K)}{\mu_2^2(K) \int \{f''(x)\}^2 \, dx} \right)^{1/5} n^{-1/5}. \tag{5.8}$$

In particular, when f is the Gaussian density function with variance σ^2, equation (5.8) becomes

$$h_I = \left(\frac{8\sqrt{\pi} R(K)}{3\mu_2^2(K)} \right)^{1/5} \sigma n^{-1/5}. \tag{5.9}$$

Various bandwidth selection rules have been proposed in the literature and a selection of them is available in XploRe, presented in Chapter 6. Therefore we focus here only on Silverman's rule of thumb, introduced in Silverman (1986). His choice of bandwidth is to replace σ by the sample standard deviation s_n. Specifically, for the popular kernel functions (5.2) to (5.4), we have

$$\hat{h}_I = \begin{cases} 1.06 \, s_n \, n^{-1/5}, & \text{for the Gaussian Kernel,} \\ 2.34 \, s_n \, n^{-1/5}, & \text{for the Epanechnikov kernel,} \\ 2.78 \, s_n \, n^{-1/5}, & \text{for the Quartic kernel.} \end{cases} \tag{5.10}$$

Note that when the underlying density is not Gaussian, (5.10) is no longer an ideal bandwidth but rather a rule of thumb. Figure 5.2 shows a kernel density estimate using $\hat{h}_I$ for net income data from the United Kingdom in 1973. The figure is obtained by the following XploRe code:

```
x=read("xagg73sh")    ; loads the data
x=x[,1]               ; extracts column 1 (net income)
library("smoother")   ; loads smoother library
fh=denauto(x)         ; automatic density estimation
                      ; with Silverman's rule of thumb
show(fh s2d)          ; displays the figure
```

While Silverman's rule works well for many data sets (in particular, Gaussian types of data), it tends to produce oversmoothed estimates. Since the bandwidth choice is crucial to the estimate, we suggest producing the estimate with bandwidth $\hat{h}_I$ along with some oversmoothed estimates and some undersmoothed estimates. In particular, we recommend using a family of estimates,

$$\{\hat{f}_h, \ h = 1.4^j \hat{h}_I, \ j = -3, -2, -1, 0, 1, 2\} \tag{5.11}$$

and then overlay them in the same plot. This way, we can examine how the estimate changes with the choice of bandwidths and we can explore some possible finer structures. The recommended choice of j is based on the considerations that we can visually differentiate six different estimates on the same plot and that $\hat{h}_I$ tends to oversmooth the data somewhat. The

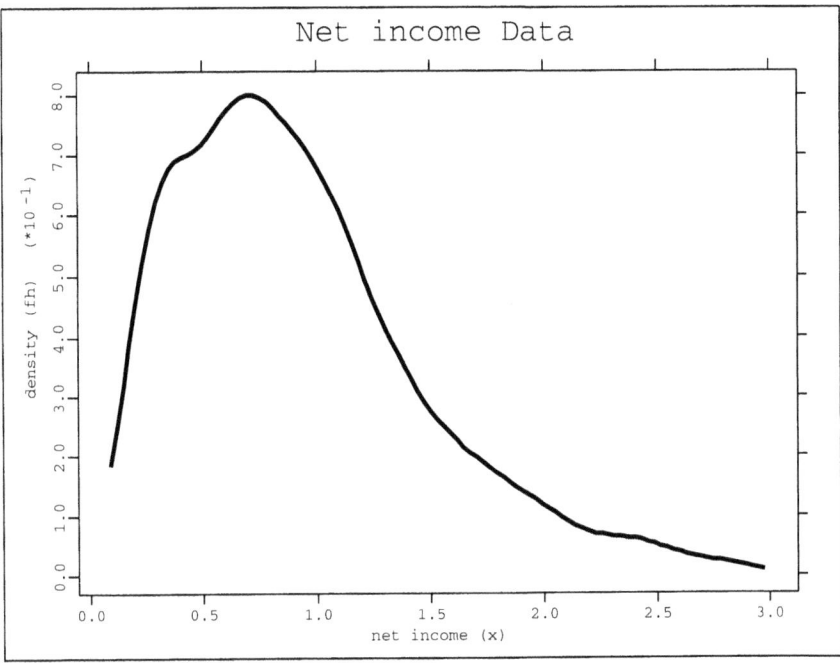

FIGURE 5.2. Kernel density estimate using the bandwidth according to the rule of thumb. The data set is the family net income in the United Kingdom in 1973.

advantage of the family-estimate method (5.11) is that output is portable so that one can present and discuss the findings with others. Figure 5.3 shows this idea by using the net income data from the United Kingdom in 1973. The family of estimates (5.11) can be computed by adding the following XploRe code:

```
sigma=sqrt(var(x))                   ; determines estimate for sigma
h=2.78*sigma*rows(x)^(-0.2)          ; determines h by rule of thumb
fh1=denest(x h*1.4^(-1))             ;
fh2=denest(x h*1.4^(-2))             ; fh1, ..., fh5 are the
fh3=denest(x h*1.4^(-3))             ; density estimates for
fh4=denest(x h*1.4)                  ; different bandwidths
fh5=denest(x h*1.4^2)                ;
show(fh fh1 fh2 fh3 fh4 fh5 s2d) ; displays estimates
```

The density estimation method is also a powerful tool for comparing the results of two experiments. The classical approaches are often based on the two-sample mean idea. The disadvantage of this method is that we have to assume that the two populations have the same family of distributions. The nonparametric density estimation method, on the other hand, compares not

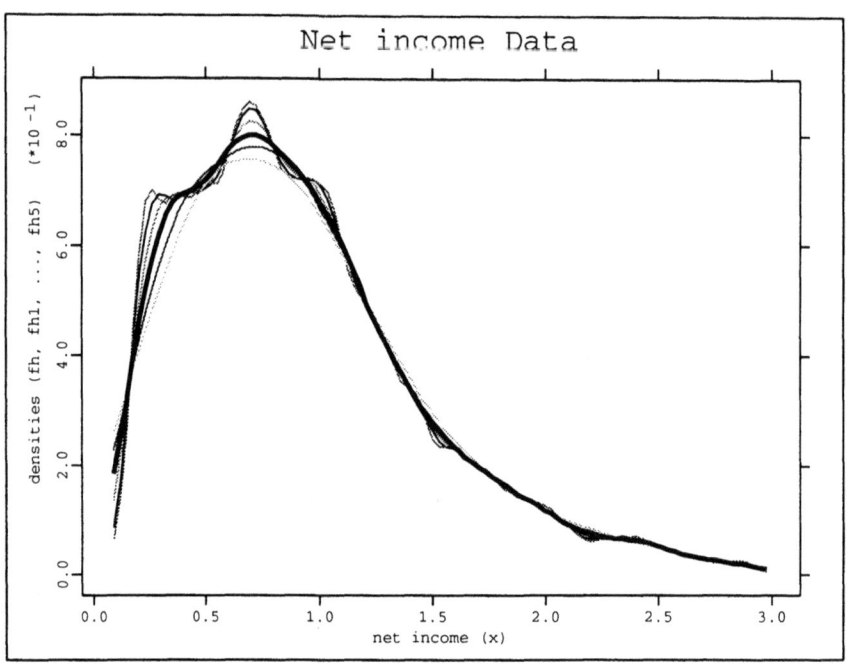

FIGURE 5.3. A family of kernel density estimates. The data set is the family net income in the United Kingdom in 1973.

only the mean, but also the overall pattern (including dispersion), of the experiments. We illustrate this idea in Figure 5.4. The data here are the diagonal lengths from genuine and forged old Swiss bank notes, see Flury and Riedwyl (1988). Clearly, the estimated densities of both groups have approximately the same shape and spread, but the forged notes on average have shorter diagonals. Here is the XploRe code that we used to produce Figure 5.4.

```
x=read("bank2")          ; loads the data
xg=x[1:100,6]            ; extracts diagonal lengths of genuine notes
xf=x[101:200,6]          ; extracts diagonal lengths of forged notes
library("smoother")      ; loads smoother library
fhg=denauto(xg)          ; automatic density estimate for xg
fhf=denauto(xf)          ; automatic density estimate for xf
show(fhg fhf s2d)        ; displays the figure
```

The bandwidth in (5.1) remains constant; it depends on neither the location x nor the datum point X_i. For the data presented in Figure 5.3, it is more desirable to use a larger bandwidth at high income regions and a smaller bandwidth at low income regions. This is due to the fact that it is more desirable (according to longitudinal studies of the family income

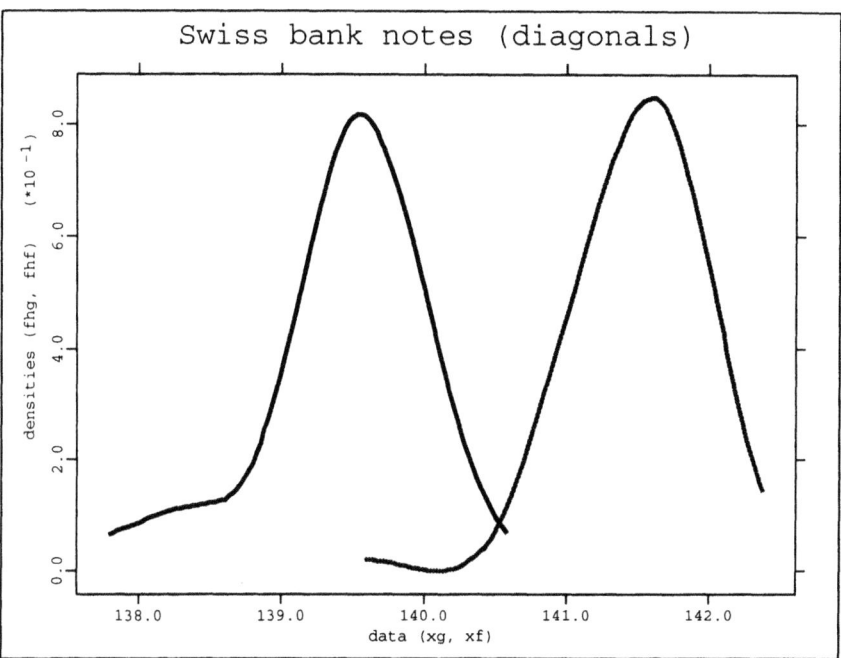

FIGURE 5.4. Comparing the diagonal lengths of genuine (red, right) and forged (blue, left) Swiss bank notes by using nonparametric density estimation. The estimates are based on the bandwidth according to the rule of thumb (see color insert).

over a number of years) to produce a sharp bimodal estimate at the low income region and a smoothed curve at the high income region. A large bandwidth produces an oversmoothed estimate at the low income region, while a small bandwidth results in an undersmoothed estimate at the high income region. Thus, a variable amount of smoothing is needed. Here we offer two approaches to accomplish this.

The first approach is to divide the data into several subgroups (for example, low, middle, and high income classes). Then, apply the kernel density estimate (5.1) with Silverman's rule to each subgroup, and glue these estimates together. Since the bandwidths are discontinuous across the boundary of the division, the resulting estimate may have discontinuities. A better idea is to use the average of the density estimates for the whole data set based on the bandwidths selected at subgroups. This would produce a continuous estimate.

Another simple method is to first transform the data by

$$Y_i = g(X_i), \ i = 1, \ldots, n,$$

where g is a given monotone increasing function. Apply the kernel density

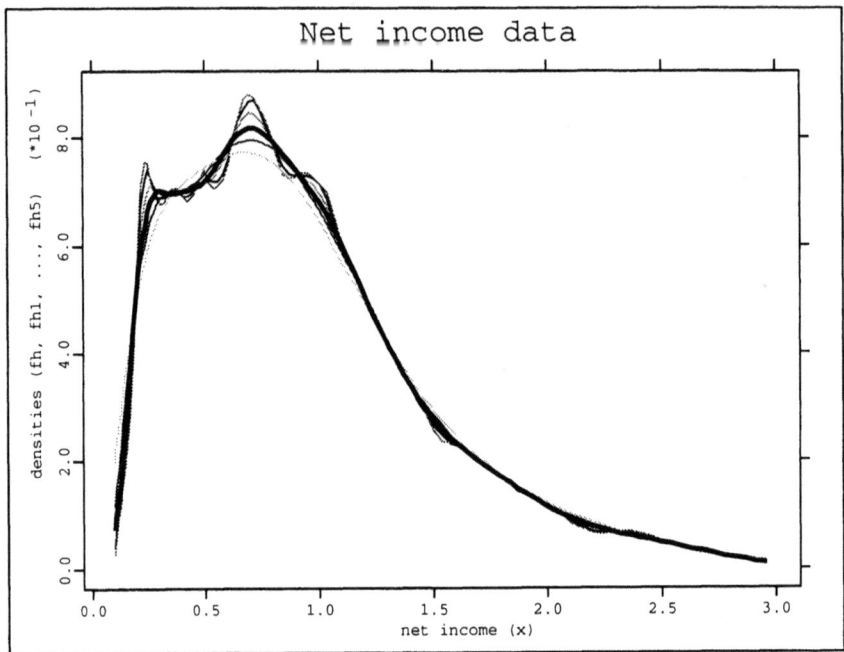

FIGURE 5.5. A family of kernel density estimates using the transformation technique. The data set is family net income in the United Kingdom
in 1973.

estimate (5.1) to this transformed data set and obtain the estimate $\hat{f}_Y(y)$.
Finally, apply the inverse transform to obtain the density of X:

$$\hat{f}_X(x) = g'(x)\hat{f}_Y(g(x)) = g'(x)\,n^{-1}\sum_{i=1}^{n} K_h(g(x) - g(X_i)).$$

Figure 5.5 illustrates this idea using the square root transform.
    A popular family of transforms is the power family indexed by α and β

$$g_{\alpha,\beta}(x) = \begin{cases} (x - \beta)^\alpha, & \alpha > 0, \\ \log(x - \beta), & \alpha = 0. \end{cases}$$

A data-driven choice of α and β is also possible (we omit details here, but
see Wand, Marron and Ruppert, 1991). The following code is used to create
the main density curve (black line) in Figure 5.5.

```
x=read("xagg73sh")          ; loads the data
x=x[,1]                     ; extracts column 1 (net income)
y=sqrt(x)                   ; transforms net income
library("smoother")         ; loads smoother library
fh=denauto(y)               ; automatic dens. est. for y
```

```
fh[,1]=fh[,1].*fh[,1]              ; retransforms column 1 of fh
fh[,2]=fh[,2]./(2.*sqrt(fh[,1]))   ; retransforms column 2 of fh
show(fh s2d)                       ; displays the figure
```

The idea of the univariate density estimation can easily be generalized to multivariate data. Two-dimensional densities can be calculated with the XploRe macro **denest2**. For the use of this macro, the computation, and visualization of higher-dimensional density estimates, we refer to Chapters 4 and 9.

5.3   Nadaraya–Watson Nonparametric Regression

Given a set of bivariate data $\{(X_1, Y_1), \ldots, (X_n, Y_n)\}$, it is often of interest to examine the association between the covariates and the responses. This is usually characterized by estimating the mean response function or its derivatives:

$$m(x) = E(Y|X = x); \quad m^{(\nu)}(x) = \frac{d^\nu m(x)}{dx^\nu}.$$

The regression $m(x)$ shows the average response for the group of individuals whose covariate is approximately x. The derivative function $m'(x)$ indicates how much the covariate variable contributes to the response variable per unit value near the point x. Take, for example, the covariate X as the years of education of individuals and Y as their associated income. The value $m(14)$ shows the average income of individuals who have 14 years education and $m'(14)$ indicates how much one additional year of education is worth after having 14 years of education. These two functions are often used in exploring useful information contained in data.

Suppose that the data $\{(X_1, Y_1), \ldots, (X_n, Y_n)\}$ are a random sample from a population that can be modeled as

$$Y = m(X) + \sigma(X)\varepsilon, \tag{5.12}$$

where $E(\varepsilon) = 0$ and $Var(\varepsilon) = 1$, and X and ε are independent. Note, that $m(x) = E(Y|X = x)$ and $\sigma^2(x) = Var(Y|X = x)$, which allows for possible heteroscedasticity. The location-scale model (5.12) is imposed simply to help us understand the methodology, and this assumption is not necessary for the development.

Since $m(x)$ is the mean response for the covariates near x, an intuitive estimator for this is the locally weighted average estimator defined as

$$\hat{m}(x) = \frac{\sum_{i=1}^n K_h(X_i - x)Y_i}{\sum_{i=1}^n K_h(X_i - x)}, \tag{5.13}$$

where K is the kernel function and h is the bandwidth. This estimator is often referred to as the Nadaraya–Watson *kernel regression estimator*. Härdle

(1990) gives extensive discussion of various statistical methods based on this estimator.

The bandwidth controls the effective size of a local neighborhood. When h is very large, the estimator is essentially the global average of the responses, producing a constant estimate; when h is very small, the estimate essentially interpolates the data. There are many interesting estimates between these two extremes. Thus, the bandwidth h effectively controls the complexity of the estimate.

The performance of the Nadaraya–Watson estimator can be summarized as

$$E\{\hat{m}(x)|X_1,\ldots,X_n\} - m(x) = \mu_2(K)\left\{\frac{m''(x)}{2} + \frac{m'(x)f'(x)}{f(x)}\right\}h^2 + o(h^2) \tag{5.14}$$

and

$$\mathrm{Var}\{\hat{m}(x)|X_1,\ldots,X_n\} = R(K)\frac{\sigma^2(x)}{f(x)nh} + o\left(\frac{1}{nh}\right), \tag{5.15}$$

where f is the marginal density of X, namely the *design density*, and x is an interior point of the support of f. Clearly, the asymptotic optimal bandwidth, which minimizes the *Asymptotic Weighted Mean Integrated Squared Error*

$$
\begin{aligned}
\mathrm{AMISE}(h) &= \mu_2^2(K)h^4 \int \left\{\frac{m''(x)}{2} + \frac{m'(x)f'(x)}{f(x)}\right\}^2 w(x)\,dx \\
&+ \frac{R(K)}{nh} \int \frac{\sigma^2(x)}{f(x)} w(x)\,dx
\end{aligned}
$$

of $\hat{m}(x)$, is given by

$$h_{opt} = \left(\frac{R(K)\int \sigma^2(x)f^{-1}(x)w(x)\,dx}{\mu_2^2(K)\int \{m''(x) + 2m'(x)f'(x)f^{-1}(x)\}^2 w(x)\,dx}\right)^{1/5} n^{-1/5}, \tag{5.16}$$

where $w(x)$ is the weight function.

An attempt to estimate h_{opt} is a cross-validatory idea with $w(x) = f(x)$. Let $\hat{m}_{h,(-i)}(x)$ be the Nadaraya–Watson estimator (5.13) without using the ith-observation (X_i, Y_i). We now validate the "goodness-of-fit" by measuring the "prediction error" $Y_i - \hat{m}_{h,(-i)}(X_i)$. The cross-validation criterion measures the overall "prediction errors", and is defined by

$$CV(h) = n^{-1}\sum_{i=1}^{n}\{Y_i - \hat{m}_{h,(-i)}(X_i)\}^2. \tag{5.17}$$

The cross-validation bandwidth selector $\hat{h}_{CV}$ chooses an h that minimizes $CV(h)$.

It is common practice to plot the function $CV(h)$ against h to examine the changes of overall prediction errors. Visualization of the function $CV(h)$

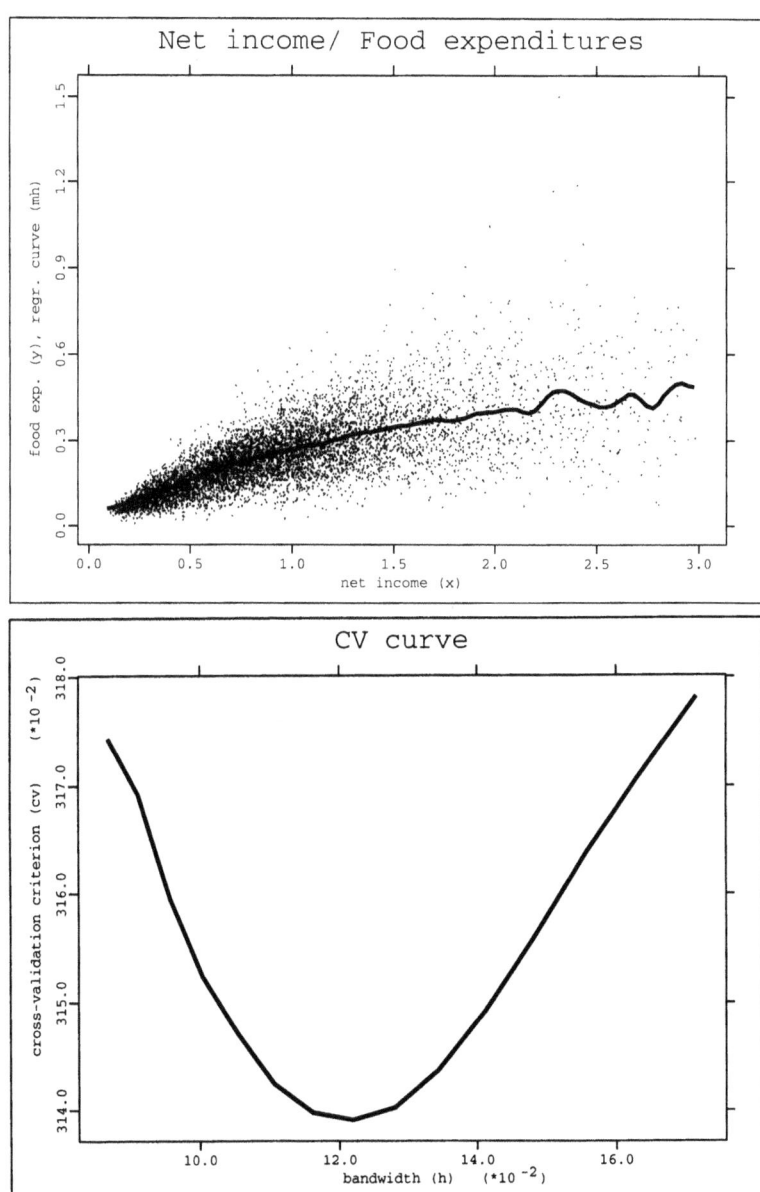

FIGURE 5.6. Nadaraya–Watson kernel regression estimate for the food expenditure data, using the cross-validated bandwidth $\hat{h}_{CV} = 0.12$. The data set is family net income and food expenditures in the United Kingdom in 1973. The top panel is the scatterplot of the data set and its estimated regression function. The bottom panel is the cross-validatory curve.

helps in the process of bandwidth selection. In some situations, if $\hat{h}_{CV}$ does not work well for a particular data set, one usually selects a bandwidth h_0 such that $CV(h_0)$ is close to the minimum and $CV'(h_0)$ is large. Figure 5.6 illustrates the Nadaraya–Watson estimator using $\hat{h}_{CV}$ for net income and food expenditures data from the United Kingdom in 1973. The XploRe code[3] is as follows:

```
x=read("xagg73sh")                      ; loads the data
x=x[,1]~x[,4]                           ; net income & food expend.
func(environ+#("\examples\regcvl"))     ; loads cv example macro
cv=regcvl(x)                            ; calculates cv function
show(cv s2d)                            ; displays cv function
cv=sort(cv 2)                           ; sorts values wrt column 2
hcv=cv[1,1]                             ; bandwidth for min. cv value
library("smoother")                     ; loads smoother library
mh=regest(x hcv)                        ; estimates regression curve
show(mh x s2d)                          ; displays curve and data
```

In view of the sensitivity of the bandwidth selection to the quality of the regression estimation, we recommend that a family of estimates be produced:

$$\{\hat{m}_h, \ h = 1.4^j \hat{h}_{CV}, \ j = -2, -1, 0, 1, 2\}. \tag{5.18}$$

This allows us to explore the different structures as the bandwidth changes. Figure 5.7 provides such a method for the food expenditure data in Figure 5.6.

The domain of the smoothing in Figures 5.6 and 5.7 is the same as that of a human's visualization. Visual inspection of estimates also helps in the process of bandwidth selection. However, the major strength of smoothing techniques is in the domain that is different from our visualization. Examples of this include estimation of regression functions (conditional probabilities) for binary data and estimation of derivative functions. Figure 5.8 presents a family of estimates for a conditional probability function. Here, the data are on women labor supply. The continuous covariate is the income of the partner, and the response is 1 or 0, depending whether the wife works. The sample consists of 1883 couples in the 1986 wave of the German Socio-Economic Panel (for details on the panel, see Projektgruppe "Das Sozio-ökonomische Panel", 1991). The previous XploRe code can be used after changing the data with x=read("xwwork86").

As with density estimation techniques, nonparametric regression methods can also be used to compare the results of experiments. We will illustrate this application in the next section.

---

[3] To get the cross-validation curve in Figure 5.6 we have modified the macro regcvl in the following way: The discretization binwidth is taken as d=(max(x[,1])-min(x[,1]))/500, the start value for the bandwidth as h=15*d, the number of bandwidths as ncv=15 and the inflation factor as h=h*1.05.

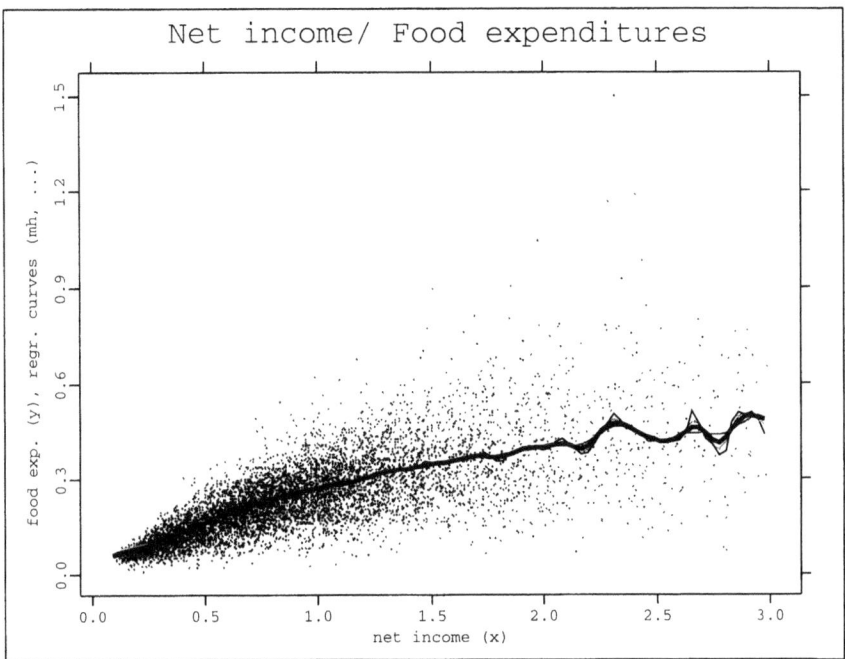

FIGURE 5.7. A family of Nadaraya–Watson kernel regression estimates. The data set is family net income and food expenditures in the United Kingdom in 1973.

5.4 Local Polynomial Fitting

From the function approximation point of view, the Nadaraya–Watson estimator uses a locally constant approximation. This approximation suffers from large bias, particularly in regions where either $m'(x)$ or $f'(x)$ is large. Also, the bias of the Nadaraya–Watson estimator at the boundary is large. One way to repair these drawbacks is to use a higher-order approximation. Theoretical and heuristical insights of this were made clear in Fan (1992) and Hastie and Loader (1993).

Suppose that *locally* the regression function $m(x)$ can be approximated by a polynomial

$$m(x) \approx m(x_0) + m'(x_0)(x - x_0) + \cdots + m^{(p)}(x_0)(x - x_0)^p / p! \quad (5.19)$$

for x in a neighborhood of x_0, by using Taylor's expansion. From the local

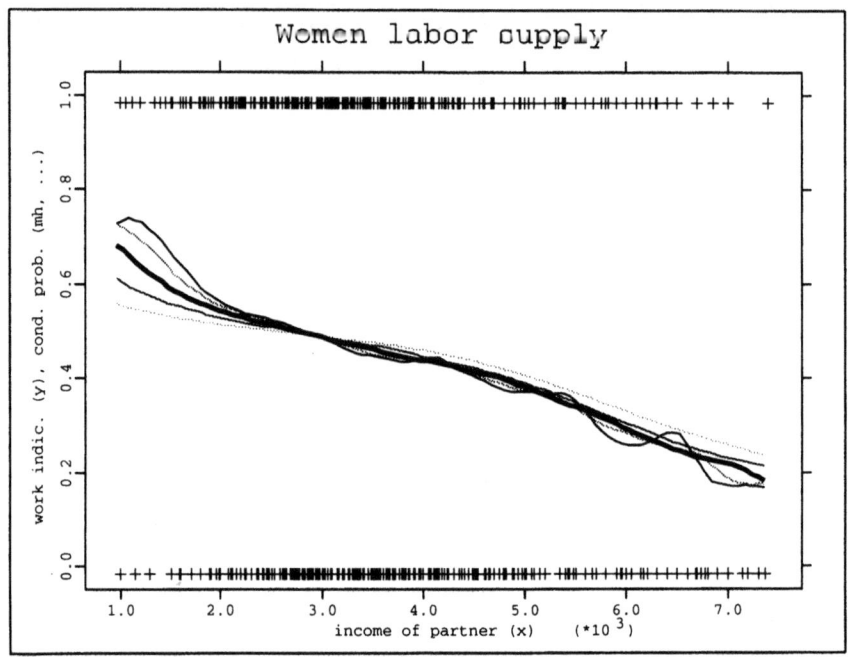

FIGURE 5.8. A family of Nadaraya–Watson kernel regression estimates for binary responses, $\hat{h}_{CV} = 1200$. The data set is partner's income, and the binary variable is 1 or 0, depending on whether the wife is employed.

modeling point of view, (5.19) models the unknown regression function by a polynomial of order p. This suggests fitting a *local polynomial regression*:

$$\min \sum_{i=1}^{n} \{Y_i - \beta_0 - \beta_1(X_i - x_0) - \cdots - \beta_p(X_i - x_0)^p\}^2 \, K_h(X_i - x_0).$$

(5.20)

Let $\hat{\beta}_0(x_0), \ldots, \hat{\beta}_p(x_0)$ be the minimizer to the problem (5.20). The *local polynomial regression estimator* of the regression function is

$$\hat{m}(x_0) = \hat{\beta}_0(x_0).$$

(5.21)

The whole curve $\hat{m}(\bullet)$ is obtained by fitting the above local polynomial regression with x_0 varying in an appropriate domain of interest.

We first note that when $p = 0$, the estimator (5.21) is the same as the Nadaraya–Watson (5.13). For estimating the regression function, the order p is usually taken to be 1 (namely, local linear fit) or occasionally 3 (namely, local cubic fit). In general, the "odd" fit outperforms the "even" order fit. For example, the local linear fit performs better than the local constant fit (that is, the Nadaraya–Watson estimator), and the local cubic fit outperforms the local quadratic fit. However, there is no direct comparison between the local linear fit and the local cubic fit. For a given bandwidth, the local linear fit has smaller variance, while the local cubic fit is expected to have a smaller bias.

The complexity of parametric polynomial model is controlled by the order p of the polynomial. By way of contrast, the complexity of the local polynomial fit is governed by the bandwidth h. Take the local linear fit as an example. When h is very small, the resulting estimate basically interpolates the data points, while when h is very large, the resulting estimate becomes the parametric linear model. There are many interesting examples that lie between these two extreme choices of bandwidth. The possible nonlinearity structure can be detected by using an appropriate bandwidth.

The simplicity, flexibility, and wide applicability of local polynomial modeling are depicted in the forthcoming monograph by Fan and Gijbels (1995b), where readers can find more references to the literature. In particular, local polynomial fitting possesses nice theoretical results such as minimax optimality, automatic boundary correction, design adaptation, and simple bias and variance expressions. The choice of bandwidth is also very simple (see, for example, Fan and Gijbels, 1995a; Ruppert, Sheather and Wand, 1996).

The *local linear regression smoother* is particularly simple to implement. Indeed, the estimator has the simple expression

$$\hat{m}_{L,h}(x) = \sum_{i=1}^{n} w_i Y_i, \tag{5.22}$$

where with $S_{n,j} = \sum_{i=1}^{n} K_h(X_i - x)(X_i - x)^j$,

$$w_i = K_h(X_i - x)\{S_{n,2} - (X_i - x)S_{n,1}\}/(S_{n,0}S_{n,2} - S_{n,1}^2). \tag{5.23}$$

The asymptotic bias and variance for this estimator are

$$E\{\hat{m}_{L,h}(x)|X_1,\ldots,X_n\} - m(x) = \mu_2(K)\frac{m''(x)}{2}h^2 + o(h^2) \tag{5.24}$$

and

$$\text{Var}\{\hat{m}_{L,h}(x)|X_1,\ldots,X_n\} = R(K)\frac{\sigma^2(x)}{f(x)nh} + o\left(\frac{1}{nh}\right), \tag{5.25}$$

where f is the marginal density of X, namely, the *design density* (Fan, 1993). By comparing (5.15) and (5.25), it is evident that the local linear fit uses locally one extra parameter without increasing its variability. But this extra parameter creates opportunities for significant bias reduction, particularly at the boundary regions and sloped regions. This is evidenced by comparing (5.14) and (5.24). Clearly, the asymptotic optimal bandwidth that minimizes the asymptotic weighted MISE of $\hat{m}_{L,h}(x)$ is given by

$$h_{opt} = \left(\frac{R(K)\int \sigma^2(x)f^{-1}(x)w(x)\,dx}{\mu_2^2(K)\int\{m''(x)\}^2 w(x)\,dx}\right)^{1/5} n^{-1/5}, \tag{5.26}$$

where $w(x)$ is the weight function. In particular, when the design density is nearly uniform in an interval $[a, b]$ and $\sigma^2(x) \sim \sigma^2$, by taking a constant uniform weighting scheme, we obtain

$$h_{opt} \approx \left(\frac{R(K)(b-a)\sigma^2}{\mu_2^2(K) \int_a^b \{m''(x)\}^2 \, dx} \right)^{1/5} n^{-1/5}. \tag{5.27}$$

Now, estimating $m''(x)$ and σ^2 from a (parametric) fourth-order polynomial fit $m(x) = \alpha_0 + \alpha_1 x + \cdots + \alpha_4 x^4$, we obtain the following rule of thumb:

$$\hat{h}_{ROT} = \left(\frac{R(K)(b-a)\hat{\sigma}^2}{\mu_2^2(K) \int_a^b \{2\hat{\alpha}_2 + 6\hat{\alpha}_3 x + 12\hat{\alpha}_4 x^2\}^2 \, dx} \right)^{1/5} n^{-1/5}. \tag{5.28}$$

This rule of thumb is a special case of the more general idea given in Chapter 4 of Fan and Gijbels (1995b).

Another simple, but much more computationally involved choice of bandwidth is the cross-validation method. Let $\hat{m}_{L,h,(-i)}(x)$ be the local linear regression estimator (5.22) without using the ith-observation (X_i, Y_i). We now validate the "goodness-of-fit" by measuring the "prediction error" $Y_i - \hat{m}_{L,h,(-i)}(X_i)$. The cross-validation criterion measures the overall "prediction error", and is defined, analogously to (5.17), by

$$CV(h) = n^{-1} \sum_{i=1}^{n} \{Y_i - \hat{m}_{L,h,(-i)}(X_i)\}^2. \tag{5.29}$$

The cross-validation bandwidth selector $\hat{h}_{CV}$ chooses the one that minimizes $CV(h)$.

The use of the local linear regression smoother is very much analogous to that of the Nadaraya–Watson estimator. Figure 5.9 shows the local linear fit and the cross-validatory curve for the net income/food expenditure data. The corresponding XploRe code[4] is the following:

```
x=read("xagg73sh")              ; loads the data
x=x[,1]~x[,4]                   ; net income & food expend.
func(environ+#("\examples\lpregcv1"));  loads cv example macro
cv=lpregcv1(x)                  ; calculates cv function
show(cv s2d)                    ; displays cv function
cv=sort(cv 2)                   ; sorts values wrt column 2
hcv=cv[1,1]                     ; bandwidth for min. cv value
library("smoother")             ; loads smoother library
```

---

[4]To get the cross-validation curve in Figure 5.9 we have modified the macro lpregcv1 in the following way: The discretization binwidth is taken to be d=(max(x[,1])-min(x[,1]))/500, the start value for the bandwidth is h=60*d, the number of bandwidths is ncv=15 and the inflation factor is h=h*1.05.

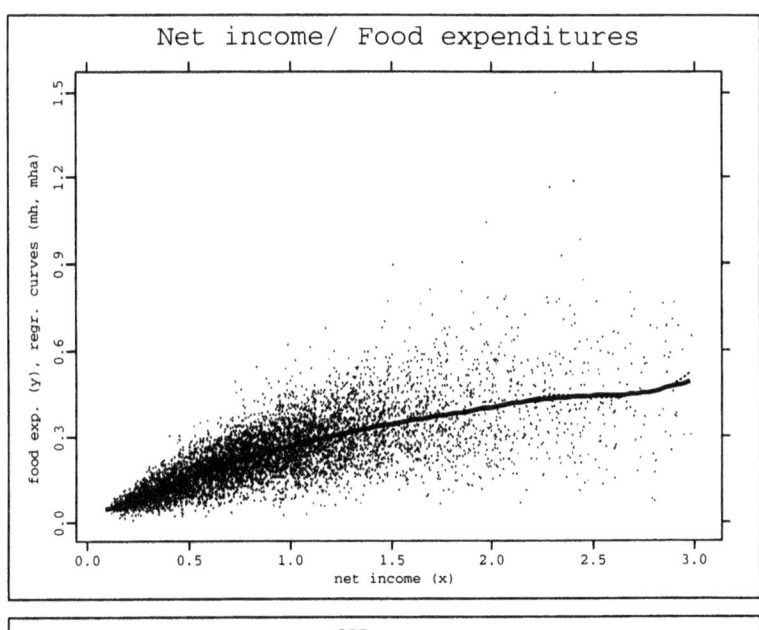

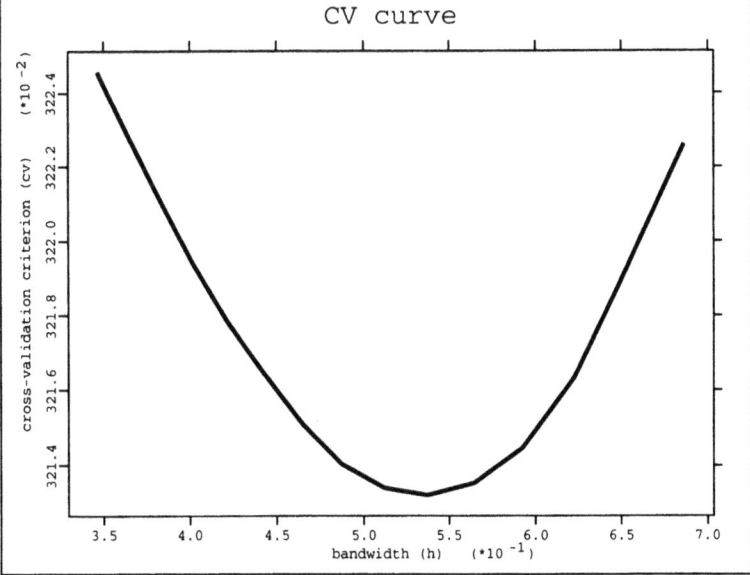

FIGURE 5.9. Local linear regression estimate for the food expenditure data. The top panel is the scatterplot of the data set and its estimated regression functions using bandwidth $\hat{h}_{CV} = 0.54$ (thick solid line) and $\hat{h}_{ROT} = 0.32$ (thin dashed line). The bottom panel is the cross-validatory curve for the local linear regression estimate.

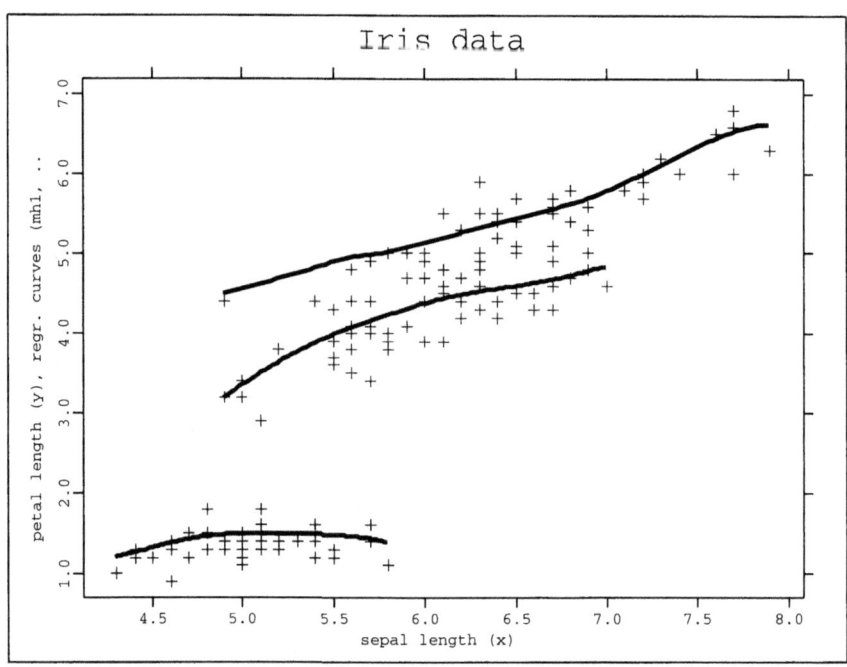

FIGURE 5.10. Comparison of three experiments using local linear regression estimates for the Iris data. The figure gives the scatterplots of sepal length vs. petal length for setosa (black, below), versicolor (red, middle), and virginica (blue, above). The estimated regression curves, using rule of thumb bandwidths 0.43, 0.65, and 0.78, are indicated in the same colors (see color insert).

```
mh=lpregest(x hcv)              ; regression using hcv
mha=lpregauto(x)                ; regression using hrot
show(mh mha x s2d)              ; displays curves and data
```

An important application of the nonparametric regression technique is to compare experiments. Figure 5.10 compares the relation between sepal and petal length for three groups of iris flowers (see Andrews and Herzberg, 1985). The following XploRe code was used:

```
(x c)=read("iris" "dd-dd-c")  ; loads columns 1 and 3 of the data
                              ; & converts flower name to integer
x1=paf(x[,1:2] x[,3].=1)      ; petal, sepal length (setosa)
x2=paf(x[,1:2] x[,3].=2)      ; petal, sepal length (versicolor)
x3=paf(x[,1:2] x[,3].=3)      ; petal, sepal length (virginica)
library("smoother")           ; loads smoother library
mh1=lpregauto(x1)             ; loc. lin. regression (setosa)
mh2=lpregauto(x2)             ; loc. lin. regression (versicolor)
mh3=lpregauto(x3)             ; loc. lin. regression (virginica)
show(mh1 mh2 mh3 x1 x2 x3 s2d); displays the figure
```

We would like to close this introduction of nonparametric kernel regression by mentioning a collection of other nonparametric univariate regression methods that are available in XploRe: spline regression (`spline`), k-nearest neighbor regression (`knn`), running median regression (`runmed`), supersmoother (`regssm`), wavelet regression (`wavereg`), and monotone regression (`monreg`). These methods can be employed similarly to the kernel regression smoother. Introduction to these techniques is scattered around in various books, in particular, Chapter 2 of Fan and Gijbels (1995b). Methods for multivariate regression, in particular generalized additive models, average derivative estimation for single index models, and sliced inverse regression, are presented in Chapter 11 of this book.

5.5   Estimation of Regression Derivatives

Derivative curves measure locally the strength of the association between the covariate variable X and the response variable Y. They also indicate the rate of growth of the regression function m. Local polynomial fitting can easily be applied to derivative estimation. Suppose that we fit the local polynomial of order p as in (5.20). Then, by comparing (5.19) and (5.20), we obtain an estimator for $m^{(\nu)}$ by

$$\hat{m}^{(\nu)}(x) = \nu!\,\hat{\beta}_\nu(x). \tag{5.30}$$

Usually, the order of polynomial p is taken to be $p = \nu + 1$ or occasionally $p = \nu + 3$, based on the consideration of the efficiency of the estimator and cost of computation. The "even" order $p = \nu + 2\ell$ (ℓ is an integer) is not admissible—it is dominated by the "odd" order fit $p = \nu + 2\ell + 1$. For example, to estimate the first derivative $m'(x)$, we use the slope of the local quadratic fit. Again, the complexity of the estimator is effectively controlled by the bandwidth h, not by the order p.

At this point, one may naturally think of estimating $m^{(\nu)}(x)$ by taking the νth derivative of the estimated regression curve. However, it has been shown that the estimator (5.30) is the most efficient one in an asymptotic minimax sense. Besides, for the recommended "odd" order fit, the estimator (5.30) possesses the properties of design adaptation and automatic boundary bias correction. In other words, the estimator (5.30) can be applied to various designs and reduce significantly the bias when one estimates the curve near the boundary of the design density (see Fan and Gijbels (1995b) for more detailed discussions and the related references).

The bias and variance of the estimator (5.30) are given (Ruppert and Wand, 1994) by

$$E\{\hat{m}^{(\nu)}(x)|X_1,\ldots,X_n\} - m^{(\nu)}(x)$$
$$= \int t^{p+1} K_\nu^*(t)\,dt\,\frac{\nu!\,m^{(p+1)}(x)}{(p+1)!}\,h^{p+1-\nu} + o(h^{p+1-\nu})$$

and

$$\text{Var}\{\hat{m}^{(\nu)}(x)|X_1,\ldots,X_n\} = \int K^{*\,2}_{\nu}(t)\,dt\,\frac{(\nu!)^2\sigma^2(x)}{f(x)nh^{2\nu+1}} + o\left(\frac{1}{nh^{2\nu+1}}\right),$$

where K^*_ν is the *equivalent kernel function*. The function K^*_ν is defined as follows. Let $\mu_j = \int t^j K(t)\,dt$ be the jth moment of K, and $S = (\mu_{i+j-2})$ be a $(p+1)\times(p+1)$ matrix. Denote by $S^{\nu\ell}$ $(\nu,\ell = 0,\ldots,p)$ the element (ν,ℓ) of the inverse matrix S^{-1}. Then,

$$K^*_\nu(t) = \left\{\sum_{\ell=0}^{p} S^{\nu\ell}t^\ell\right\}K(t).$$

The asymptotic optimal bandwidth for the estimator (5.30) is

$$h_{opt} = C_{p,\nu}(K)\left[\frac{\int \sigma^2(x)w(x)f^{-1}(x)\,dx}{\int \{m^{(p+1)}(x)\}^2 w(x)\,dx}\right]^{1/(2p+3)} n^{-1/(2p+3)}, \qquad (5.31)$$

which minimizes the asymptotic weighted MISE with the weight function w. Here,

$$C_{p,\nu}(K) = \left\{\frac{(p+1)!^2(2\nu+1)\int K^{*\,2}_{\nu}(t)\,dt}{2(p+1-\nu)\{\int t^{p+1}K^*_\nu(t)\,dt\}^2}\right\}^{1/(2p+3)}.$$

Using the same heuristic as in Section 5.4, we obtain the following rule of thumb bandwidth selector:

$$\hat{h}_{ROT} = C_{p,\nu}(K)\left(\frac{(b-a)\hat{\sigma}^2}{\int_a^b \{\tilde{m}^{(p+1)}(x)\}^2\,dx}\right)^{1/(2p+3)} n^{-1/(2p+3)}, \qquad (5.32)$$

where $\tilde{m}^{(p+1)}(x) = (p+1)!\hat{\alpha}_{p+1} + (p+2)!\hat{\alpha}_{p+2}x + \frac{1}{2}(p+3)!\hat{\alpha}_{p+3}x^2$ and $\hat{\sigma}$ are obtained by fitting a (parametric) polynomial $m(x) = \alpha_0 + \alpha_1 x + \cdots + \alpha_{p+3}x^{p+3}$. This is a special case of the rule of thumb given in Fan and Gijbels (1995b).

While the asymptotic optimal bandwidth (5.31) looks complicated, it is indeed very easy to estimate. This reflects another advantage of local polynomial fitting. A bandwidth selector that converges very fast to h_{opt} in (5.31) is proposed by Fan and Gijbels (1995a).

We now illustrate how to use XploRe to estimate derivative curves. The data consist of 133 observations of the time (in milliseconds) after a simulated impact with motorcycles and the head acceleration (in g) of a test object. We take the local quadratic fit, using the Quartic kernel and bandwidth $\hat{h}_{ROT}$. Note that in this case the constant $C_{2,1}(K) = 2.586$. Figure 5.11 depicts the result. The code is as follows:

```
x=read("motcyc")        ; loads the data
library("smoother")     ; loads smoother library
mh=lpregauto(x)         ; automatic regression curve
mh1=lpderauto(x)        ; automatic derivative estimate
show(mh1 mh x s2d)      ; displays curves and data
```

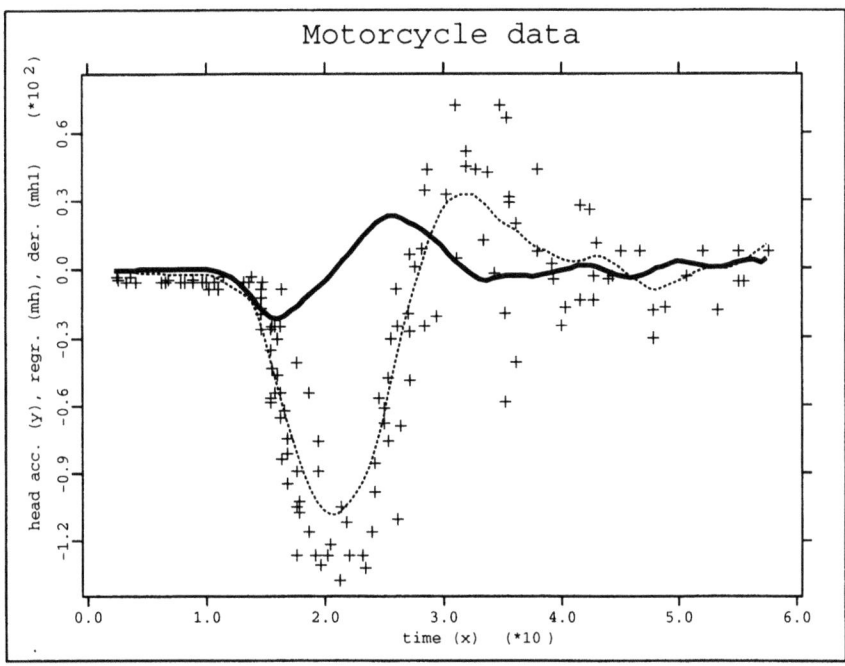

FIGURE 5.11. Local quadratic estimation of the derivative curve for the motorcycle data (thick solid line), $\hat{h}_{ROT} = 5.04$, and local linear regression curve (thin dashed line), $\hat{h}_{ROT} = 4.50$ (see color insert).

5.6   Variable Amount of Smoothing

In Sections 5.3 to 5.5, we use a constant amount of smoothing h. While this works for many practical cases, it might not work very well for *spatially inhomogeneous curves*, namely curves with different degrees of smoothness at different locations. In this case, a variable amount of smoothing is more desirable: In smooth regions, a larger bandwidth should be used to reduce variance, while in nonsmooth regions, a smaller bandwidth is needed to reduce the bias.

The automatic choice of the variable bandwidth is difficult. Here, we describe a general principle for a variable amount of smoothing. Suppose that we have a smoother $\hat{g}_h(x)$ with an automatic global bandwidth selector $\hat{h}$. This can be in one case the local linear regression with $\hat{h}_{CV}$ and in another case the local quadratic estimation of $m'(x)$ with $\hat{h}_{ROT}$. The recipe of variable amount of smoothing reads as follows:

- partition the interval $[a, b]$, where the curve is to be estimated, into subintervals I_ℓ $(\ell = 1, \ldots, L)$;

- apply the global bandwidth selector $\hat{h}$ for the data in each subinterval and obtain the bandwidth step function (on each interval I_ℓ, the bandwidth is a constant);

- smooth the resulting bandwidth step function and obtain a continuous variable amount of bandwidth $\hat{h}(x)$; and

- calculate the variable amount of smoothing estimator $\hat{g}_{\hat{h}(x)}(x)$.

This idea was suggested in Fan and Gijbels (1995a). Their experience suggests the choice of $L = [n/(10 \log n)]$ equally long subintervals I_ℓ. They also used the Nadaraya–Watson estimator with window size of the length of I_ℓ to smooth the bandwidth step function.

We close this section with the remark that a sharpened tool such as a variable amount of smoothing is indeed available to us when it is needed. This can be accomplished by using our old toolbox introduced in Sections 5.3 to 5.5, along with the above recipe.

REFERENCES

Andrews, D.F. and Herzberg, A.M. (1985). *Data: A Collection of Problems from Many Fields for the Student and Research Worker*, Springer Series in Statistics, Springer-Verlag, New York.

Fan, J. (1992). Design-adaptive nonparametric regression, *Journal of the American Statistical Association* **87**(420): 998–1004.

Fan, J. (1993). Local linear regression smoothers and their minimax efficiency, *Annals of Statistics* **21**(1): 196–216.

Fan, J. and Gijbels, I. (1995a). Data-driven bandwidth selection in local polynomial fitting: Variable bandwidth and spatial adaptation, *Journal of the Royal Statistical Society, Series B*. To appear.

Fan, J. and Gijbels, I. (1995b). *Local Polynomial Modeling and Its Application—Theory and Methodologies*, Chapman and Hall, New York.

Flury, B. and Riedwyl, H. (1988). *Multivariate Statistics, A Practical Approach*, Cambridge University Press, New York.

Härdle, W. (1990). *Applied Nonparametric Regression*, Econometric Society Monographs No. 19, Cambridge University Press, New York.

Hastie, T.J. and Loader, C.R. (1993). Local regression: Automatic kernel carpentry (with discussion), *Statistical Science* **8**(2): 120–143.

Marron, J.S. and Nolan, D. (1988). Canonical kernels for density estimation, *Statistics & Probability Letters* **7**(3): 195–199.

Projektgruppe "Das Sozio-ökonomische Panel" (1991). *Das Sozio-ökonomische Panel (SOEP) im Jahre 1990/91*, Deutsches Institut für Wirtschaftsforschung. Vierteljahreshefte zur Wirtschaftsforschung, pp. 146–155.

Ruppert, D., Sheather, S.J. and Wand, M.P. (1996). An effective bandwidth selector for local least squares regression, *Journal of the American Statistical Association*. To appear.

Ruppert, D. and Wand, M.P. (1994). Multivariate weighted least squares regression, *Annals of Statistics*. To appear.

Scott, D.W. (1992). *Multivariate Density Estimation: Theory, Practice, and Visualization*, John Wiley & Sons, New York, Chichester.

Silverman, B.W. (1986). *Density Estimation for Statistics and Data Analysis*, Vol. 26 of *Monographs on Statistics and Applied Probability*, Chapman and Hall, London.

Wand, M.P., Marron, J.S. and Ruppert, D. (1991). Transformations in density estimation (with discussion), *Journal of the American Statistical Association* **86**(414): 343–361.

6
Bandwidth Selection in Density Estimation

Marco Bianchi[1]

6.1 Introduction

The motivation for density estimation in statistics and data analysis is to realize where observations occur more frequently in a sample. The aim of density estimation is to approximate a "true" probability density function $f(x)$ from a sample information $\{X_i\}_{i=1}^n$ of independent and identically distributed observations. The estimated density is constructed by centering around each observation X_i a kernel function $K_h(u) = K(u/h)/h$, with $u = x - X_i$, and averaging the values of this function at any given x. The estimator in its general form is defined as

$$\hat{f}_h(x) = n^{-1} \sum_{i=1}^n K_h(u),$$

where $h > 0$ is the *bandwidth* or *window width*. Comprehensive monographs on density estimation such as Silverman (1986) or Härdle (1991) point out to what extent there is little to choose in practice about the form of the kernel function, at least for certain given general properties of the kernel, whereas much more important is the *choice of the bandwidth*. The Gaussian kernel

$$K(u) = \frac{1}{\sqrt{2\pi}} \exp\left(-\frac{1}{2}u^2\right)$$

is commonly preferred for its computational advantages (Turlach, 1993, section 6), but the estimated density doesn't virtually change its shape if other kernels (Epanechnikov, Quartic, etc.) are used with comparable bandwidths.[2] Whatever the kernel function $K_h(u)$, however, when h is too small, the resulting curve is too wiggly, reflecting too much of the sampling variability, with a tendency to show spurious structure in the data; when h is too large, on the other hand, the resulting estimate may tend to smooth

---

[1]Bank of England, London SE 1, UK.
[2]Comparable bandwidths are here synonymous with *canonical bandwidths*, as defined, among others, in Härdle (1991, pp. 72–76).

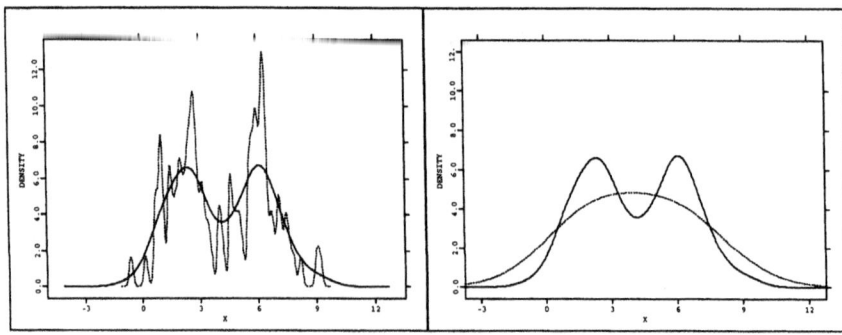

FIGURE 6.1. Density estimate for a mixture of two normals with different values of h. Undersmoothing with $h = 0.1$ (left), and oversmoothing with $h = 2.0$ (right). Solid line is the true density.

away important features of the underlying density. This situation is shown in Figure 6.1. A reasonable choice of the bandwidth really constitutes the key aspect of density estimation.

6.2   Choosing the Smoothing Parameter

There are mainly two approaches to bandwidth selection in practice. These are:

- subjective choice, and
- automatic or *data-driven* methods.

The former approach is suggested by Silverman (1986) as a practical tool for density estimation in data analysis because it is extremely easy to understand and implement. It consists of estimating the density several times by using a wide range of different bandwidths with the purpose of being aware of all possible structures (number of modes) in the density. The data-driven approach is a more theoretical one; an automatic selection of the bandwidth is performed, in fact, by identifying that h which minimizes some "distance" between the true and estimated density. Turlach (1993) provides an excellent up-to-date review about the "state of the art" in data-driven bandwidth selection. In the following, the main fundamental ideas are summarized.

The first question of interest in data-driven bandwidth selection is how an optimal bandwidth is defined. An optimal bandwidth is the minimizer of distances between the true and the estimated density, such as, for example (Turlach, 1993, section 2), the *Integrated Squared Error (ISE)*,

$$ISE(h) = \int [\hat{f}_h(x) - f(x)]^2 \, dx,$$

its expected value, the *Mean Integrated Squared Error (MISE)*,

$$MISE(h) = \mathrm{E}\left(\int [\hat{f}_h(x) - f(x)]^2 \, dx\right),$$

and the asymptotic (Taylor expansion of f) $MISE$, $(AMISE)$,

$$AMISE(h) = \frac{1}{nh}R(K) + h^4 \left(\frac{\mu_2(K)}{2!}\right)^2 R(f^{(2)}),$$

where $R(L) = \int L^2(x) \, dx$, $\mu_j(L) = \int x^j L(x) \, dx$, and $f^{(j)}$ is the jth derivative of f. For a Gaussian kernel, $R(K) = \int K^2(x) \, dx = 1/(2\sqrt{\pi})$, and
$\mu_2(K) = \int x^2 K(x) \, dx = 1$.

To make clear the trade-off existing between the bias and the variance
of the estimator for different values of h, the $MISE$ is rewritten as

$$
\begin{aligned}
MISE(h) &= \int \mathrm{E}\{\hat{f}_h(x) - \mathrm{E}[\hat{f}_h(x)] + \mathrm{E}[\hat{f}_h(x)] - f(x)\}^2 \, dx \\
&= \int \mathrm{E}\{\hat{f}_h(x) - \mathrm{E}[\hat{f}_h(x)]\}^2 + \{\mathrm{E}[\hat{f}_h(x)] - f(x)\}^2 \, dx \\
&= \int \mathrm{Var}\,\hat{f}_h(x) \, dx + \int \mathrm{bias}^2 \hat{f}_h(x) \, dx.
\end{aligned}
$$

By denoting the first integral as IV (integrated variance), and the second
integral as IB (integrated bias), we have

$$
\begin{aligned}
IV(h) &= (nh)^{-1}R(K) + n^{-1}R(f) + O(n^{-1}h^2), \\
IB(h) &= h^4/4 \cdot \mu_2^2(K)R(f^{(2)}) + O(h^8)
\end{aligned}
$$

(Turlach, 1993, page 7). Asymptotically, big values of h reduce the variance
of the estimator $\hat{f}_h(x)$, but at the cost of increasing the bias.

Indicating by $\hat{h}_0$ the minimizer of ISE, by h_0 the minimizer of $MISE$,
and by h_∞ the minimizer of the $AMISE$, all above optimal bandwidths
depend on the unknown density f. Estimates of $\hat{h}_0$, h_0, and h_∞ are needed
through the estimation of f or derivatives of f. With this respect, most
data-driven methods are classified into the following three categories:

- "quick and dirty" methods,
- "cross-validation" methods, and
- "plug-in" methods.

These categories are separately discussed in the following.

6.2.1 Silverman's Rule of Thumb

The most well-known among quick and dirty methods is the *Rule of thumb*,
proposed by Silverman (1986, section 3.4.2), in which the aim is to minimize

the $AMISE$. The best trade-off between asymptotic variance and bias is given by

$$h_\infty = \left(\frac{R(K)}{\mu_2^2(K)R(f^{(2)})}\right)^{\frac{1}{5}} n^{-\frac{1}{5}}, \tag{6.1}$$

where $R(f^{(2)})$ is the only unknown. The idea of the *Rule of thumb* is to refer to a family of distributions as a candidate for f, such as, for example, the standard normal distribution. This allows us to derive, for a Gaussian kernel, the following estimate of h_∞:

$$h_{\text{rot}} = 1.06 \cdot \hat{\sigma} \cdot n^{-\frac{1}{5}},$$

where $\hat{\sigma}$ is the sample standard deviation. A more robust version of the rule of thumb with respect to the presence of outliers is obtained by eventually replacing the interquartile range R with the standard deviation, as a measure of the dispersion of the data. This leads to

$$h_{\text{rot}} = 1.06 \min\left(\hat{\sigma}, \frac{\hat{R}}{1.34}\right) n^{-\frac{1}{5}}.$$

The advantage of the rule is, of course, the fact that it provides a very practical method of bandwidth selection; the disadvantage, however, is that the smoothing parameter is wrong if the population is not normally distributed (if the true density is a mixture of normals, for example). In this eventuality, the estimated density obtained through h_{rot} hides a lot of structure by smoothing away modes.

6.2.2   Least-Squares and Bandwidth-Factorized CV

From the class of cross-validation bandwidth selectors, we will discuss here two methods, *Least Squares Cross-Validation* (*LSCV*) and *Bandwidth-Factorized Cross-Validation* (*BFCV*). Other methods, such as *biased* and *smoothed* cross-validation, which have also been included in XploRe, are not discussed here but can be found in Turlach (1993).

The least-squares CV function is defined by

$$LSCV(h) = R(\hat{f}_h(x)) - 2\sum_{i=1}^{n} \hat{f}_{h,i}(x_i),$$

where $\hat{f}_{h,i}(x)$ is the density estimate obtained by all data points except for the ith-observation. The $LSCV$ function can be viewed as an estimator of $ISE(h) - R(f)$ (Turlach, 1993, p. 13); that is, it represents in fact an attempt to estimate $\hat{h}_0$.

The starting point for $BFCV$ is the representation $MISE(h) = IV(h) + IB(h)$. As far as $(nh)^{-1}R(K)$ has been proved to be a good estimator of

$IV(h)$, the main task of $BFCV$ is to search for a "good" estimate of $IB(h)$. The integrated bias can be written as

$$IB(h) = \int (K_h * f - f)^2(x)\, dx,$$

where $*$ denotes the convolution of two functions K and L, $(K * L)(x) = \int K(x-u)L(u)\, du = \int K(u)L(x-u)\, du$. Indicating by K_0 the Dirac function, $IB(h)$ is estimated by (Turlach, 1993, p. 16)

$$\widehat{IB}(h) = n^{-2} \sum_{i=1}^{n} \sum_{j=1}^{n} (K_h * K_h - 2K_h + K_0) * K_g * K_g(x_i - x_j),$$

that is, by combining the density estimates obtained with *two* different bandwidths, h and g, where g is a defined function of h. The $BFCV$ function is obtained as

$$BFCV(h) = (nh)^{-1} R(K) + \widehat{IB}(h).$$

6.2.3   Plug-In Methods

Plug-in methods target the $AMISE$ as the distance to be minimized, that is, an estimate for h_∞ is looked for. h_∞ is a function of $R(f^{(2)})$, which is estimated, according to plug-in methods, through a *sequence* of bandwidths h_1, h_2, $\ldots$. The first bandwidth h_1 serves initially to estimate the density $\hat{f}_{h_1}(x)$, which is used to compute $\hat{R}(f^{(2)}) = R(\hat{f}_{h_1}(x))$; subsequently, $\hat{R}(f^{(2)})$ is plugged into (6.1) to calculate the second bandwidth h_2 in the sequence. Thus a new estimate of $R(f^{(2)})$ is obtained as $\hat{R}(f^{(2)}) = R(\hat{f}_{h_2}(x))$, and again plugged into (6.1) to derive h_3, etc. The process iterates until convergence of the bandwidths is reached. Some well-known plug-in methods have been developed by Park and Marron (1992) and Sheather and Jones (1991) (although those do not iterate until convergence). In both cases, a score function of h is defined, and the zero of the function is searched for. Exact formulas of the methods are not reported here, but can be found, for example, in Turlach (1993, p. 28).

6.2.4   Simulation Studies

As far as the performance of the different data-driven bandwidth selectors is concerned, Sheather (1992) and Park and Turlach (1992) contain some practical indications. In Park and Turlach (1992), the relative performance of different selectors is tested over a wide range of structures of densities through Monte Carlo simulations. According to their findings, Sheather and Jones' plug-in method appears one of the best performing, especially when dealing with multimodal densities.

6.3 Density Estimation in Action

Bandwidth selection and density estimation are available in XploRe through the BWSEL module included in the SMOOTHER library. The methods currently available in the routine are

1. Least-Squares Cross-Validation (Rudemo, 1982),

2. Biased Cross-Validation (Scott and Terrell, 1987),

3. Smoothed Cross-Validation (Hall, Marron and Park, 1992),

4. Bandwidth-Factorized Smoothed Cross Validation (Jones, Marron and Park, 1991),

5. Park and Marron plug-in (Park and Marron, 1992),

6. Sheather and Jones plug-in (Sheather and Jones, 1991), and

7. Silverman's rule of thumb (Silverman, 1986, p. 47).

The bandwidths are obtained for Gaussian kernels, and in order to hasten computations, the method of WARPing (Härdle, 1991, pp. 27–36) has been used—this means that either the cross-validation or the score functions are evaluated at gridpoints. The number of gridpoints (the dimension of the h vector) is set by default equal to 10, but can be changed by expert users.

For an application, the data set CARKILL.DAT in the \XPLORE3\DATA directory, concerning the logs of the number of persons killed or seriously injured in car accidents, monthly observations from January 1969 to December 1984, is considered.[3] The raw data are shown in Figure 6.2. A feature of interest with the series is to what extent the seat belt law, which went into effect in February 1983, had a significant impact in reducing the number of persons killed in car accidents. It is shown in the following how kernel density estimation may serve to investigate such a feature.

The XploRe routine is activated by typing the commands at the XploRe prompt:

```
x=read("carkill")
library(smoother)
bwsel(x)
```

Through the second command, the SMOOTHER library (in which the DENBWS module has been included) is loaded before the main program for bandwidth selection is called by the command BWSEL(x).

---

[3]The car data set has been analyzed by Harvey (1989) and Balke (1993), who transformed the original series by taking natural logs and adjusting for seasonality by regressing the logs against twelve seasonal dummies.

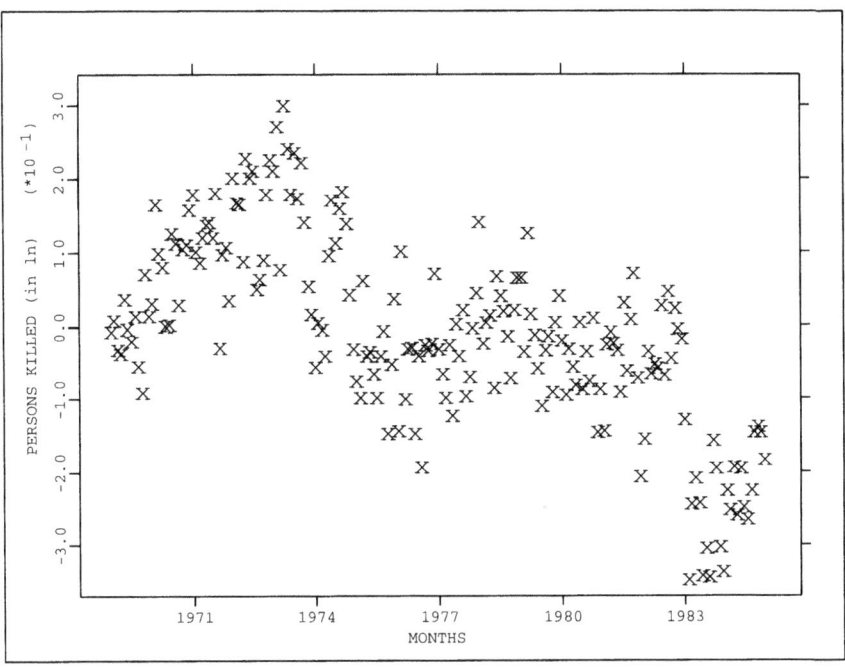

FIGURE 6.2. Number of persons killed or seriously injured in car accidents (in logs), January 1969 to December 1984.

After calling the BWSEL macro, XploRe displays the picture shown in Figure 6.3, where the screen is divided into four regions: at the bottom is the menu which allows the user to select the desired method, or to change the range of the bandwidths being investigated. On the right, the current range of $h \in [\underline{h} \ \bar{h}]$ is reported; by default, $\bar{h} = h^* + 0.10h^*$, $\underline{h} = \bar{h}/5$, with h^* being Silverman's rule of thumb bandwidth.[4] In the lower corner, the optimal bandwidth will appear after selecting any of the menu-available methods. The central area is finally reserved to plot the cross-validation or the score function and, subsequently, the associated density estimate.

---

[4]Note in general that such a range should cover the "right" support of both the cross-validation and the score functions. In fact, there are, broadly speaking, two possibilities: either the density is unimodal or multimodal. If the density is unimodal, then it is known that the optimal bandwidth is approximately h^*, whereas if the underlying density is multimodal, h^* greatly oversmooths the density; in this case, Silverman's h^* is an upper bound, and has to be reduced by a scale factor (5 in our case). If no minimum value of the cross-validation function or, alternatively, no zero in the score function is found on the default support, the user can modify $\bar{h}$ and $\underline{h}$ by pressing the <HOME> key.

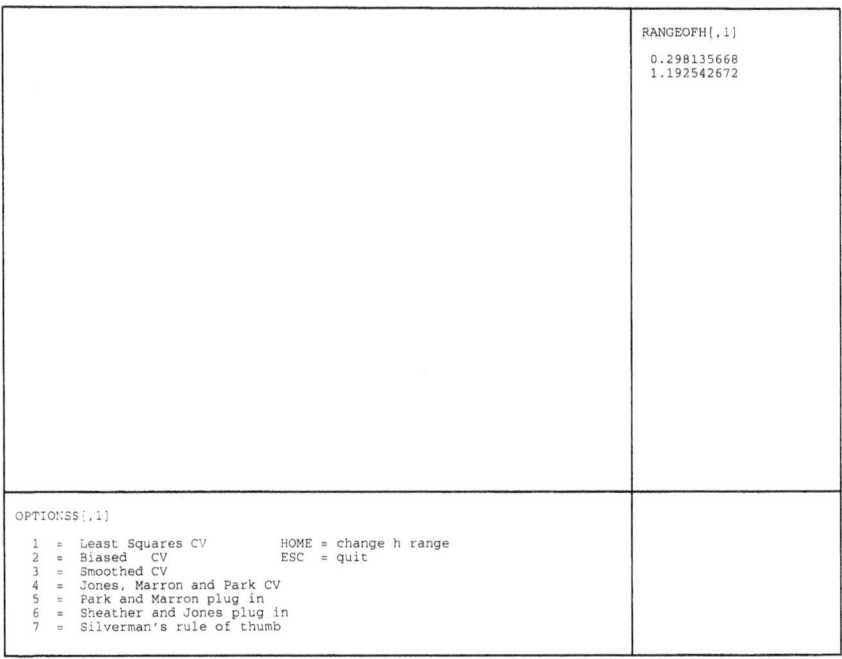

FIGURE 6.3. XploRe display after calling the macro **BWSEL**.

If the user selects Sheather and Jones' plug-in method by typing `<5>` at the prompt, then XploRe starts computing until the result appears on the screen, as reported in Figure 6.4. It can be seen in the right lower corner that the optimal h suggested by the method is now reported. Subsequently (that is, once the `<ESC>` key is pressed), XploRe automatically uses the optimal bandwidth to estimate the density using a Gaussian kernel. The result of the density is reported in Figure 6.5.

From the density estimate reported in Figure 6.5, a discriminant analysis may be implemented by placing cut-points in the density at $c_1 = -0.11$ and $c_2 = 0.09$ (vertical lines in Figure 6.5). The cut-points divide the values of x into three regions or intervals: $R_0 = \{-\infty, c_1\}$, $R_1 = \{c_1, c_2\}$, and $R_2 = \{c_2, \infty\}$, and the series is defined to be in state $S = j$ at time $t = i$ if $X_i \in R_j$, $j = 0, 1, 2$. The result of this discrimination analysis is summarized in Figures 6.6 and 6.7 and Table 6.1.

It is evident from the table that the seat belt law of February 1983 had a clear impact on the reduction of the number of drivers killed or seriously injured in car accidents. The state-labeled "Low Mortality" starts, in fact, in January 1983, which is one month before the effective introduction of the law (in order to prevent bills from the police, English drivers regulated their position *before* the law went into effect!).

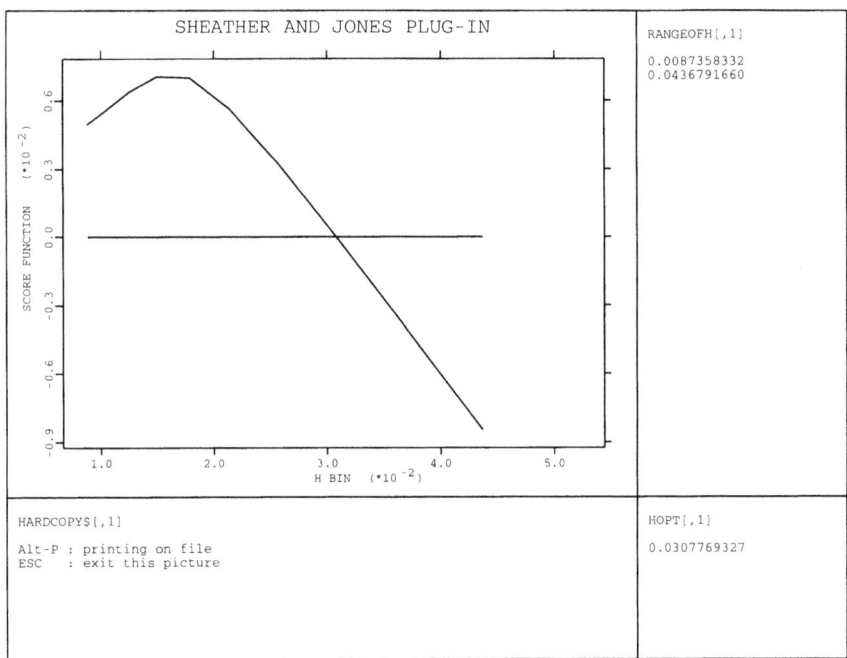

FIGURE 6.4. XploRe display after selecting SJ (option number 5).

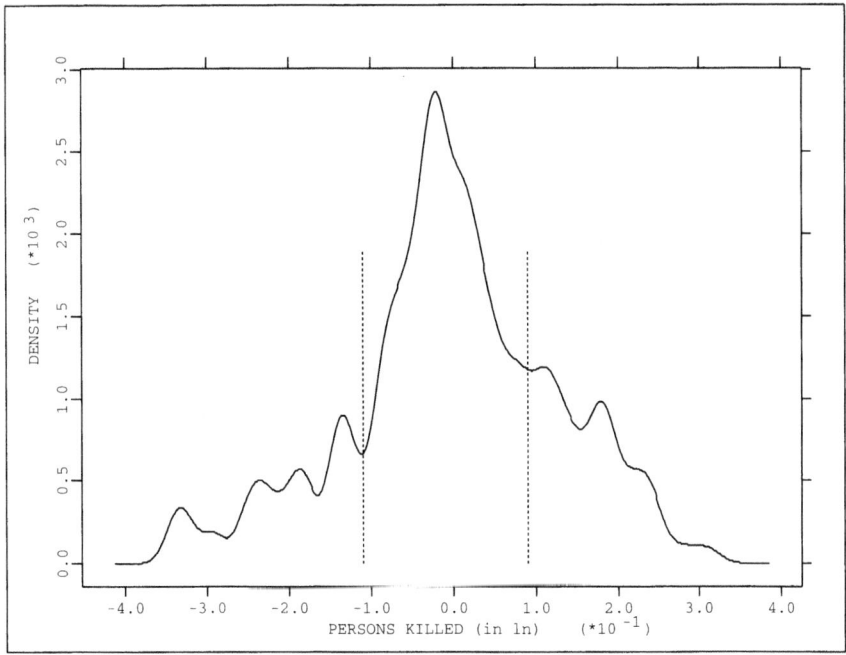

FIGURE 6.5. Density estimate after SJ bandwidth selection.

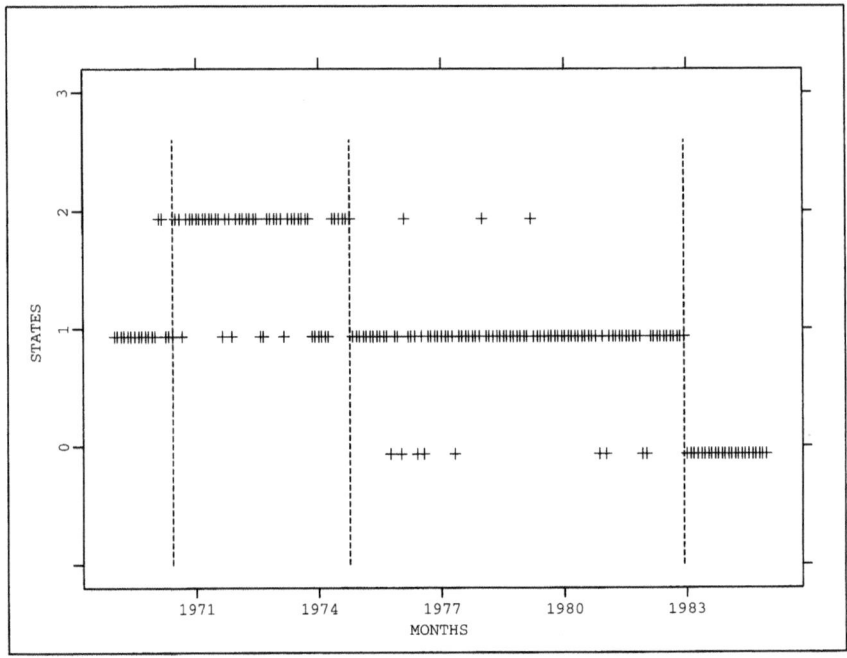

FIGURE 6.6. Allocating actual observations to different states.

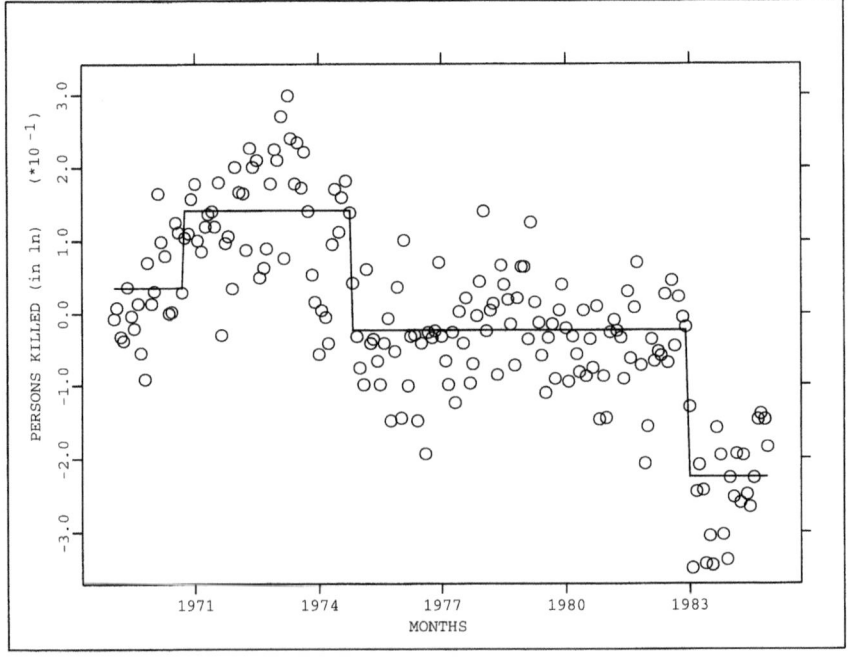

FIGURE 6.7. Actual and fitted data according to discrimination analysis.

State	Period	Mean	St.err.
Medium mortality	Jan. 1969 to Sep. 1970	1725	167
High mortality	Oct. 1970 to Oct. 1974	1960	126
Medium mortality	Nov. 1974 to Dec. 1982	1562	158
Low mortality	Jan. 1983 to Dec. 1984	1153	67

TABLE 6.1. Allocating observations to states. Cut-points at −0.11 and 0.09.

REFERENCES

Balke, N.S. (1993). Detecting level shifts in time series, *Journal of Business & Economic Statistics* **11**(1): 61–85.

Hall, P., Marron, J.S. and Park, B.U. (1992). Smoothed cross-validation, *Probability Theory and Related Fields* **92**: 1–20.

Härdle, W. (1991). *Smoothing Techniques, With Implementations in S*, Springer-Verlag, New York.

Harvey, A.C. (1989). *Forecasting, structural time series models and the Kalman filter*, Cambridge University Press, New York.

Jones, M.C., Marron, J.S. and Park, B.U. (1991). A simple root *n* bandwidth selector, *Annals of Statistics* **19**(4): 1919–1932.

Park, B.U. and Marron, J.S. (1992). On the use of pilot estimators in bandwidth selection, *Nonparametric Statistics* **1**: 231–240.

Park, B.U. and Turlach, B.A. (1992). Practical performance of several data driven bandwidth selectors, *Computational Statistics* **7**(3): 251–270.

Rudemo, M. (1982). Empirical choice of histograms and kernel density estimators, *Scandinavian Journal of Statistics* **9**: 65–78.

Scott, D.W. and Terrell, G.R. (1987). Biased and unbiased cross-validation in density estimation, *Journal of the American Statistical Association* **82**(400): 1131–1146.

Sheather, S.J. (1992). The performance of six popular bandwidth selection methods on some real data sets, *Computational Statistics* **7**(3): 225–250.

Sheather, S.J. and Jones, M.C. (1991). A reliable data-based bandwidth selection method for kernel density estimation, *Journal of the Royal Statistical Society, Series B* **53**(3): 683–690.

Silverman, B.W. (1986). *Density Estimation for Statistics and Data Analysis*, Vol. 26 of *Monographs on Statistics and Applied Probability*, Chapman and Hall, London.

Turlach, B.A. (1993). Bandwidth selection in kernel density estimation: A review, *Discussion Paper 9317*, Institut de Statistique, Université Catholique de Louvain, Louvain-la-Neuve, Belgium.

7
Interactive Graphics for Teaching Simple Statistics

Isabel Proença[1]

7.1 Introduction

The progress of computer science, in association with the progress of computer hardware, is making computers a more accessible and important tool in assisting individuals in processing and computing information. This fact enhances the use of computers for educational and teaching purposes. That is, the current sophisticated potential of computers can be used to provide a means to acquire concepts and to develop reasoning and problem-solving skills. Computer-aided learning (CAL) and computer-aided instruction (CAI) have become established fields in computer science; they utilize the computer as a tool for learning and presenting instruction that is individualized, interactive, and guided.

In this chapter, a teachware (or courseware) module will be presented that uses modern computer capabilities for the purpose of improving the quality of learning and the instruction of statistics. In the conception of this teachware, the following basic ideas were taken into consideration:

1. The learning process should be interactive, exploratory and directed by the student (and not by the program).

2. The system should be fully menu driven and easily usable by students without the need for supervision. In other words, it must be simple and very user-friendly, designed to be used essentially by persons who may have little knowledge of programming or computer handling.

3. Because some of the statistical and econometric techniques are highly graphical in nature, a highly interactive graphical interface routine is needed to give good graphical representations of the performance of the procedures.

---

[1]Instituto Superior de Economia e Gestao, Universidade Técnica de Lisboa, P-1200 Lisboa, Portugal. This work was developed within a project financed by the Fonds de Dévelopement Scientifique at the Université Catholique de Louvain. I would like to thank L. Simar and M. de Wolf for valuable suggestions.

Interactive CAL systems have appreciable benefits: they save time; they provide fast manipulation of statistical procedures and give immediate feedback; they reduce computational effort and therefore allow students to concentrate on concepts and to understand principles; and finally, they make learning more attractive and pleasant, therefore better motivating student attentiveness.

7.2   The General Structure of the Teachware System

This section aims to describe the main characteristics of this teachware implementation, such as the computer environment, the user interface, and the teaching/learning setting.

7.2.1   The Computer Environment

Software can be broadly divided in two categories: closed software and open software. Closed software systems are those where the user cannot modify or manipulate procedures at will; he must follow the options offered by the system in an automatic and direct way. With open software, the user has to learn a macrolanguage and build the procedures necessary to achieve the desired results. Usually, teachware is closed software. The system proposed here aims to be more flexible than the so-called closed software while simultaneously maintaining their user-friendly properties. Thus, it should allow the teacher to manipulate the procedures and tasks in order to suit particular teaching needs, and it should be direct and very easy to use for students. For these reasons, the system was written using a macro language. It was also required that the teachware run on IBM-compatible microcomputers.

The system is written in the macrolanguage of XploRe, mainly because of the highly interactive capabilities and useful graphic interface properties of XploRe. XploRe is a high-performance computing environment specially designed for statistical/econometric problems. It is written entirely in the C language and is a completely integrated system that includes graphics, programmable macros, an interpreter, and many analytical commands, collected in libraries. These libraries are reasonably exhaustive and make some tasks very easy and fast to implement (namely, tasks related with smoothing techniques). It uses the matrix as a basic data element which does not require dimensioning. This allows many numerical problems to be solved more easily using XploRe than by using a language like Pascal, Fortran, or C. The results can be readily displayed as graphics, with good-quality output, to plotters and printers.

The teachware presented here is a module comprising a set of XploRe macros. Each teaching subject is attached to a particular macro. However, to perform some specific tasks other macros have to be called within these main macros. Nevertheless, those will be effectively "hidden" to the user.

7.2.2   The User Interface

The teachware system under analysis is highly interactive: the computer communicates with the user via a system of tree-structured menus. Users communicate by pressing a key that identifies their option. Menus describe the tasks and actions that can be pursued, and each action is identified by a specific key on the computer keyboard. The user has an active role in the process of performing a task and getting a result. This seems to be more attractive for the student and more effective for learning than simply having the solution (and the particular action that produces it) predetermined by the program. The goal is to allow the student to carry out iteratively a variety of actions and see the respective consequence of those actions until he or she reaches the "best" solution. Criteria are shown in order to evaluate the actual solution and to guide the student to achieve the "best" answer. The results of an action are shown graphically and numerically.

7.2.3   The Teaching/Learning Setting

The target population of this teachware system is undergraduate students of statistics and/or econometric courses. These students do not need to know the XploRe language, and they do not need special knowledge about computers to be able to use the teachware module. However, they do need a preinstruction on the subjects that the module is concerned with. This prior instruction should be presented by the teacher in lectures or should be obtained by the student by reading the appropriate course textbooks.

The teachware module can provide a rich interaction between the teacher and the student. It can be used in the classroom by the teacher (using the appropriate technology for projection of the computer screen) as a more efficient method than the blackboard (more efficient in the sense that not only are the computations much easier and faster, but it also avoids the sometimes painful attempts of the teacher to draw graphics using chalk). The teachware module can also be used by the student alone, whether at home (self-instruction) or in class under the tutoring of the teacher.

7.3   Description of the Main Macros in the Module

When the implementation of a teachware system for statistics and econometrics was planned, it was decided that the focus should initially be on core subjects of those fields traditionally taught to undergraduate students. The list of these subjects is given below:

1. The thinking behind the ordinary least squares rule (OLS).

2. Impact of influential observations on the OLS rule.

3. Principal component analysis.

4. Density estimation.

5. Nonparametric regression.

The reasons of these choices are the following: The OLS rule is the most popular technique used in basic econometrics, mainly because of its simplicity and wide use. OLS is the first rule learnt by students for fitting a model to a data set. Outliers in data can be a serious problem, interfering with the quality of a fit, and they are not uncommon when dealing with real data. It is important to know how utilitarian methods like the OLS react (concerning robustness) when outliers are present. Principal component analysis is a useful technique in a basic approach for dimension reduction with many applications. Density estimation is often needed in applied statistics. Here, we are concerned mainly with the kernel method, given its well-known superior performance related to other methods like the histogram. Today computer packages can automatically provide a kernel density estimate if the user chooses the bandwidth value (and sometimes also the kernel). The choice of bandwidth is the most interesting problem related to kernel estimation, in the sense that the shape of the estimate is influenced by the bandwidth (and is practically the same for all kernels). It is important for the student to know the meaning and consequences of undersmoothing and oversmoothing the data. Nonparametric regression has recently become more usable due to the increasing speed of modern computers. It is a valuable tool for exploratory analysis of data in order to guide (or even to validate) the choice of a parametric model.

The macros that constitute the teachware presented here are stored in the XploRe library named **TEACHWAR**. To gain access to this tutoring system, the user has to call the library **TEACHWAR** and the macro **TWARE** by typing (from the command line):

```
library(TEACHWAR)
TWARE(0)
```

A menu will appear describing all the tasks that can be performed. Each task corresponds to a certain macro. To perform a task, the user has only to position the cursor in the line that contains the task description and press return. The corresponding macro starts to run automatically. In some situations another menu may appear showing several options related to this task. An option is selected by pressing the indicated key. Then the screen will be divided into several windows. One window shows a graphic display of the object(s) related to the task, representing an initial situation. After analyzing this situation, the user can take an action to produce a desired modification. The specific action is taken by pressing a matching key. In another window a help menu is shown which presents all the possible actions that can be pursued and indicates the appropriate key. In the last window, some information will be provided concerning numerical results or other graphic representations that describe the state of the actual situation or show criteria to evaluate it. When the user carries out an action by pressing the matching key, all windows will automatically be updated showing the situation resulting from that action. The user can end the task (exit the macro) whenever he wants. In almost all the tasks, the user can manipulate the input objects by manipulating the parameters of the simulated data or by introducing a data set of his choice. The module includes the macros named TWARE, TWLINREG, TWINFOBS, TWPCOMPO, TWDENES1, TWDENES2, TWREGSMO, and TWREGEST, each one corresponding to one of the subjects (tasks) mentioned previously. TWARE gives the interactive initial menu and calls the macro corresponding to the task selected by the user. The other macros require a more detailed analysis, presented below.

7.3.1   Linear Regression: TWLINREG

OLS provides a simple tool to summarize a data set by a linear model. The students learn easily how to compute the coefficient estimates of the regression line, but they should also understand the performance in practice of this technique; that is, how the fitted line is determined for a given data set. The aim is to show that the data are projected in the direction that minimizes the L_2 distance from the observations, the sum of squared residuals (SSR). To illustrate this purpose, students are confronted with a "good" data set (in the sense that it is adequate to perform on OLS) and a line (direction) that projects the data and represents a linear fit. The data are generated from the linear model with random error $Y = \alpha + \beta X + \sigma U$ verifying the classic conditions. Students can move the line in any direction and visually measure the distance between the line and the data. They can also see numerically the importance of this distance evaluated by the SSR. Moreover, they can perceive that any other linear fit besides the OLS produces larger residuals, or in other words, is more apart from the bulk of the data. The student can manipulate the values of the parameters in the model in order to generate different data sets. This can be especially interesting in illustrating how the accuracy of the OLS estimates depends on the variance of the error term.

When the macro starts to run, a graphic with the simulated data set and an initial fit are shown, together with the parameter values of this actual fit and the respective SSR (Figure 7.1). The student should move the line (according to the actions shown in the help menu) and simultaneously check the changes in SSR. For instance, the student can move the line from the initial position to the position shown in Figure 7.2 (by successively

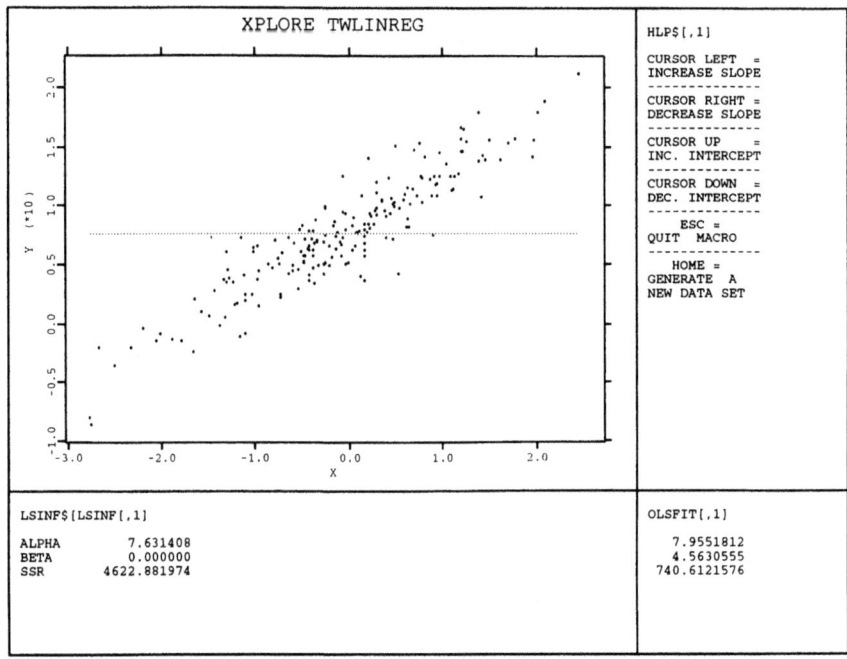

FIGURE 7.1. The initial plot of **TWLINREG**. In the upper left window a data set is displayed together with a line that gives a first fit to the data. In the window below, the values of the parameters of this line are shown together with the SSR of the corresponding fit. The smaller window gives the OLS results, while the upper right window shows a help menu.

pressing **<←>**), making an improvement in the SSR. However, OLS results (presented in the smaller window) are still better.

The actions that can be performed are

- increase/decrease the slope of the line on the screen;

- increase/decrease the intercept of the line on the screen;

- enter new values for the parameters α, β, σ; and

- quit the macro and return to initial menu.

7.3.2   Influential Observations: TWINFOBS

OLS is a useful rule but it does not perform equally well for all data sets. A common problem when analyzing real data is the presence of outliers (as was already mentioned). Outliers are observations with a clearly distinct value compared to the rest of the data, and they can occur in the dependent variable or in the explanatory variables, with different consequences

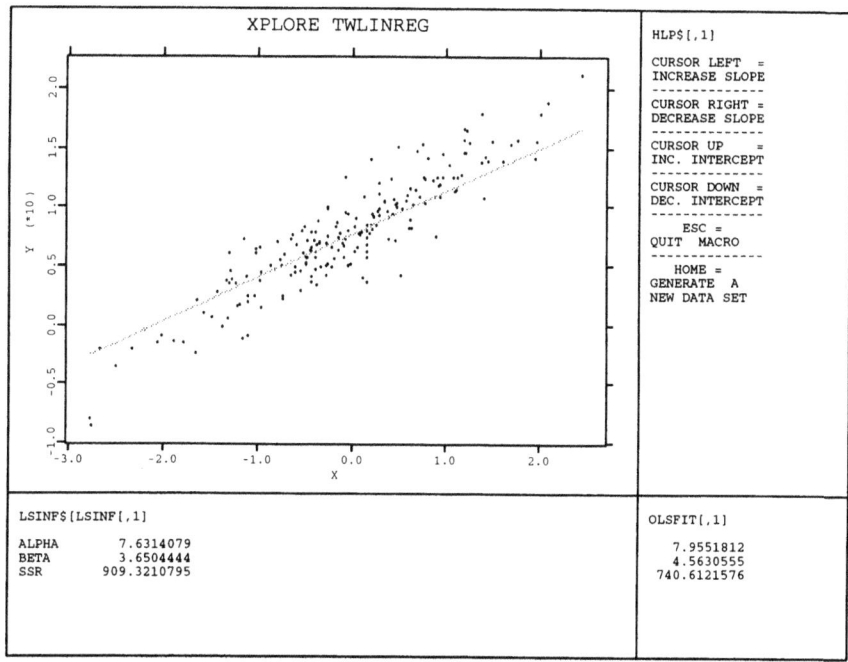

FIGURE 7.2. The result obtained after moving the line displayed before by increasing the slope. The SSR has diminished but OLS results are still better.

in estimation. Outliers in the explanatories can be quickly identified by a high influence index measured by the elements in the diagonal of the "hat" matrix, the matrix $X(X^T X)^{-1} X^T$. In most cases, unusual values in the observations are due to measurement errors. In this circumstance it is adequate to use an estimation method that will restrict the influence of this observation on the resulting estimate, and better reflect the behavior of the bulk of the data, that is, a robust method. OLS is not a method robust to outliers and students should be aware of this.

The aim of **TWINFOBS** is to show how sensitive the OLS rule is to changes in the observation value, whether the change is in the explanatory or in the dependent variable. This setup is useful for illustrating how OLS behaves in the presence of outliers in the data. The data are simulated from a classical linear model with random errors and is displayed, together with the OLS fit, as can be seen in Figure 7.3. In the upper right window the elements of the diagonal of the "hat" matrix are shown, while the smaller window contains the maximal value of these elements and the value corresponding to the fifth observation (the one represented by a circle).

The user has the option of modifying the position of one observation (the one represented by the circle in Figure 7.3) by changing the value

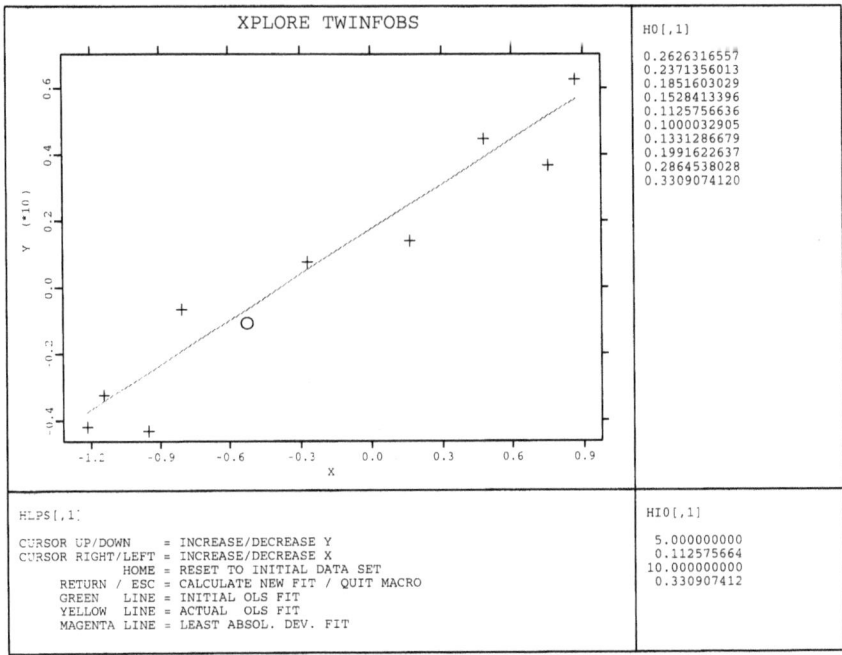

FIGURE 7.3. The initial plot of **TWINFOBS**. The upper left window shows a data set together with the OLS fit. The fifth observation is identified by a circle and can be manipulated in order to simulate an outlier. In the upper right window, the diagonal of the "hat" matrix is displayed, while in the smaller window, the influence value of the fifth observation and the most influential observation are displayed together with the observation number. In the lower left, a help menu is shown.

of X and/or Y within the graphic display (as much as he wants) to see how the OLS fit is changed. The L1-line fit (resulting from minimizing the sum of the absolute value of residuals) is also shown as a more robust alternative to OLS. However, this fit is not completely robust to outliers in X. Figure 7.3 shows one situation where the L1-line is more robust than the OLS in the presence of an outlier in the explanatory variable. Note that after having been manipulated, the fifth observation is now represented by a yellow star.

The possible actions that can be undertaken are

- increase/decrease the value of the observation in X;

- increase/decrease the value of the observation in Y;

- calculate the new fit for the modified data set;

- reset to initial data set; and

- quit macro and return to initial menu.

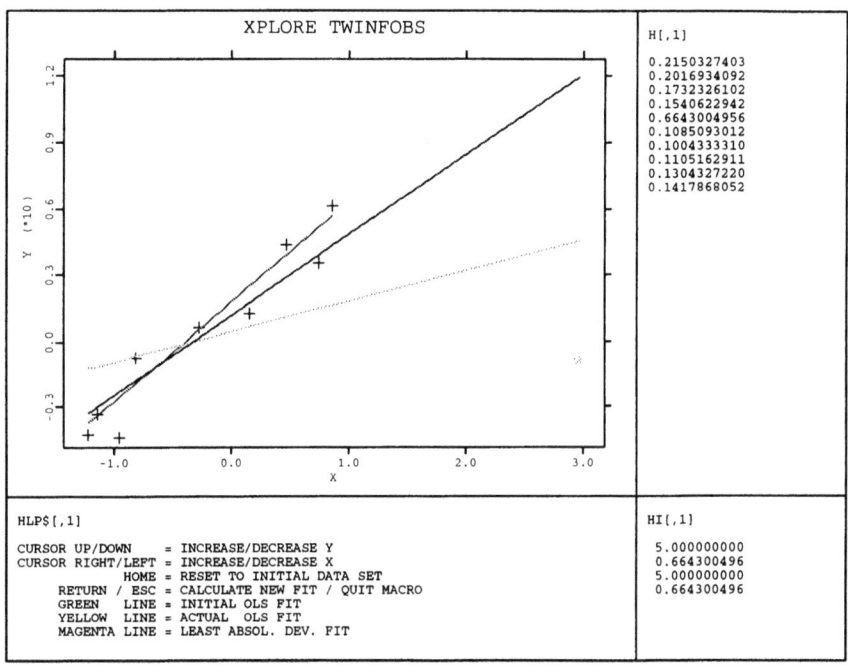

FIGURE 7.4. After moving the fifth observation (now represented by the yellow star), a situation was reached where the OLS of the manipulated data (yellow line) moved from the bulk of the data in the direction of the outlier. The original OLS fit (green line) is also shown together with the L1-line (magenta line). Here, the L1-line is more robust than the OLS (see color insert).

7.3.3 *Principal Component:* TWPCOMPO

The aim of principal component analysis is to find appropriate linear combinations of the variables in the data (projections) that best summarize the general behavior of the data. The goal of this technique is the reduction of the dimension of the problem by substituting for the original variables a smaller number of projections, new variables, achieving effectively the same descriptive power as the original data. The principal components give the best projections in the sense that they incorporate better the variability of the original data. Consequently, these projections have maximal variance (among all linear combinations of the original variables where the square of the weights sum to one).

When showing the performance of principal component analysis graphically, we are severely restricted regarding the number of variables that can be used. XploRe provides 2-D and 3-D plots. Three-dimensional plots are generated by projecting the third variable into a plane generated by the

122   I. Proença

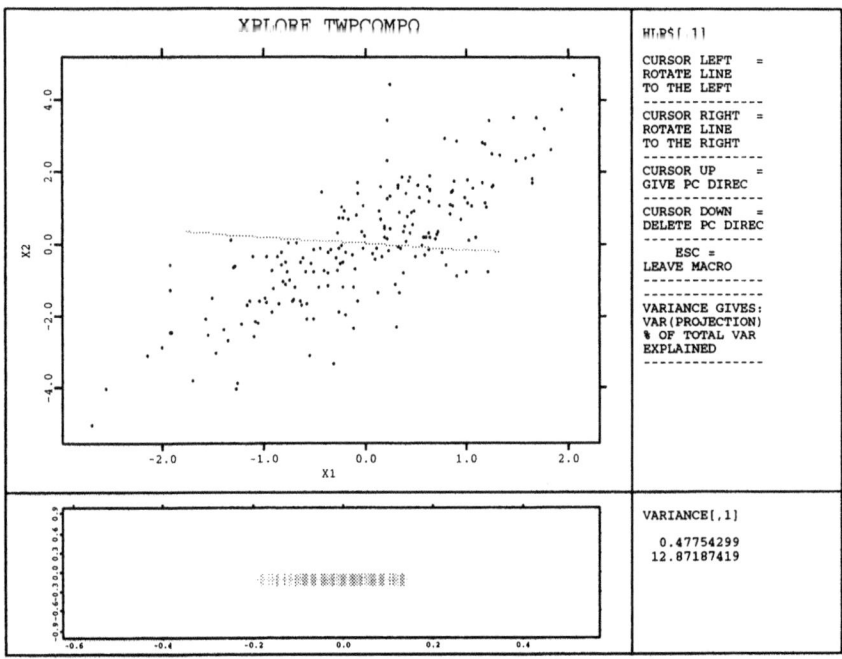

FIGURE 7.5. The initial plot of TWPCOMPO. In the upper left window
a data set is displayed together with a line representing a direction to
project the data. This projection is shown below. The smaller window
displays the variance of the projection and the percentage of total variance explained, while in the upper right, a help menu is shown.

other two. This plane can be rotated interactively (generating new projections), allowing us to see the data set from several angles. In the framework
of this TEACHWAR module, it was shown to be most convenient to use two-
dimensional plots. Even if the principal component analysis is used for
problems in higher dimensions in practice, a graphical example with two
variables can illustrate the performance of this method well.

The data used in TWPCOMPO are simulated from a bivariate normal distribution. The task of this macro consists of projecting the data into one
direction, given by a linear combination of the two variables (where the sum
of the weights is one). In Figure 7.5 the data set and an initial direction are
shown in the upper left window. In the window below, the data projected
in this direction are also displayed. The smaller window gives the variance
of this projection and the percentage of the total variance explained.

The student can manipulate this direction in order to find the best projection of the data by rotating iteratively the line in the upper graphic;
the changes on the projected data can be seen in the window below. One
situation that can be achieved is shown in Figure 7.6. Here, the principal
component is also given, corresponding to the red line. The smaller window

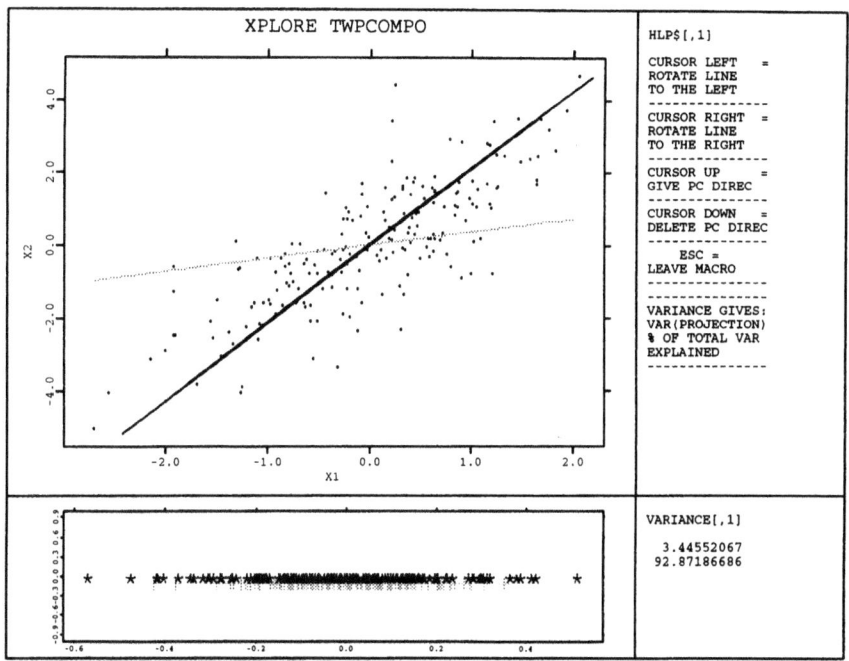

FIGURE 7.6. The upper left window shows a direction (yellow line) resulting from moving the initial line. Below, the data projected in this direction are shown with crosses; it can be seen that this projection has more spread than the initial in Figure 7.5. The principal component is also plotted in the upper left (red line). The data projected in the principal component are shown with red stars in the lower left; this projection has the biggest variance. The smaller window shows the results concerning the principal component (see color insert).

shows that the principal component explains 93.28% of the total variance. The other direction in the graph (the yellow line) gives a projection of the data with less variance, as can be seen in the lower left window, but it is better than the initial projection.

After some trials, the student will see that there is no other direction that gives a projection with higher variance than the principal component.

The possible actions are

- rotate line to the right/left;

- give principal component results;

- delete principal component; and

- quit macro and return to initial menu.

7.3.4   Density Estimation: TWDENES1 and TWDENES2

These macros perform density estimation of a data set. As mentioned previously, they are focused on the kernel method. The kernel density estimator of a univariate data set X_i, $i = 1, \ldots, n$ is defined by

$$\hat{f}_h(x) = \frac{1}{nh} \sum_{i=1}^{n} K\left(\frac{x - X_i}{h}\right),$$

where $K(\bullet)$ is the kernel and h is the bandwidth. The kernel must be a function symmetric around zero with integral equal to one. Härdle (1991) gives a good survey of kernel density estimation. To obtain a kernel estimate, the kernel function and the bandwidth value must be chosen. It is proved that the particular kernel used does not have much effect on the shape of the density estimate (Härdle, 1991, p. 78). Thus, the choice of the kernel is not important. One of the most widely used kernels in practice is the Gaussian. On the other hand, the bandwidth value does affect the resulting estimate in terms of its bias and its variance. The bias of the kernel estimate increases with h, while its variance decreases with h. Bandwidth choice is a crucial problem for kernel density estimation which is not simple, given that it is necessary to balance the trade-off between the bias and the variance.

A "good" bandwidth can be defined as one that minimizes the mean integrated square error (MISE) of the kernel estimate. However, to calculate the MISE, the true density of the data needs to be known. In real applications, this is not the case and other criteria should be used. One popular criterion suggests choosing the bandwidth that minimizes the least squares cross-validation (LSCV) function. For details about this method, see Härdle (1991). On the other hand, the following rule of thumb gives a quick method of choosing the bandwidth. This rule consists of using

$$\hat{h} \cong 1.06\hat{\sigma}n^{-1/5}.$$

There is also a more robust variant of the rule of thumb, a "better rule of thumb", which determines h according to

$$\hat{h} \cong 1.06 \min\left(\hat{\sigma}n^{-1/5}, \frac{\hat{R}}{1.34}\right).$$

In the formulas above, $\hat{\sigma}$ is the sample estimate of the standard deviation of X, and $\hat{R}$ is the interquartile range of the data. These rules were deduced by assuming that the true distribution of X is Gaussian, so that, whenever the data comes from a distribution which is not too far from Gaussian the rule of thumb (or the better rule of thumb) should provide a good value for h.

The kernel density estimates calculated in the this teachware system were obtained using a Gaussian kernel and the technique of binning (considering

250 bins) to speed up calculations. The binning technique is explained in detail in Härdle (1991). The criteria used to guide the bandwidth choice were the minimum of MISE for simulated data and the minimum of LSCV for real data. The macros work as follows: TWDENES1 reads by default the data stored in the file chron.dat in the XploRe data directory. However, the macro may read any other data set indicated by the user (provided that the file specification where these data are stored is correctly given to the program). TWDENES2 simulates a data set from a density function. The user can choose this density from three different options. The options are a symmetric density (which is a standard normal), a bimodal density (which is a mixture of normals), and an asymmetric density (also a mixture of normals). The sample size is also defined by the user.

After the data have been read or simulated (depending on the macro used), a first kernel density estimate appears in the screen. This estimate is calculated for a bandwidth h given by the empirical rule $h = 5d$ where d is the binwidth. The user may change the bandwidth and see the effects of oversmoothing and undersmoothing the data and the behavior of the criterion function. In the lower left window, a graphic of the criterion function for all bandwidths used until this moment is displayed. The point corresponding to the actual bandwidth is identified by a different color to help the user easily locate the present criterion state. The actual bandwidth, the increment value to change it, and the numerical value of the criterion function are also shown in another window. A histogram can also be obtained by pressing the key h. Figures 7.7 and 7.8 show a situation produced with TWDENES1. The red points correspond to the initial density estimate of the data set, and the yellow line corresponds to a smoother estimate which has the minimal MISE (for the bandwidths tried) as can be seen in the graphic in the smaller window. The second figure shows the same situation with the addition of a histogram.

Figure 7.9 was given by TWDENES2. The graphic shows the true simulated density, which is asymmetric. The cyan line is the density obtained with the better rule of thumb, which is quite close to the true density with an MISE very near to the minimum presented in the graphic in the lower right window. An oversmooth density estimate is given by the yellow line. The MISE of this estimate is represented by the yellow point in the smaller window.

The possible actions are

- increase/decrease the bandwidth by an increment;

- increase/decrease the increment;

- calculate a histogram;

- calculate the kernel estimate with the *better rule of thumb*;

- read a new data set (for TWDENES1);

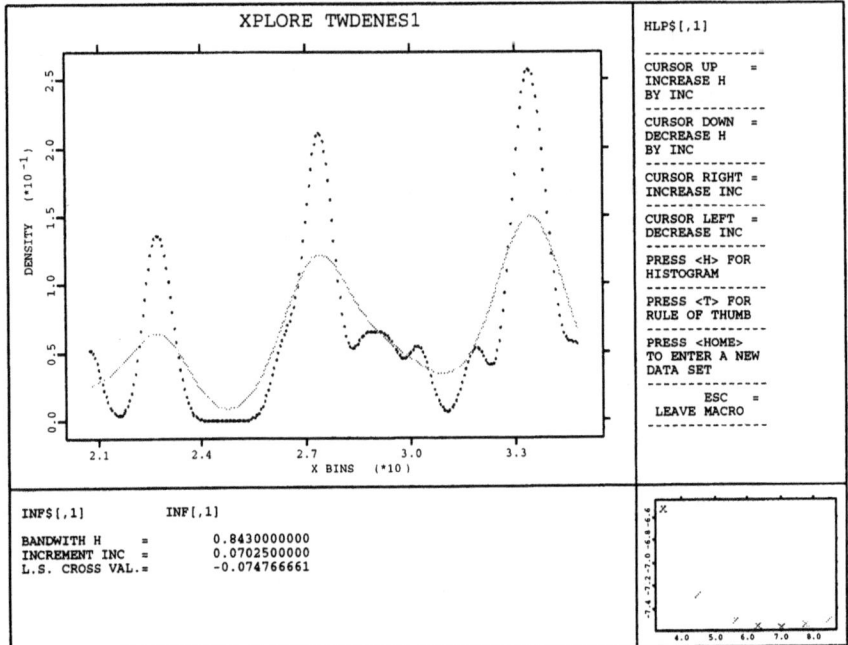

FIGURE 7.7. The upper left window shows two kernel density estimates of the **chron** data for different bandwidths. The red points indicate the initial estimate obtained for a small bandwidth, while the yellow line gives the actual estimate, which is smoother. Below, values for the bandwidth and LSCV concerning the actual estimate are shown, together with the increment to change the bandwidth in order to replace the actual estimate. In the lower right, the LSCV values for the different bandwidths tried before are shown. The upper right shows a help menu (see color insert).

- return to last menu in order to choose the density to simulate the data (for **TWDENES2**); and

- quit macro and return to initial menu.

7.3.5 *Kernel regression:* **TWREGEST**

Kernel estimation is a well-known technique for nonparametric regression. Assume that a variable Y is related to an exogenous variable X by an unknown model. Suppose that the expectation of Y conditional on X is $E[Y|X] = m(X)$, where $m(X)$ is an unknown function. An estimate of $m(X)$, the regression curve, can be determined nonparametrically using

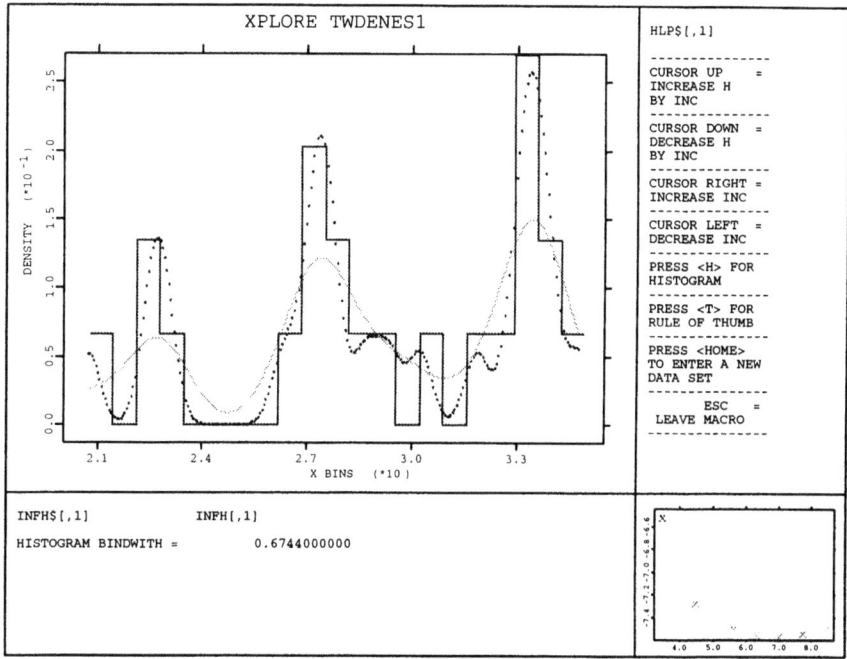

FIGURE 7.8. Here we have the same situation as in the previous figure, with the addition of a histogram (see color insert).

the Nadaraya–Watson estimator defined by

$$\hat{m}_h(x) = \frac{\sum\limits_{i=1}^{n} K\left(\dfrac{x - X_i}{h}\right) Y_i}{\sum\limits_{i=1}^{n} K\left(\dfrac{x - X_i}{h}\right)},$$

where $K(\bullet)$ is the kernel. Details about kernel regression smoothing can be seen in Härdle (1990). Here, as in density estimation, the choice of the bandwidth is crucial (while the choice of the kernel plays no important role). Least squares cross-validation (LSCV) is a common method of bandwidth choice in kernel regression. The "best" bandwidth is the one that minimizes the LSCV function.

TWREGEST performs a kernel estimate of the regression of the one-dimensional variable Y on X. The main purpose is to show the flexibility of a kernel regression of a given data set and the effects of bandwidth choice on undersmoothing or oversmoothing. This macro evolves in the same way as TWDENES1. By default, the data are read from the file motcyc.dat that

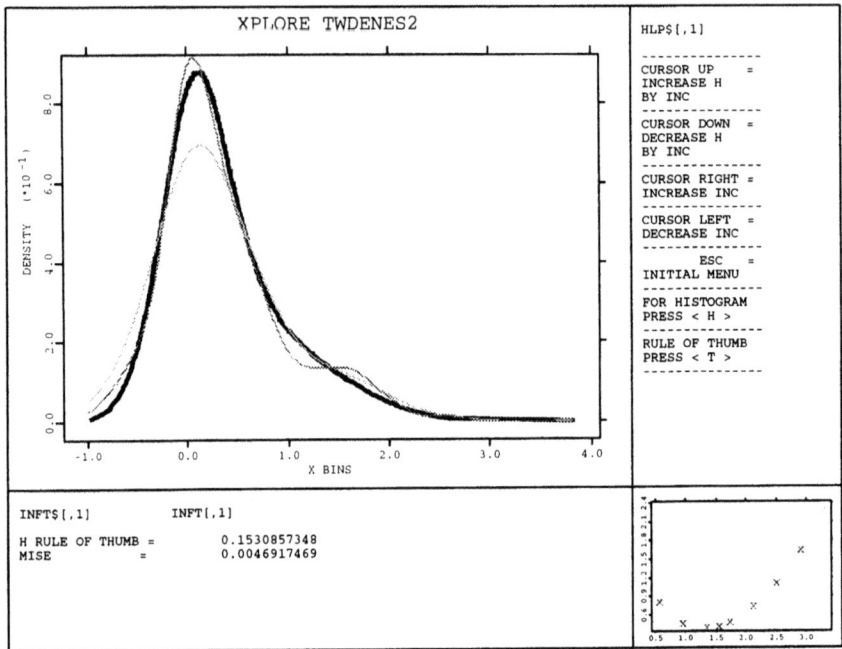

FIGURE 7.9. In the upper left window, a simulated asymmetric density (thick red line) is shown, together with two kernel density estimates. One was obtained for the bandwidth given by the better rule of thumb (cyan line) and the other for a bandwidth chosen by the user. The estimate for the better rule of thumb is pretty close to the true density with a very small MISE, while the other estimate is oversmoothing the data. The plot of the MISE for the different bandwidths used before is shown in the lower right. The yellow point gives the MISE for the estimate represented by the yellow line.

is in the XploRe data directory. Any other data set can be read by the program provided that the user correctly inputs the file specification where the data are stored. The estimates are calculated with binning using 250 bins. The initial bandwidth is calculated in the same way as in TWDENES1. For every bandwidth selected by the user, the corresponding value in the LSCV function is given numerically and graphically. Figures 7.10 and 7.11 were produced by TWREGEST. They show the kernel regression estimate that minimizes the LSCV (for the bandwidths tried) given by the yellow line together with the data set. Figure 7.10 refers to the motcyc data, and Figure 7.11 refers to the geyser data. The red line corresponds to a rough kernel estimate.

The actions that can be pursued are

- increase/decrease the bandwidth by an increment;

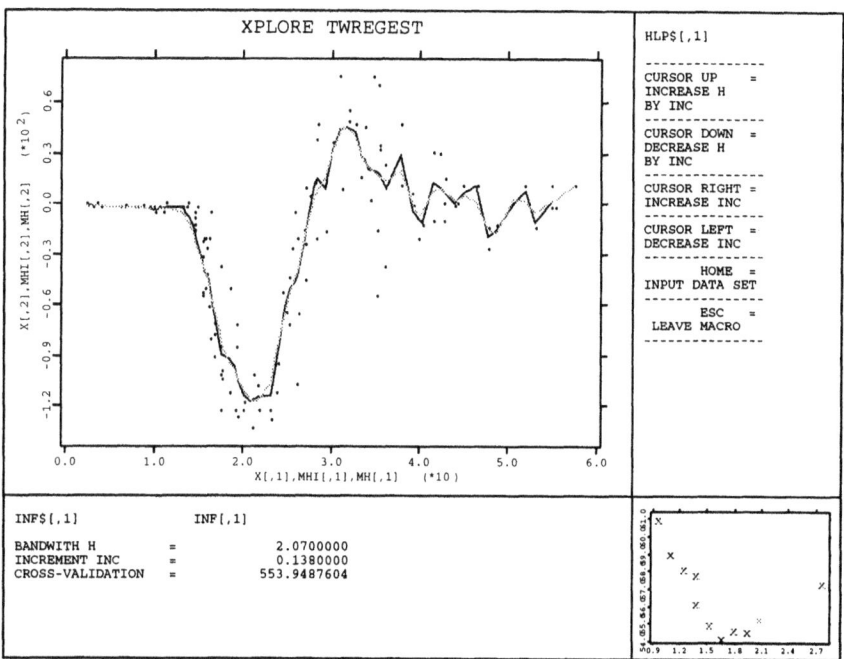

FIGURE 7.10. The plot in the upper left shows two kernel regression estimates for different bandwidths of the motcyc data. The initial estimate is given by the red line, while the smoother estimate (the yellow line) was obtained for a bandwidth chosen by the user. Below, values for the bandwidth and LSCV of the actual estimate are shown together with the actual increment to change the bandwidth in order to obtain another estimate. In the lower right, a graphic with the LSCV for the bandwidths tried is also shown. In the upper right, a help menu is displayed.

- increase/decrease the increment;

- read a new data set; and

- quit macro and return to initial menu.

7.3.6   Regression smoothers: TWREGSMO

There are many more methods of nonparametric regression in addition to the kernel regression by the Nadaraya–Watson estimator. The choice of one method depends mainly on the data in question and the purpose of the estimation. These methods may differ in robustness, smoothness, calculation complexity, and speed, among others. Härdle (1990) provides a good survey of nonparametric regression methods.

The aim of this procedure is to show the performance of different regres-

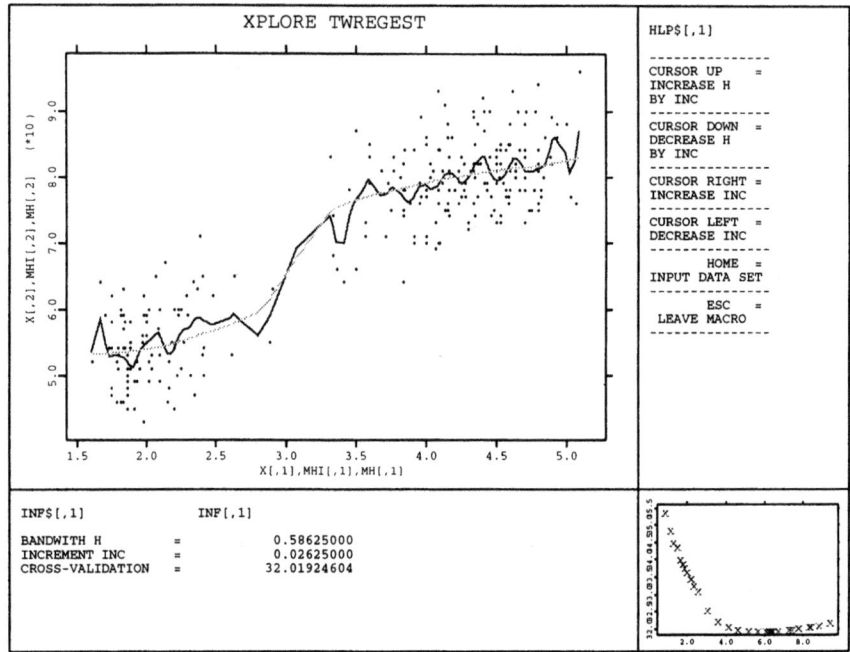

FIGURE 7.11. The plot in the upper left shows two kernel regression estimates for different bandwidths of the **geyser** data. The initial estimate is given by the line with points, while the smoother estimate (the yellow line) was obtained for a bandwidth chosen by the user. Below, values for the bandwidth and LSCV of the actual estimate are shown, together with the actual increment to change the bandwidth in order to obtain another estimate. In the lower right, a graphic with the LSCV for the bandwidths tried is also shown. In the upper right, a help menu is displayed (see color insert).

sion smoothers for the same data set. The program reads the previously mentioned **motcyc.dat** data and displays it on the screen. As before, any other data set can be read. The user must perform an action indicating which method should produce the smoothed fit. This fit will then be displayed on the same graphic as the data. For the smoothing methods that depend on a bandwidth value, the user has the option of changing this value. Figures 7.12 and 7.13 show, respectively, a running median and a spline fit.

The available actions are

- calculate the supersmoother;

- calculate a running median smoother;

- calculate a spline regression smoother;

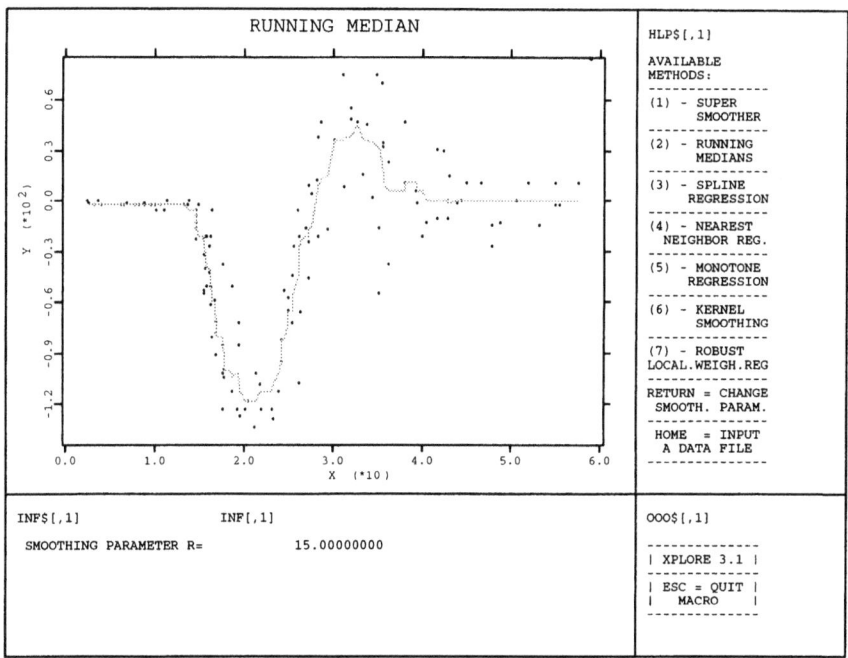

FIGURE 7.12. In the upper left window, the `motcyc` data are plotted, together with a running median fit. The smoothing parameter used in this fit is shown below. The upper right window shows a help menu.

- calculate a k-nearest neighbor smoother;
- calculate the isotonic regression;
- calculate a kernel smoother;
- calculate a locally weighted regression smoother;
- input a new smoothing parameter;
- read a new data set; and
- quit macro and return to initial menu.

7.3.7 Remarks

Before concluding this section, a remark should be made. Procedures like `TWDENES1`, `TWREGSMO`, and `TWREGEST` can also be used for other than instructional purposes; they can also be used as quick and easy exploratory tools to analyze a given data set. For example, the user may use `TWREGSMO`

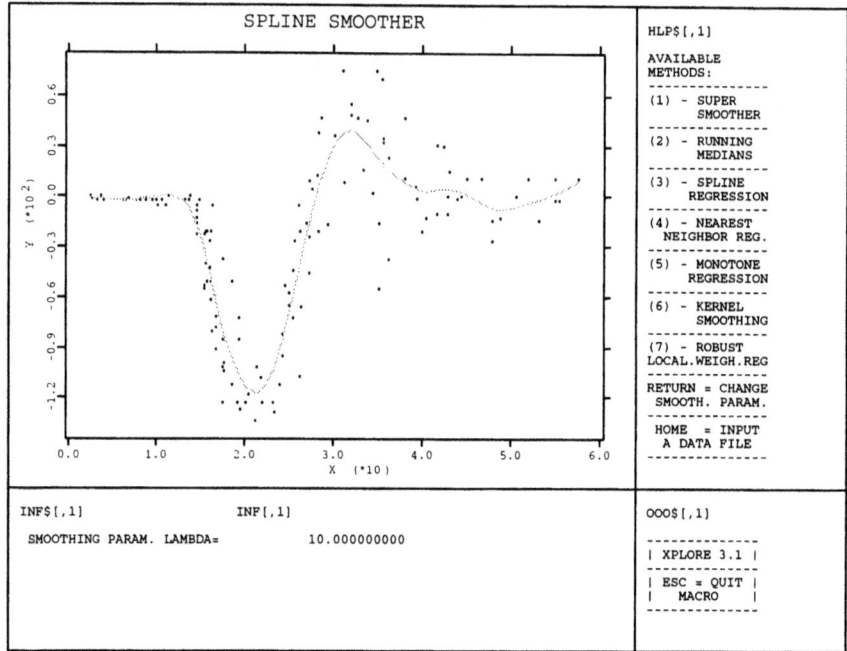

FIGURE 7.13. In the upper left window, the `motcyc` data are plotted together with a spline fit. The smoothing parameter used in this fit is shown below. The upper right window shows a help menu.

to see how the data are fitted by different smoothers (for different bandwidth values) in order to understand the data set better and to help to choose the method that best suits his or her purpose. With `TWDENES1` and `TWREGEST` the user has a quick and easy way to see the effects of bandwidth selection in density or kernel regression evaluated by the LSCV rule.

7.4   Details on XploRe Language

In this section some of the XploRe commands that have been used to write the teachware macros will be explained in some detail. This allows users to either build their own instruction macros or modify those already existent in order to suit some designed purpose.

7.4.1   Creating Windows and Showing Objects

One important feature of this teachware system is the simultaneously display in the screen of more than one graphic together with numeric results and text strings such as the help menu. In order to do this, the computer screen has to be divided into several windows of different type.

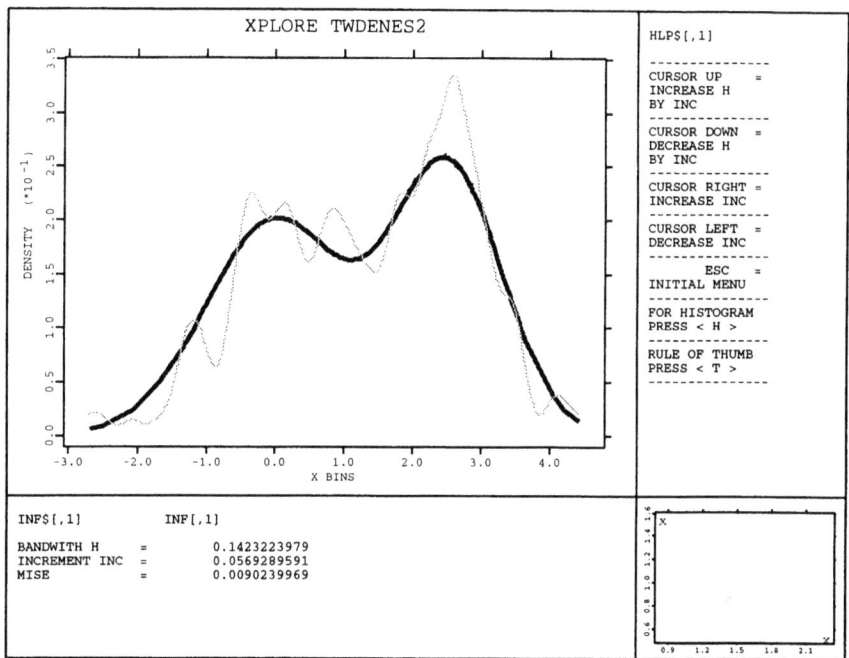

FIGURE 7.14. The initial display of **TWDENES2**. The screen is divided into four asymmetric windows using the XploRe command **createdisplay**.

The particular XploRe command for creating windows is **createdisplay**. To see an example of how to work with **createdisplay**, consider the following display created by the macro **TWDENES2** (see Figure 7.14).

The four windows of the above display were generated with the command

```
createdisplay(displayname , (-2) (-2) , s2d text text s2d).
```

The parameter **displayname** gives the name of the display, which identifies it as an XploRe object. The numbers in parentheses refer to the numbers of vertical and horizontal windows, respectively. Negative numbers create windows of asymmetric size. The last parameters indicate the type of window. In this case the first and fourth windows show static 2-D graphics (**s2d**) while the second and third are designed to show matrices (float or string matrices).

In the XploRe environment, objects are displayed using the command **show**. As an example, to generate the display from **TWDENES2** shown in Figure 7.14, the following **show** command was given (note that the *style* of the graphic was manipulated after the call of this command, as will be explained below):

```
show(tru fh s2d1, inf text1, hlp text2, misemsk s2d2)
```

The objects **tru** and **fh** are shown in the first window of type **s2d**. In the first window of type text, the object **inf** (which is a matrix where the first column is a string vector and the second is a float vector) is displayed,

134   I. Proença

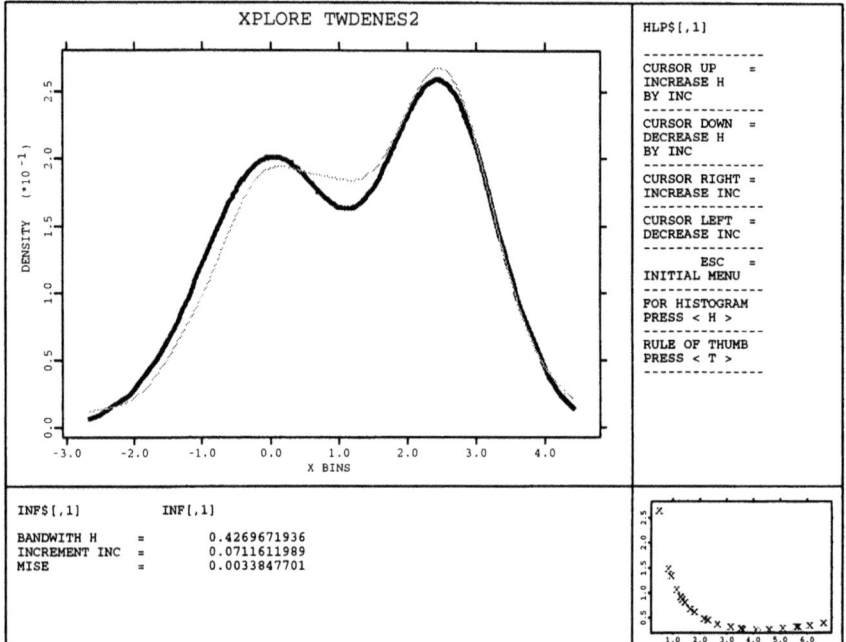

FIGURE 7.15. The yellow line in the upper left window was obtained by pressing <↑>. It has immediately replaced the line given in Figure 7.14. This process is carried out with the XploRe command **update** (see color insert).

while in the second window of type text, the string matrix **hlp** is exhibited. Finally, the graphic in the lower right window was produced by showing the object **misemsk** in the second **s2d** window.

7.4.2   Updating Displays

A fundamental feature of the teachware system is its ability to replace displayed graphics and matrices easily and quickly. That is, as soon as the user takes an action, the display has to be updated showing the resulting situation. The XploRe environment allows this procedure to be done easily using the command **update**.

Updating an Object

Suppose that in the situation presented in the display shown above, the user decides to increase the bandwidth in order to produce a smoother density estimate. After the action is communicated to the computer (by pressing the adequate keys), the display in Figure 7.15 appears as a result.

The replacement of the old display with the new situation was produced with the commands

```
upd = update(fh 2 s2d1 solid line newproj)
upd = update(misemsk 1 s2d2 newproj)
upd = update(inf 1 text1).
```

The first **update** command replaces the new object **fh** (after it has been modified according to the action taken by the user) in the first **s2d** window. When the first **show** command was called, XploRe retained the order with which each of the objects were given. Returning to the example in Figure 7.14, XploRe puts the object **tru** in the first position of the graphic and the object **fh** in the second position. This order is important for the update command because it is necessary to indicate which object is going to be replaced, that is, the first, second, etc., by giving the order that the new one is going to take in the graphic. Eventually, objects can also be added if this order is larger than the number of objects previously displayed. Thus, in the previous example, it is necessary to indicate that the new **fh** is going to take the position 2 (the same as before); it is also going to be drawn as a solid line. The parameter **newproj** is very useful because it allows the scale of the graphic to change for the new values assumed by the objects updated. If this parameter is not given the scale remains the same as before. With **update**, it is also possible to replace or actualize matrices in text windows. In this example, the matrix **inf** was also updated after being changed to incorporate the new bandwidth (and increment) values.

Updating the *style*

The command **update** is also useful to manipulate the *style* of the graphic, for example, the title, the axes name, the drawstyle of the data (line, points, etc.) that is usually undertaken in XploRe interactively by the user. In order to obtain the display in Figure 7.14, after the **show** command, the following update commands have to be given:

```
upd = update(tru 1 s2d1 title "XPLORE TWDENES2" xaxis "x bins")
upd = update(tru 1 s2d1 solid line thick yaxis "density")
upd = update(fh  2 s2d1 solid line).
```

7.4.3 Mask Operations

Mask operations manipulate the color and outfit of a data point. XploRe has a special type of matrices, mask matrices, that allow the user to define a specific color and a specific style (point, star, etc.) for each data point. The mask matrix defined for a particular data set must have the same number of rows as the data matrix and must be vertically concatenated with it. That is, having a data matrix (float type) x and defining a mask, the mask type columns should be "glued" to the float columns, resulting in a bigger matrix x. However, when x is displayed, the mask columns are

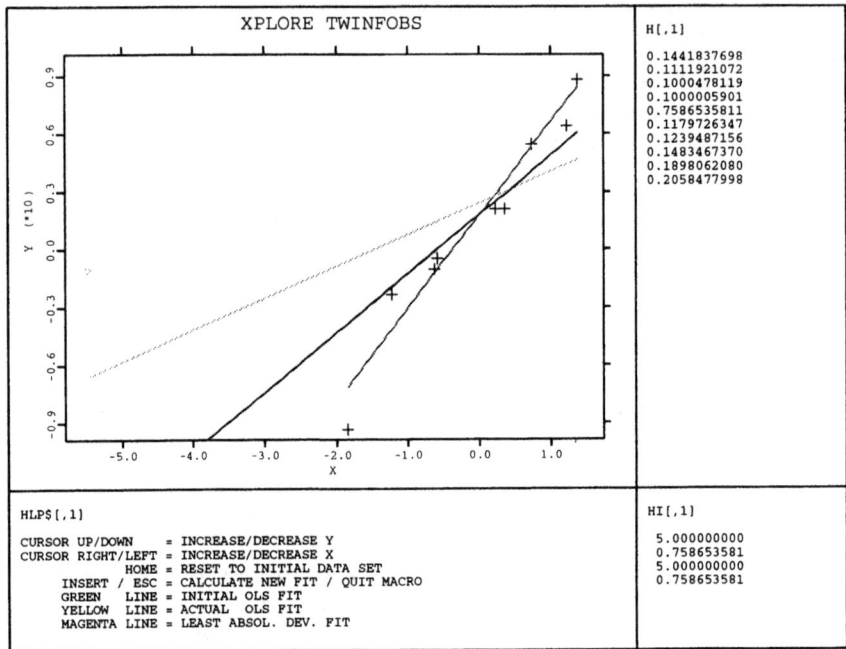

FIGURE 7.16. This display is an example of the XploRe features on masking. The different colors and styles can be assigned using the XploRe commands **mask** or vtocc (see color insert).

ignored. They are ignored as well in all mathematical operations applied to x. A mask matrix can be created using either the command **mask** or vtocc. The command vtocc may be preferable when different configurations in color or style are assigned to the data points, because the input matrix is numeric (with **mask**, the color and style are given by text strings). That is, using vtocc, the particular color combined with a particular style required for a data point is identified by a specific number.

Consider the display in Figure 7.16 produced by the macro TWINFOBS. Here, a data set with ten points is displayed where one of the points (the one corresponding to the fifth observation) has a different color (yellow) and style (star). This is the observation whose value has been manipulated by the user. Suppose that the data are stored in the float matrix **dat** and the OLS and L1 line-fitted data are in the float matrices **aols** and **al1**, respectively. The results of the OLS fit for the original data are in matrix **ools**. The mask operations to produce the outfit in the display shown in Figure 7.16 are obtained with the commands

```
ools = ools~mask(10 1 lgreen)
aols = aols~mask(10 1 lyellow)
al1  = al1~mask(10 1 lmagenta)
```

```
stco = 31*matrix(10)
stco[5,.] = 95
dat  =  dat~vtocc(stco)
```

The first three lines define mask matrices with 10 rows and 1 column and with style by default *point*; color is, respectively, light green, light yellow, and light magenta. Then a vector of 10 rows with all elements equal to 31 was generated. The number 31 identifies *point* with the color white. Then the fifth observation is labeled with 95, which identifies *star* with color yellow. Finally, the mask vector corresponding to `dat` is created with the vtocc command that interprets the numbers in `stco`. To obtain the display, it is necessary to give a `show` or `update` command to those variables.

7.4.4   Producing Interactiveness

Within the XploRe environment, the computer understands the particular action undertaken by the user by means of the following commands:

- `readcon`,

- `readstr`, and

- `readval`.

Reading a Key: `readcon`

XploRe assigns to each key in the keyboard a specific code number (for example, the key corresponding to number 1 has the code number 49). The command **readcon** returns to the program the code number of the key that has been pressed. It has useful optional parameters that allow the program, for example, to pause until a key is pressed or until a key belonging only to a prespecified subset of keys is pressed.

When the user chooses an action and transmits it by pressing a designated key, the program can identify this key with the **readcon** command and then select the action from the set of all possible actions by a conditional branching structure. This procedure can be illustrated by analyzing the XploRe code in the macro TWLINREG.

In TWLINREG, the user may implement one of the following actions:

- increase/decrease the intercept **alp** by pressing <↑>/<↓>;

- increase/decrease the slope **bet** by pressing <←>/<→>;

- insert new data;

- generate a new data set by pressing <HOME>; and

- quit the macro by pressing <ESC>.

The conditional branching structure that allows the program to select and run one of these actions is presented below.

```
k = readcon(-1 usekeys) ; waits until a key which code number
                        ; is in the vector usekeys is pressed
exit((k.=27).|(k.=327)  ; exits the loop if <ESC> or Home are
                        ; pressed
if      (k.=328)        ; cursor up
  alp = alp + dalp
elseif (k.=336)         ; cursor down
  alp = alp - dalp
elseif (k.=331)         ; cursor left
  bet = bet + dbet
elseif (k.=333)         ; cursor right
  bet = bet - dbet
endif
yfit   = x * (alp|bet)  ; fitted y in the actual direction
datfit = x[.,2]~yfit
ssr    = sum((y - yfit)^2)
tdat   = alp|bet|ssr1
upd    = update(zfit 2 s2d1 line solid)
lsinf  = nam~tdat
upd    = update(lsinf 1 text1)
```

Reading a String: `readstr`

Sometimes an action has to be transmitted as text. This is the case when the user wants to analyze a particular data set and has to input the name specification of the data file. XploRe enables the program to read a text typed by the user via the command **readstr**. When **readstr** is called, a box appears in the screen where the user can input the text. It is possible to send a message to the user inside this box. A string matrix is then created with the text typed by the user. In the macro TWDENES1, this command is used in the following way:

```
dat = readstr("input the name of the data file")
x = read(dat)     ; reads the file inputed by the user
```

If the data file is not stored in the XploRe data directory, then the full path name has to be given.

Reading a Value: `readval`

When the action has to be transmitted as a numeric value (usually a value that the user wishes to assign to some parameter), XploRe uses the command **readval** to retrieve it. The command **readval** works very similarly to **readstr** but creates a float matrix with the user input instead of a string. TWREGSMO is one of the macros that uses this command. For certain smoothing procedures, the user may change the smoothing parameter and input a value of his choice. This is the case for the running median smoother where the smoother parameter is identified by r. The command used is as follows:

```
r = readval("input new r (r has to be odd)")
```

A box will appear in the screen with the message given in the command and the user must insert a number into it. The smoothing parameter r is automatically set to this value.

Others: `message` and `writecon`

Sometimes the program needs to send a message to the user. This can be done using the command `message`. An example is provided by the macro `TWDENES1`. When the user asks for the histogram or the estimate given by the better rule of thumb, the program should pause as long as the user wishes to analyze the resulting picture because when it continues, the display will be set to the previous situation. With `message`, this is done by giving the directive

```
message("press any key" WAIT)
```

After `message` is called, the text between the quotation marks will appear on the computer screen (in the command line). The optional parameter `wait` obliges the program to hold in pause until a key is pressed.

Another useful command in creating interactiveness is `writecon`. This command inserts an XploRe code into the keybuffer; this has the same result as if the user would press the key corresponding to the code inserted. In the teachware macros, `writecon` is used very often to simulate the pressing of the key `<ESC>`. Within the XploRe environment, when the `show` command is invoked and an object is displayed in a window, the program automatically stays in the window in order to allow the user to manipulate the graphic outfit interactively. To leave the window and continue the program, the user must press `<ESC>`. In the teachware macros, the graphic outfit is always controlled by the program and not by the user. Thus, after a `show` call, the program automatically leaves the window. This is done by typing

```
writecon(27)      ; 27 is the code of <ESC> key
show(tru fh s2d1)
```

Note that `writecon` has to be given before `show`.

7.4.5 Calling External Procedures

Some of the teachware macros already analyzed need to call external procedures that are not included in any library. Procedures to generate the distributions in `TWDENES2` or a sample from those distributions, and procedures to calculate the MISE function or the LSCV for density estimation or kernel regression are examples. They are sufficiently complex to be kept independent in order not to burden the code of the interactive macros.

These procedures have to be stored in a file with a name equal to the procedure name and they are all stored in the same directory as the teachware macros (usually in the subdirectory lib of the \XPLORE3 directory). Before they are called, their existence and localization has to be recognized by the program using the command func. However, the whole path has to be given to func and this path depends on where XploRe is installed in each personal computer. Here, the command environ can be very useful. This command returns a text string containing the DOS environment variable xplore3, that is, the path where the main executable files of XploRe are stored. To illustrate how the external procedure twmise can be called from TWDENES2, consider the following code segment:

```
ftwnam = environ + #("\lib\twmise")
func(ftwnam)
mise    = twmise(n va mu w h)
```

In the first code line, the text matrix ftwnam is created containing the whole path where the file twmise.xpl is stored. Then this file is loaded into the program by the command func. Finally, the procedure may be invoked.

7.5   Conclusions

The potential of the computer as a tool to provide instruction is enormous and is being recognized in all fields of education. The accessibility of computers and their sophistication has stimulated the development of software for educational purposes (teachware or courseware).

This chapter presents a teachware module to support the instruction of basic statistics and econometrics. This teachware system is fully interactive and can be easily learned and used by those who do not have a special knowledge about computers or programming languages. It is written in a macrolanguage, XploRe, in order to allow teachers to manipulate the program without much effort by introducing changes or building additional procedures that suit their teaching needs.

The system presented is still in development. However, the basic structure involving the programming language has been achieved and can be used to inspire and help others in making their own extensions.

REFERENCES

Härdle, W. (1990). *Applied Nonparametric Regression*, Econometric Society Monographs No. 19, Cambridge University Press, New York.

Härdle, W. (1991). *Smoothing Techniques, With Implementations in S*, Springer-Verlag, New York.

8
XClust: Clustering in an Interactive Way

Hans-Joachim Mucha[1]

8.1 Introduction

Cluster analysis attempts to detect structures in the data. Some of the most important and widely applicable clustering techniques are partitioning methods and hierarchical clustering algorithms. Well-known methods from both these families are available simply by commands in the interactive statistical computing environment XploRe. Moreover, new adaptive clustering methods (which are often much more stable against random selection or small random disturbance of the data, and which seem to be a little bit intelligent because of their ability for learning the appropriate distance measures) can be carried out by macros. Additionally, the importance of each variable involved in clustering can be evaluated by taking into account its adaptive weight. The adaptive techniques are based on adaptive distances which should also be used in order to obtain multivariate plots (Mucha, 1992).

Usually, a cluster analysis should be accompanied by multivariate graphics in order to provide a better interpretation of the clustering results. Both interactive multivariate graphics based on the (generalized) principal component analysis (PCA) and dynamic graphics are recommended. Furthermore, linear discriminant analysis can be used either to describe cluster analysis results or for to check an a priori given class membership of the observations. In both cases, multivariate graphics based on the canonical discriminant variables can be recommended as an appropriate additional way of displaying the class structure of the data. In contrast to cluster analysis, previous information about classes is taken into account in order to determine decision rules. Last but not least, CART (classification and regression trees; see Breiman, Friedman, Olshen and Stone, 1984) aims at an optimum prediction of a given independent variable by decision rules as well. The results are presented in an extra graphical output, the so-called binary decision tree.

---

[1]Weierstraß Institut für Angewandte Analysis und Stochastik, D-10117 Berlin, Germany.

XploRe offers several possibilities for interactive use of the XClust-functions and macros. For example, one can cut the dendrogram obtained by a hierarchical cluster analysis or a recursive partitioning regression at an "interesting" cluster distance level which corresponds to a partition of all clustering objects into, say, K clusters. Another highly interactive action is clustering, as well as regrouping, the observations by visual inspection of a multivariate display. In this way, one can get a new or modified partition which can be improved in an objective manner by partitioning clustering algorithms. Throughout this chapter, applications to empirical data are presented. Their practical attractions and limitations are discussed.

8.2   Cluster Analysis and Classification

Clustering techniques are frequently used statistical tools in ecology, biological sciences, marketing, chemistry, geology, social sciences, economics, archaeology, ornithology, etc. Cluster analysis provides a useful reduction or description of the data at hand. Unlike classification (for example, discriminant analysis or classification tree algorithms), no previous information is available about classes, that is, neither the number of clusters nor the rules of assignment into clusters is known; they have to be discovered exclusively from the given data set without any reference to a training set. There are also several synonyms, like segmentation in the field of market research, or taxonomy in the biological sciences. Generally speaking, clustering techniques divide a set of points into subsets (clusters) in such a manner that similar points belong to the same cluster, whereas dissimilar ones are allocated to different clusters.

The structure of row points (and column points, respectively) of an $(I*J)$-data table $\mathbf{X} = (x_{ij})$, $i = 1, 2, \ldots, I$, $j = 1, 2, \ldots, J$, can be investigated. $\mathbf{X}$ may contain various types of values; for example, 0-1-values, measurements, frequencies, percentages, and a mixture of them. Without loss of generality, we shall concentrate mainly on clustering of row points (observations). By, analyzing the row points of $\mathbf{X}^T = (x_{ji})$ instead of $\mathbf{X}$, cluster analysis of variables is often practicable in the same way as clustering the observations (without any further considerations in the case of a contingency table $\mathbf{X}$, or a 0-1-matrix $\mathbf{X}$, that is, a matrix containing ones and zeros only). Because of the XploRe matrix language, the transposition of a matrix (or other data management) is quite easy to do.

As a result of cluster analysis we get either a single partition $P(I, K)$ of the I row points into K clusters ($K \ll I$), or a sequence of partitions, that is, a so-called hierarchy. A partition is a categorical variable $\mathbf{p}$ which allocates an integer value (state) $p_i \in \{1, 2, \ldots, K\}$ to every row point $\mathbf{x}_i$. The distances between points play an important role in both hierarchical and nonhierarchical clustering. There are several distance measures available using the XploRe command **distance**. A detailed description will be given later in this chapter. Moreover, additional distance measures can be computed by using the XploRe matrix language.

Let us focus first on metric-scaled variables. The squared weighted Euclidean distance

$$d_Q^2(\mathbf{x}_i, \mathbf{x}_{i'}) = (\mathbf{x}_i - \mathbf{x}_{i'})^T \mathbf{Q}(\mathbf{x}_i - \mathbf{x}_{i'}) = \|\mathbf{x}_i - \mathbf{x}_{i'}\|_Q^2 \qquad (8.1)$$

between two observations $\mathbf{x}_i$ and $\mathbf{x}_{i'}$ is a well-known dissimilarity measure used in cluster analysis and PCA. Here, $\mathbf{Q}$ is diagonal. Usually the weights

$$q_{jj} = 1/s_j^2, \qquad (8.2)$$

called standard weights, or

$$q_{jj} = 1, \qquad (8.3)$$

called trivial weights, are used, where s_j^2 is the sample estimate of the total variance of the variable j:

$$s_j^2 = \frac{1}{M} \sum_{i=1}^{I} m_i(x_{ij} - \bar{x}_j)^2. \qquad (8.4)$$

Here, $\bar{x}_j$ is the sample estimate of the total mean value of the variable j, that is,

$$\bar{x}_j = \frac{1}{M} \sum_{i=1}^{I} m_i x_{ij}, \qquad (8.5)$$

and m_i is the mass of the observation $\mathbf{x}_i$. Accordingly M is the total mass of all I observations. I focus here mainly on the well-known K-means clustering (Mucha, 1992) using the so-called exchange algorithm (Späth, 1985). Other clustering techniques and distance measures are described in detail by Mucha (1992). This method tries to minimize the sum of the within-cluster variances

$$V_K = \sum_{k=1}^{K} \sum_{i=1}^{I} \delta_{ik} m_i d_Q^2(\mathbf{x}_i, \bar{\mathbf{x}}_k). \qquad (8.6)$$

In practice, however, the minimum is sometimes a local one (see Figure 8.3 for a bad result). The indicator function δ_{ik} equals 1 if the observation $\mathbf{x}_i$ comes from cluster k, or 0 otherwise. Furthermore, the element $\bar{x}_{kj}$ of the vector $\bar{\mathbf{x}}_k$ is the mean value of the variable j in the cluster k:

$$\bar{x}_{kj} = \frac{1}{n_k} \sum_{i=1}^{I} \delta_{ik} m_i x_{ij}. \qquad (8.7)$$

We denote by n_k the mass of the cluster k, which is equal to the sum of the masses of all observations belonging to the cluster k.

A graphical representation in a low-dimensional space is a helpful tool in order to interpret the clustering results. In the case of the standard

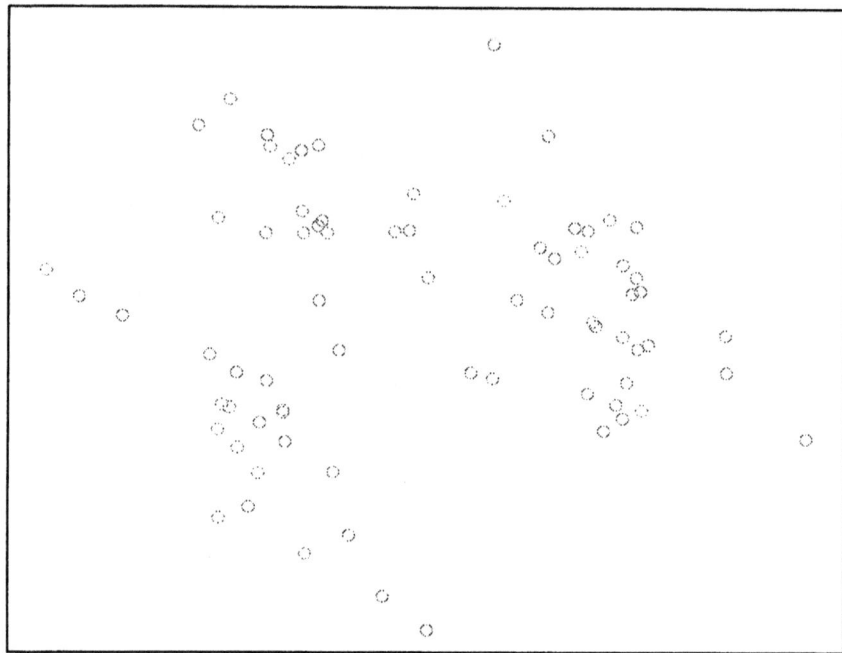

FIGURE 8.1. Plot of the observations of the "Lubischew data" in the plane of the first two principal components (54%+26%=80% of the total variance). The PCA is based on the correlation matrix.

weights $q_{jj} = 1/s_j^2$ in (8.1), the PCA of the correlation matrix (instead of the covariance matrix) is computed. As a result, each variable gets equal importance in cluster analysis as well as in PCA. Moreover, the use of the inverse total variances (8.2) in formula (8.1) as weights in a cluster analysis contradicts the fundamental assumption that there are several populations (clusters) with different parameters that we look for. As a final consequence of using these standardizations, detection of an underlying cluster structure in $\mathbf{X}$ becomes difficult or even impossible.

The Lubischew data (Lubischew, 1962) consists of six measurements on individual male flea beetles which fall into three (a priori known) species of the genus *Chaetocnema*. Jones and Sibson (1987) use this data to test the performance of projection pursuit algorithms. There are 74 observations (cases, or row points) in total (Figure 8.1): 21 observations in class (or species) 1, marked by the symbol y and blue color (Figure 8.2); 31 observations in class 2, marked by the symbol o and green color; and finally, 22 observations in class 3, marked by the symbol * and red color.

With the exception of discriminant analysis, the clustering techniques and multivariate graphics do not assume groups and make no use of the species labeling, but it will be interesting to compare the a priori given

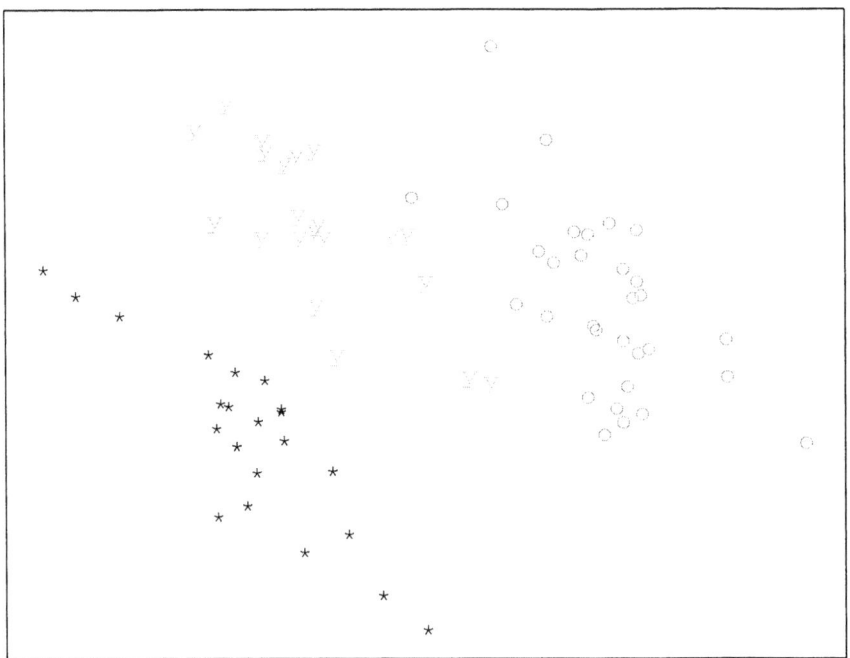

FIGURE 8.2. Same plot as in Figure 8.1 but the observations are marked by their given species labels and their species colors (see color insert).

groups with any structure found in the data by clustering methods or PCA. Figure 8.1 is obtained by the first **show** command (see the stream of commands below). It gives an initial idea about this data set, and it shows projections onto the plane of the first two principal axes (PCA on the basis of the correlation matrix). This is a "usual" multivariate display. In Figure 8.2, created by the second **show** command below, the species labels y, o, and * were added after carrying out the PCA using the **vtocc** function in order to make clear that there is some structure there. Additionally, the classes are colored. The following commands are executed (it is assumed that the data are already stored as an ASCII file named `lubisch.dat`):

```
library("xclust")
x = read("lubisch")          ; read the raw data
p = x[,8]                    ; p is the true partition
x = x[,2:7]                  ; x is the data matrix
z = x.*trn(1./sqrt(var(x)))  ; standardize the data
r = pca(z)                   ; r = PCA coordinates
show(r[,1:2] s2d)
r = r[,1:2]~vtocc(63+16.*p)  ; join the true partition
show(r s2d)
randomize(1239)              ; seed value of random numbers
s = ceil(uniform(rows(z)).*3); set the initial partition
```

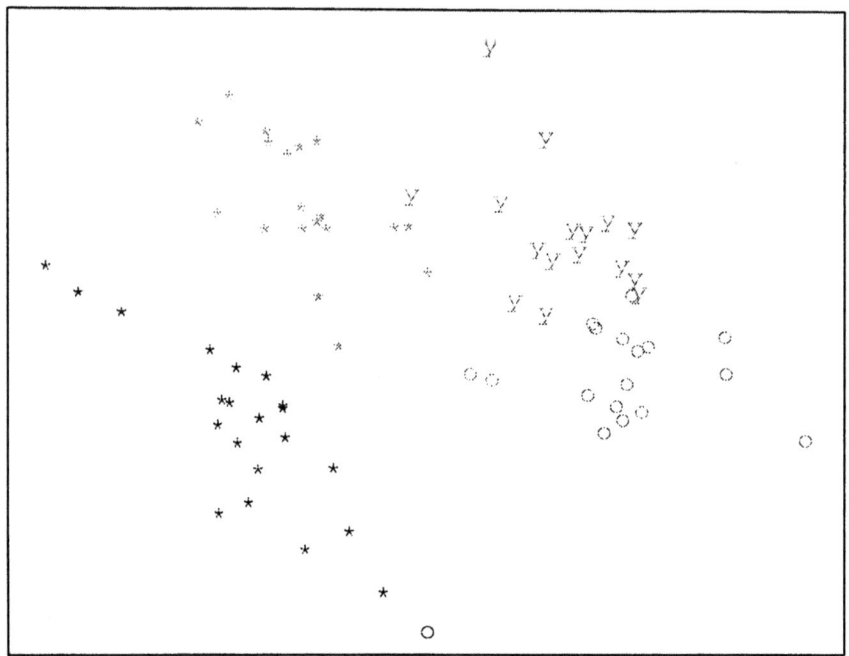

FIGURE 8.3. A bad local minimum of the K-means clustering. The resulting clusters are marked by the symbols y, o and *, whereas the given classes are colored in the same manner as shown in Figure 8.2.

```
(t c v n) = kmeans(z s 0)     ; clustering
r = r[,1:2]~vtocc(63+16.*t)   ; join the K-means partition
show(r s2d)
```

A cluster structure is not easy to identify by looking at Figure 8.1, where the species labels are erased. The PCA of the covariance/variance matrix (that is, using (8.3) in formula (8.1)) gives a plot similar to the one in Figure 8.1. The third **show**-command creates Figure 8.3, containing a "bad" result of the K-means cluster analysis (see later in this chapter for details).

Another multivariate graphic can be obtained by applying discriminant analysis. Here, we take the knowledge about the species labels of the observations into account. Consider here only a simple model: K populations (classes) are given, and suppose that associated with each population k, $k = 1, 2, \ldots, K$, there is a multinormal density $N(\mu_k, \boldsymbol{\Sigma})$ on R^J, where the K densities have the same covariance matrix $\boldsymbol{\Sigma} > 0$ in common. Given a partition $\mathbf{p}$ of I observations into K clusters, the eigenvalue problem

$$\mathbf{Te} = \lambda \ \mathbf{\bar{S}e} \qquad (8.8)$$

has to be solved, where the $(J * J)$-matrix

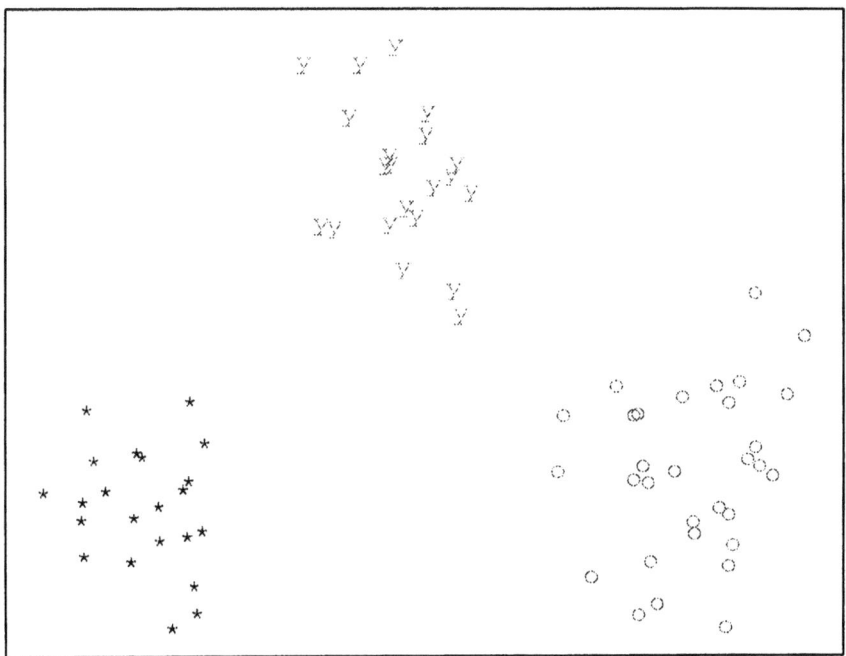

FIGURE 8.4. Plot of the species labels of the observations of the "Lubischew data" in the plane of the first two discriminant scores (82%+18%=100% of the total variability in the discriminant space).

$$T = \sum_{k=1}^{K} n_k(\overline{\mathbf{x}}_k - \overline{\mathbf{x}}) (\overline{\mathbf{x}}_k - \overline{\mathbf{x}})^T \tag{8.9}$$

has a rank $L = \min(J, K - 1)$ at most, and

$$\overline{\mathbf{S}} = \sum_{k=1}^{K} \sum_{i=1}^{I} \delta_{ik} m_i (\mathbf{x}_i - \overline{\mathbf{x}}_k) (\mathbf{x}_i - \overline{\mathbf{x}}_k)^T \tag{8.10}$$

is the $(J * J)$ matrix of the pooled within-cluster covariances and variances, that is, $\overline{\mathbf{S}}$ is an estimator of the true but unknown Σ of each class. Obviously, m_i and n_k are the weight of the observation $\mathbf{x}_i$ and the weight of the class $\overline{\mathbf{x}}_k$, respectively. The indicator function δ_{ik}, the total means $\overline{\mathbf{x}}$, and the cluster means $\overline{\mathbf{x}}_k$ are defined as above given the partition $\mathbf{p}$ of the I observations into K clusters. The well-known results of (8.8) are the L eigenvalues $\lambda_1 \geq \lambda_2 \geq \ldots \geq \lambda_L \geq 0$, and the matrix of corresponding eigenvectors $\mathbf{E} = (\mathbf{e}_1, \mathbf{e}_2, \ldots, \mathbf{e}_L)$. Afterwards, the matrix $\mathbf{G}$ of coordinates of the discriminant variables (or discriminant scores, see Figure 8.4) can be computed by the formula

$$\mathbf{G} = \mathbf{XE}. \tag{8.11}$$

Additionally, the discriminant variables can be used in order to discriminate the observations of the training sample (or some new ones) by statistical tests. For instance, the observation $\mathbf{x}_i$ is assigned to class k if

$$\frac{M - K - L + 1}{L(I - K)} \frac{n_k}{n_k + 1} (\mathbf{g}_i - \bar{\mathbf{g}}_k)^T (\mathbf{g}_i - \bar{\mathbf{g}}_k) \leq F_{L, M - K - L + 1; \alpha} \tag{8.12}$$

holds for a chosen significance level α of the F-distribution, where the vector

$$\bar{\mathbf{g}}_\mathbf{k} = \mathbf{E}\bar{\mathbf{x}}_\mathbf{k}^\mathbf{T} \tag{8.13}$$

contains the coordinates of the centroids of the class k in the discriminant space. As previously mentioned, M is the total mass over all observations, that is,

$$M = \sum_{i=1}^{I} m_i. \tag{8.14}$$

Usually the masses of the observations are each equal to 1 so that $M = I$ holds in this case. The discriminant analysis, including the display (see Figure 8.4), requires the following XploRe commands

```
(h E S T A B) = discrim(X p 2)          ; discriminant analysis
G = X * E ~ vtocc(63+16.* p)            ; discriminant scores
show(G s2d)
F =sum(sum(unit(cols(x)).*inv(T*S.*(I-K)))) ; total F-value
```

where, as usual, the input data matrix is denoted by $\mathbf{X}$. The vector $\mathbf{p}$ contains the given partition of the I points into the K clusters. Furthermore, the number n of discriminant variables used for the discrimination step can be specified (in the case given above $n = 2$ is specified), and an optional input parameter $\mathbf{m}$ can be given which contains the weights (masses) of the observations. The output parameters are the vector $\mathbf{h}$ containing the revised partition after discriminating the observations according to n dimensions of $\mathbf{G}$ in (8.11), the matrix $\mathbf{E}$ of weights of the discriminant functions, the matrix $\mathbf{S}$ of the pooled within-cluster covariance/variances with the elements according to (8.10), and $\mathbf{T}$, which is given by formula (8.9). The matrix $\mathbf{A}$ contains the univariate/bivariate f-values (lower triangle/diagonal) and the coefficients of affinity (upper triangle). The last output parameter $\mathbf{B}$ is the matrix of f-values (lower triangle) and t-values (upper triangle) between clusters. The matrix $\mathbf{G}$ contains the discriminant scores (which were used in order to draw Figure 8.4). The last XploRe statement above makes use of the two last matrices in order to determine the overall F-value:

$$F = \frac{1}{M - K} \sum_{k=1}^{K} n_k (\bar{\mathbf{x}}_k - \bar{\mathbf{x}})^T \bar{\mathbf{S}}^{-1} (\mathbf{x}_i - \bar{\mathbf{x}}_k) , \tag{8.15}$$

which is a well-known measure for the discriminant power. For further details concerning discriminant analysis, the books by Mardia, Kent and Bibby (1979) and Dillon and Goldstein (1984) are recommended. Figure 8.4 shows very well-separated groups of observations. These a priori given groups (species) correspond precisely to the discriminant analysis result **h**. In contrast to Figure 8.4, Figure 8.3 shows the failure of the usual K-means clustering. The question arises: how can we get a similar multivariate display (showing large distances between classes) with the help of clustering algorithms, that is, without making use of the species labels?

8.3   The K-Means Method in XploRe

The criterion (8.6) of the K-means method can be derived directly by using the maximum likelihood approach, assuming that the populations have the multinormal densities $N(\mu_k, \sigma\mathbf{I})$ with expected values $\mu_k \neq \mu_{k'}$, $k \neq k'$, and independent variables, that is, a simple identical covariance structure is assumed. Other approaches using density estimation are given, for example, by Silverman (1986).

Several of variants of the K-means clustering method are called simply by the command

```
(p C V n) = kmeans (X b r { w { m }})
```

where the vectors **w** (column weights $w_j = q_{jj}$) and **m** (row weights, or masses) are optional input parameters. As usual, the input data matrix is denoted by **X**; it consists of I rows (row points) and J columns (column points). With the help of the parameter r, one can select first another version of algorithms based on the K-means method and secondly, one determines stopping rules of the iteration process. For instance, $r = 0$ means that the standard K-means algorithm, the so-called exchange algorithm, is used and it stops after an iteration cycle without any shift of an observation from one cluster into another. An iteration cycle covers checking the class membership of all I observations in order to minimize the criterion (8.6). An initial partition **b** of the I points into K clusters can be a categorical variable containing randomly generated integers. It could be a vector computed by the XploRe command

```
b = ceil(uniform(rows(X)).*K)
```

or

```
b = 1 + floor(K*uniform(rows(X)))
```

where K denotes the number of clusters that have to be chosen in advance. Otherwise, for example, **b** can contain $I - K$ zeros (that means the corresponding points do not affect the initial clusters) and the K values $1, 2, \ldots, K$, which define the clusters (so-called seed points). Or **b** can be the categorized first principal component from a PCA.

The output of the **kmeans** command consists of the final partition **p** of the I points into K clusters (which minimizes the sum of within-cluster variances), the matrix **C** of the cluster mean values (the so-called centroids computed by (8.7); **C** has K rows and J columns), the matrix **V** of the within-cluster variances divided by the weight (mass) of the corresponding cluster (**V** consists of K rows and J columns)

$$v_{kj} = \frac{1}{n_k} \sum_{i=1}^{I} \delta_{ik} m_i q_{jj} (x_{ij} - \overline{x}_{kj})^2, \tag{8.16}$$

and the vector **n**, containing the weights (masses) of the K clusters. In the case of row weights $m_i = 1, (i = 1, 2, \ldots, I)$, the vector **n** contains the number of observations per cluster. These matrices and vectors can be used in the matrix language of XploRe in an easy and convenient way. The partition **p** is useful for multivariate and dynamic XploRe graphics. One can either transform **p** into a so-called mask vector **f** (which contains the codes of K different colors) using the **vtocc** command as, for example,

```
f = vtocc(80.+ p)
```

or into a vector **f** (which contains the codes of K different symbols) by using

```
f = vtocc(63+16.* p)
```

These mask vectors are used by the XploRe command **show**. Otherwise one can specify both symbols and colors (see, for example, Figure 8.3).

8.3.1 Example 1

Suppose that there are 1000 randomly generated observations in total (with 9 variables each): 300 in class 1, 150 in 2, 350 in 3, and 200 in 4. The classes differ in their location, variance, and sample size, respectively, and the variables give different contributions to the distinction of the classes. Additionally, quite different scales of the variables occur. These XploRe commands are used:

```
library("xclust")
randomize(11111)      ; in order to rerun with the same data
z = normal(350 9)                ; 350*9 standard normals
x1 = z[1:300,] .*1.2-#(0,0,0.5,3.5,(-1),(-1), 0 , 0 ,0)
x2 = z[201:350,]    +#(0,0,0.5, 2 , 0.5, 0.5,0.3,0.5,0)
x3 = z[1:350,] .*1.5+#(0,0, 0 , 1 , 5.5,  2 ,0.5,0.5,0)
x4 = z[101:300,].*0.9-#(0,0,1.5,1.5,  2 ,  0 , 0 ,0.5,0)
x = x1 | x2 | x3 | x4          ; join together the classes
x = x.*trn(#(1000 10 100 .001 .01 .1 1 100 1)) give scales
q = 1./var(x)                 ; standard weights
randomize(1111) ; in order to rerun with the same partition
r = ceil(uniform(rows(x)).*4)  ; get an initial partition
(p C V n) = kmeans(x r 0 q)
```

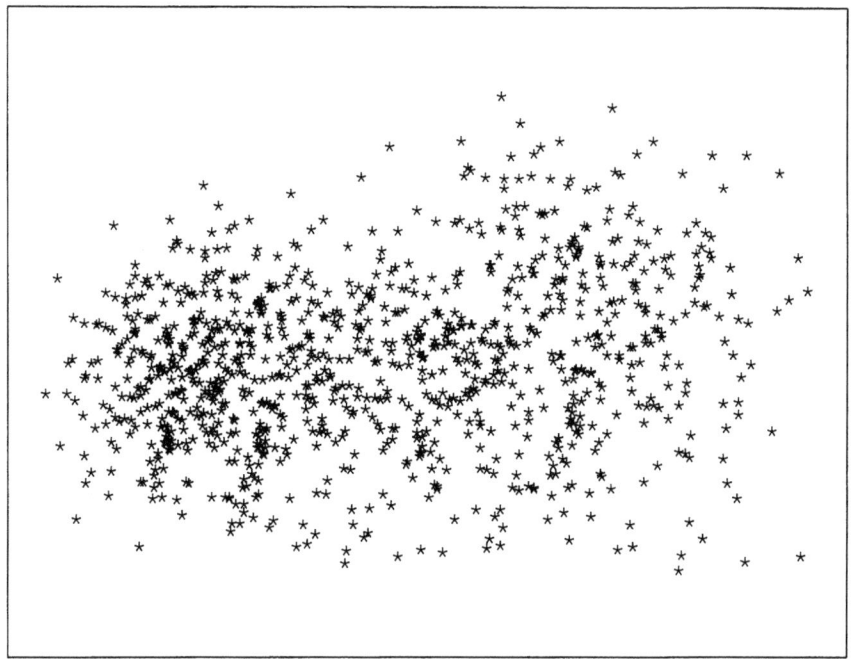

FIGURE 8.5. Principal component analysis (PCA) plot of random generated nine-dimensional data with four groups: about 35.6% (23% + 12.6%)
of the variance is explained by the first two principal components.

Figure 8.5 gives you an initial idea about the data of our example; it
shows projections onto the first PCA plane. The PCA is based on the
correlation matrix, which is a standard option of statistical software. The
class labels are ignored, but even in the case when the given class labels
are added into the plot, it does not become clearer that there is some
structure in the data. Further on that means that an interactive clustering
by human eyes on the basis of this multivariate graphic will probably fail.
Here, we are interested in a K-means cluster analysis of the 1000 points
into 4 clusters. Because of the known true class membership, we are able
to assess the performance of this clustering technique simply by the error
rate. Figure 8.6 shows the cluster membership marked by different symbols.
The K-means method based on the standard weights $q_{jj} = 1/s_j^2$ in (8.1)
obviously fails.

8.3.2   Example 2

The Iris data (see, for example, Andrews and Herzberg, 1985) consist of
three classes which are known beforehand. As a result of this knowledge, the
performance of the K-means clustering can readily be evaluated. For further

152   H.-J. Mucha

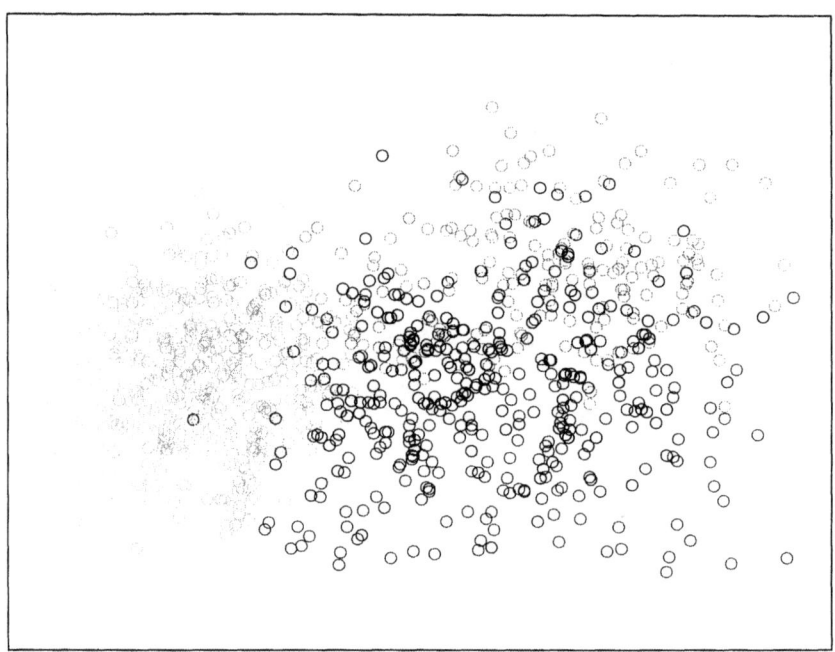

FIGURE 8.6. PCA plot of the cluster labels of the observations after a
K-means cluster analysis using the weights inverse total variances (8.2);
about 46% errors are counted. This is by no means the worst result of
the K-means method obtained by using different random initial partitions
(see color insert).

details about this data set, for example, concerning simulation results, see
Mucha (1992). Figure 8.7 shows the final result of the K-means clustering
of $I = 150$ observations into $K = 3$ clusters in the plane of the first two
principal components (here, again, the correlation matrix was used in the
PCA). In contrast to the data set shown in Figure 8.1, a well-separated
class occurs here. It can be visualized even when the observations are not
marked by the symbols of the classes. First the data are read from an ASCII
file named iris.dat. The following XploRe commands must be executed:

```
library("xclust")
X = read("iris")
X = X[,1:4]
y = matrix(50)|matrix(50).+1|matrix(50).+2 ; true classes
z = uniform(rows(X) 1)
X = X~y~z
X = sort(X 6)      ; randomizes the sequence of the rows
y = X[,5:5]        ; true partition
X = X[,1:4]
w = 1./var(X)
```

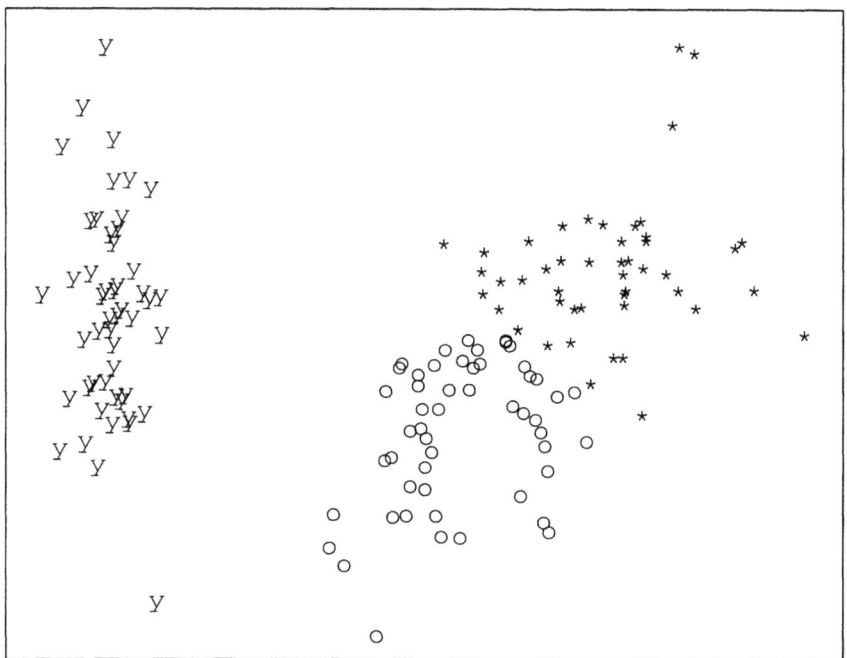

FIGURE 8.7. PCA plot of the "Fisher *Iris* data" after a usual K-means clustering using the weights inverse total variances (8.2).

```
b = ceil(uniform(rows(X)).*3)
(p C V n) = kmeans(X b 0 w)
U = X .* sqrt(w') ; prepares the weighted PCA
Z = pca(U)
Z = Z[,1:2] ~ vtocc(63+16.* p)
show(Z s2d)
(yb pb) = bindata(y ~ p 1 0.5)
yb ~ pb
```

The vector **y** contains the a priori known cluster membership of each observation. The input value 0 of the **kmeans** command fixes the following usual stopping rule: the iteration will be finished if no shift of an object is counted during an iteration cycle. One can get the nonempty elements of the contingency table, obtained by crossing the true partition **y** with the computed partition **p**, using the XploRe command **bindata**. Instead of **bindata**, the macro **conting** calculates the full contingency table **H**:

```
H = conting(y p)
```

The number of misclassifications e can be computed by using the macro **contmax**

```
(e cp) = contmax(H)
```

In the first column, the $(K*2)$-matrix **cp** contains the category labels k, $k \in \{1, 2, \ldots, K\}$, of the rows of **H** which correspond at most to the category labels of the columns of **H**. The latter ones are saved in the second column of **cp**. Generally, **cp** contains those K pairs of indices of the contingency table **H** whose corresponding elements are maximum under the restriction that the indices within each column of **cp** are different from one another. The sum of the remaining $K*(K - 1)$ elements of **H** gives the number of misclassifications.

The error rate is 16.7% (that is, 25 errors are counted). The reason for this failure which is also responsible for Figure 8.3 or Figure 8.6 is the use of the standard weights $1/s_j^2$ in the squared weighted Euclidean distance (8.1). Because of this, each variable contributes nearly the same amount of variability to the clustering task, that is, the contributions are rather independent of the shape (number of modes, etc.) of the univariate distributions of the underlying variables.

In the case of the Lubischew data, the K-means method based on the weights (8.2) does not recognize the given true classes correctly: varying numbers of errors (35 at most, see Figure 8.3, and 4 at least) are counted depending on different initial partitions. Figure 8.3 is the result of using the initial partition based on the seed value 1239 (command: `randomize(1239)`). The K-means method results are highly dependent on both the initial partition and the sequence of observations. There is a very wide range of results depending on these circumstances/parameters. Therefore, it is recommended that the user should try several different initial partitions and/or different random orderings of the observations.

The K-means method is available in different variants depending on the choice of the parameter r: exchange technique, minimum distance method, and gradient technique. For a discussion of advantages and disadvantages of the variants, the reader is referred to Mucha (1992). All algorithms start from an initial partition of the objects into a fixed number of clusters, and their results are all dependent on the initial partition. However, the techniques differ in the ways they minimize the sum of within-cluster variances. The exchange technique minimizes this criterion directly. The centroids of the clusters are corrected immediately after shifting a point from one cluster to another. On the other hand, the minimum distance method allocates an observation to that cluster whose centroid has the shortest (weighted) Euclidean distance to this observation. Therefore, it minimizes the sum of within-cluster variances (8.6) indirectly. The centroids of the clusters are corrected either immediately or after each iteration cycle. The latter can lead to empty clusters, but otherwise this technique gives results that are independent of the sequence of the observations. This advantage is valid for the gradient technique, too. Only one shift per iteration cycle is carried out here, namely the one that gives the best improvement of the criterion.

8.4   The Adaptive K-Means Method

The usual weights used in the Euclidean distance often lead to a bad result in cluster analysis and PCA. A better one can be obtained in almost every case by using specific or adaptive weights of the variables (Mucha, 1992). Which is the best among the weighted Euclidean distances? For a contingency table $\mathbf{X} = (x_{ij})$, the chi-squares distance (χ^2-distance)

$$d_W^2(\mathbf{y}_i, \mathbf{y}_{i'}) = (\mathbf{y}_i - \mathbf{y}_{i'})^T \mathbf{W}(\mathbf{y}_i - \mathbf{y}_{i'}) = \|\mathbf{y}_i - \mathbf{y}_{i'}\|_W^2 \qquad (8.17)$$

is the appropriate dissimilarity measure between the row profiles $\mathbf{y}_i = \mathbf{x}_i z_{i+}$ and $\mathbf{y}_{i'} = \mathbf{x}_{i'} z_{i'+}$. Here $z_{i+} = 1/x_{i+}$ and $z_{i'+} = 1/x_{i'+}$ are the inverses of the row total x_{i+} of the row point i and the inverse of the row total $x_{i'+}$ of the point i', respectively. The special weights are given by $w_{jj} = x_{++}/x_{+j}$, with the grand total x_{++} and the column total $x_{+j}, j = 1, 2, \ldots, J$. The decomposition of the total inertia

$$T = \sum_{i=1}^{I} d_W^2(\mathbf{y}_i, \mathbf{a}) x_{i+}/x_{++} \qquad (8.18)$$

has to be performed subject to the sum of within-cluster inertia becoming minimal. Here, the vector $\mathbf{a}$ is the average row profile with the elements $a_j = x_{+j}/x_{++}, j = 1, 2, \ldots, J$. For example, the K-means method (introduced above) minimizes the sum of the within-cluster inertia

$$V_K = \sum_{k=1}^{K} \sum_{i=1}^{I} \delta_{ik} d_W^2(\mathbf{y}_i, \mathbf{b}_k) x_{i+}/x_{++} \qquad (8.19)$$

for K clusters. The indicator function δ_{ik} is 1 if the row profile $\mathbf{y}_i$ comes from cluster k, or 0 otherwise. Obviously the vector $\mathbf{b}_k$ is the average row profile of all row points within the cluster k. The ith row point is weighted by its respective mass $x_{i+}/x_{++}, i = 1, 2, \ldots, I$.

Suppose that a contingency table $\mathbf{X}$ is read in, or $\mathbf{X}$ is computed by crossing two categorical variables (by using the macro conting, which is provided with XploRe). With the help of the matrix language of XploRe, one can use the command kmeans to carry out a partitioning cluster analysis of the rows (as well as of the columns) of a contingency table $\mathbf{X}$ (for example, into five clusters):

```
library("xclust")
w = sum(sum(X))./sum(X)
m = sum(trn(X))/sum(sum(X))
Y = X./sumr(X)                    ; row profiles
b = ceil(uniform(rows(Y)).*5)
(p C V e) = kmeans(Y b 0 w m)
```

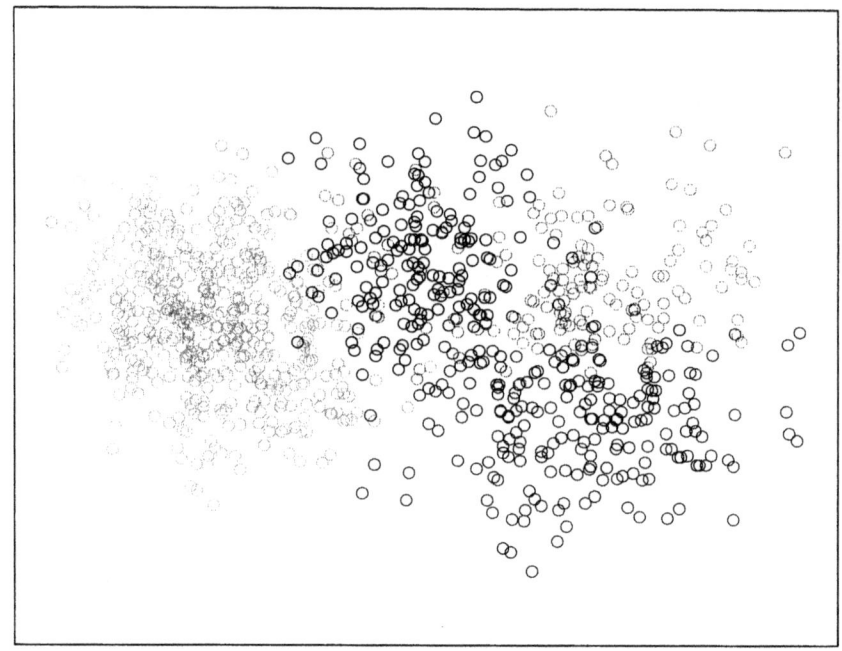

FIGURE 8.8. PCA plot of the cluster labels after the second step of the adaptive K-means method; 39% errors are counted, and about 42.8% of the variance is explained here (see color insert).

where the vectors $\mathbf{w}$ (with the elements $w_j = x_{++}/x_{+j}, j = 1, 2, \ldots, J$) and $\mathbf{m}$ (with the elements $m_i = x_{i+}/x_{++}, i = 1, 2, \ldots, I$) have to be computed before calling the procedure **kmeans**. Since there is an additional plot by correspondence analysis, I recommend the macro **kmcont** for partitioning cluster analysis of contingency tables. This macro looks like the **kmeans** function; here the number of clusters k is a necessary input parameter.

```
(p C V e a) = kmcont(X k)
```

Moreover, Greenacre (1988) describes the hierarchical clustering of row points (as well as column points) of a contingency table using Ward's clustering method on the basis of the χ^2-distance. The clustering method by Ward (1963) provides an optimum decomposition of the total inertia (8.18) within the meaning that the sum of within-cluster inertia becomes minimal. Usually the K-means method provides a better partition of I row points into K clusters in the sense of minimum V_K than the Ward clustering method. The latter is available in XploRe too. Once more, a macro is recommended: **wardcont** performs hierarchical cluster analysis of the rows as well as of the columns. In addition to the output of dendogram (see later: Figure 8.13 and Figure 8.14) a simultaneous graphic of both row points and column points by means of correspondence analysis is created.

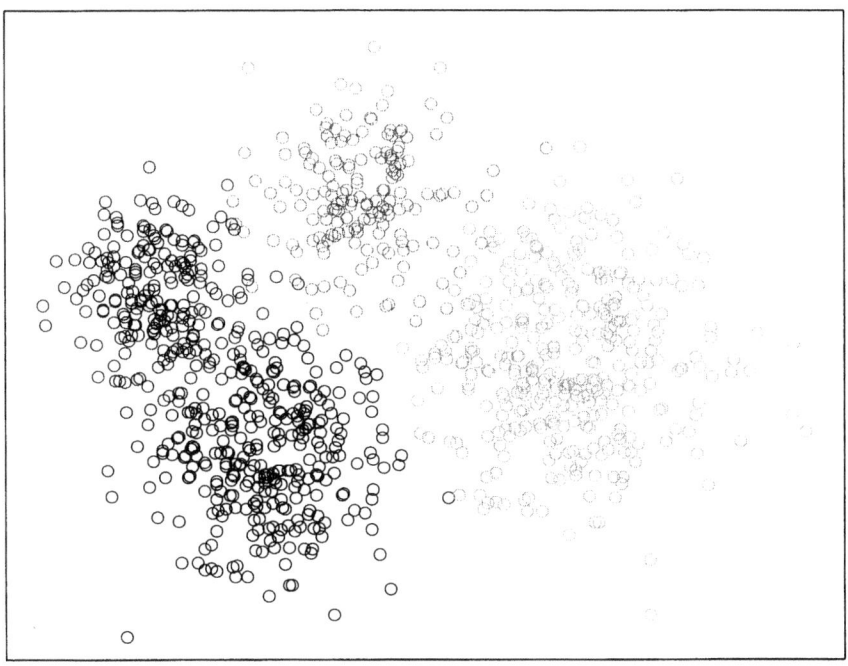

FIGURE 8.9. PCA plot of the final results of the adaptive K-means method; 4.2% errors are counted. The PCA is based on the corresponding adaptive distances; the first PCA plane accounts for about 60.8% (45.0% + 15.7%) of the total variance (see color insert).

The partition $\mathbf{p}$ of the row points into k clusters and the partition $\mathbf{b}$ of the column points into l clusters are computed:

(p b) = wardcont(X k l)

Analogously to the χ^2-distance, correspondence analysis (named dual scaling by Nishisato, 1980, 1994) provides an appropriate graphical representation (plot) of the row and column points of a contingency table $\mathbf{X}$. Moreover, the clusters (represented by their average row profiles $\mathbf{b}_k$) should be shown in the graphic.

Let us return to metric-scaled data values. To increase the stability in cluster analysis, specific weights or adaptive weights in the distance formula can often be applied, rather than the usual weights $q_{jj} = 1/s_j^2$ or $q_{jj} = 1$. For example, the simple adaptive weights

$$q_{jj} = 1/\bar{s}_j^2 \qquad (8.20)$$

can be used in the squared weighted Euclidean distance, where $\bar{s}_j$ is the

158   H.-J. Mucha

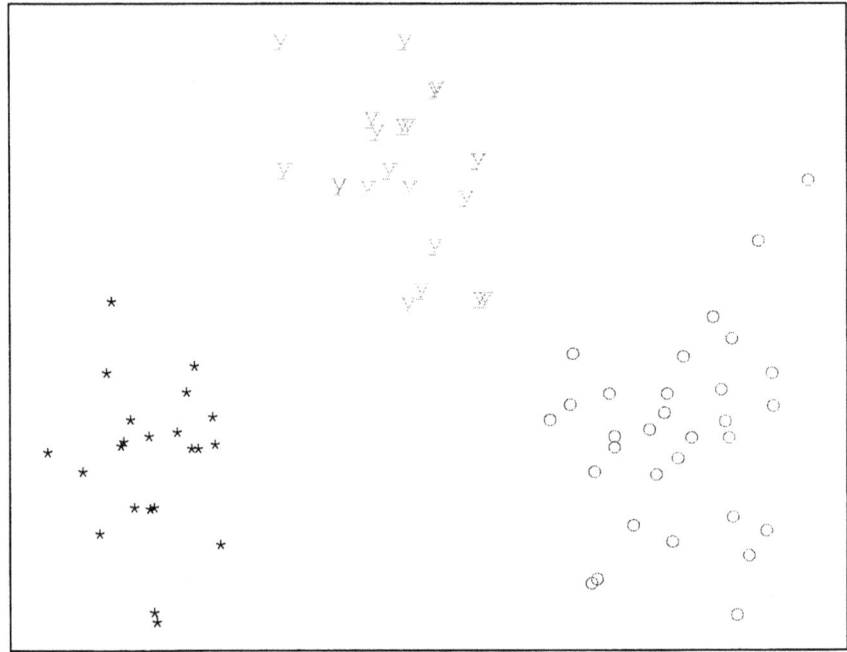

FIGURE 8.10. PCA plot of the "Lubischew data" based on the final adaptive weights (8.23). Here, as well as with the weights (8.20), all observations are grouped correctly (see color insert).

pooled standard deviation of the variable j

$$\bar{s}_j^2 = 1/M \sum_{k=1}^{K} \sum_{i=1}^{I} \delta_{ik} m_i (x_{ij} - \bar{x}_{kj})^2. \tag{8.21}$$

The indicator function δ_{ik} is defined in the usual way. For simplicity, one can use M, the sum of all weights m_i of the observations i, $i = 1, 2, \ldots, I$, that is, M becomes independent of the number of clusters K.

The "true" pooled standard deviations cannot be computed in cluster analysis in advance because the cluster structure is usually unknown. Otherwise it is known that the pooled standard deviations concerning a random partition are nearly equal to the total standard deviations. Therefore, starting with the weights $q_{jj} = 1/s_j^2$ and a random initial partition $P^0(I, K)$, the K-means method computes a (local) optimum partition $P^1(I, K)$ of I observations into K clusters. In a repeated K-means clustering (with a new random initial partition or another arbitrary partition), the above weights $q_{jj}^{(1)} = 1/\bar{s}_j^2$ are used, where $\bar{s}_j$ is the pooled standard deviation over K clusters for the new partition $P^1(I, K)$. After carrying out the second K-means run, we get a new partition $P^2(I, K)$. The new weights $q^{(2)}$, which

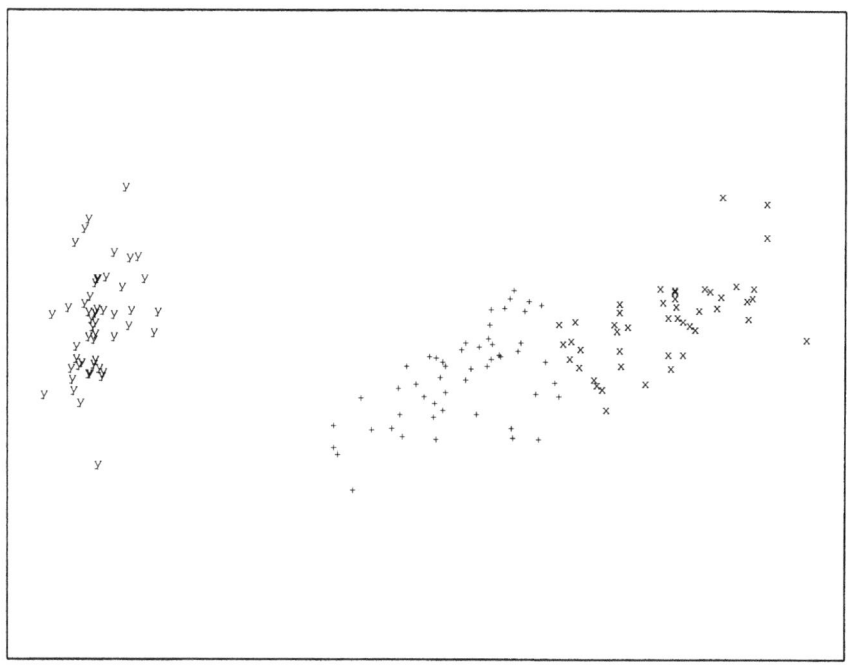

FIGURE 8.11. Scaled PCA plot of the Iris data based on the final adaptive weights (8.20). Here, the number of misclassifications decreases from 25 (Figure 8.7) to 6.

correspond to the partition $P^2(I, K)$, are used in the next K-means clustering, and so on. We repeat this procedure as long as no changes in the partition or the adaptive weights are detected.

In Figure 8.9, the final partition of the adaptive K-means method applied on the random generated data (Figure 8.5) is shown in the first PCA plane. Six iterations are necessary for the convergence (Figure 8.8 shows the result of the second step of this fitting process). Obviously, the clusters become visible and the error rate decreases rapidly. Moreover, the importance of each variable for clustering can be assessed. For instance, the following final weights can be computed by

$$g_j = s_j^2 \, / \overline{s}_j^2. \qquad (8.22)$$

In the case of the random generated data of Figure 8.9, we obtain the values g_j 1.0006, 1.0010, 1.3755, 3.8546, 6.6682, 1.3247, 1.0352, 1.1080, and 1.0005.

The result of the adaptive K-means clustering of the Lubischew data is presented in Figure 8.10. Here, the adaptive weights

$$q_{jj} = s_j^2 \, / \overline{s}_j^4 \qquad (8.23)$$

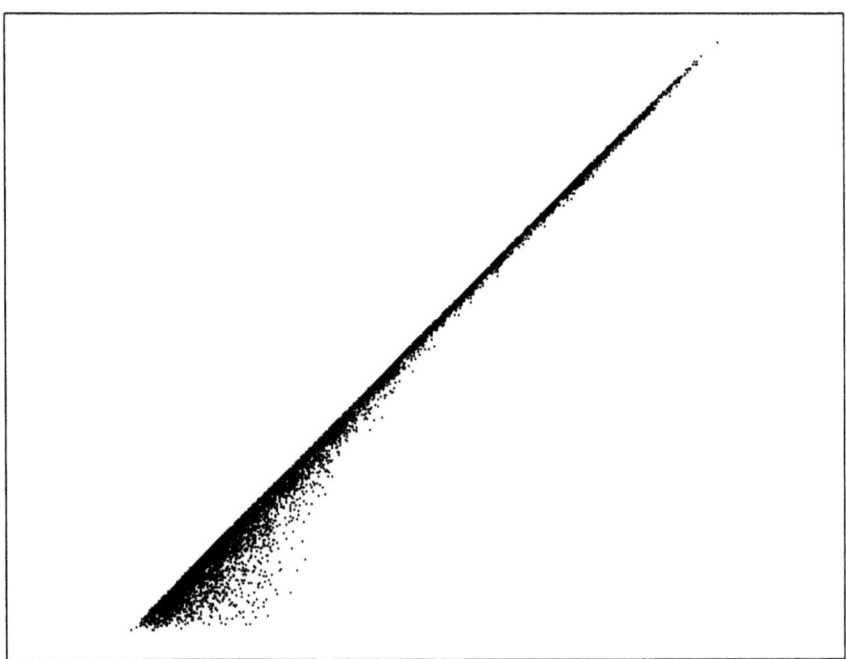

FIGURE 8.12. Scatter plot of the true weighted Euclidean distances of the observations of the "Lubischew data" versus the Euclidean distances based on the first two principal components of Figure 8.11. The final adaptive weights of the variables are considered here.

are used in the squared weighted Euclidean distance. The adaptive method becomes stable after two cycles. That is also true in the case of using the weights (8.20).

Figure 8.11 shows the final result of the adaptive K-means clustering of the Iris data in the scaled plane of the first two principal components (here, the covariance matrix of the weighted variables was used in the PCA). The weights (8.20) are used. Figure 8.12 shows the high correspondence between the distances of the "Lubischew data" used in the adaptive K-means clustering and the distances presented by the first two principal components only (Figure 8.10).

The adaptive K-means method is performed by the macro **adaptive** which looks like the command **kmeans**. Additionally to the output of **kmeans**, the final adaptive weights **q** of the variables are prepared for further use. The number of clusters K replaces the **kmeans** input **b**:

```
(p C V e q) = adaptive(X K w m)
```

Furthermore, the macro **adaptive** shows the adaptive (dynamic) graphic using the original weighted variables as well as the scores from the PCA.

In order to improve the cluster analysis results and their stability (for instance, against small random errors or random sampling) adaptive techniques are recommended. In consequence the specific or adaptive metrics should be used in principal component analysis (Figure 8.9, Figure 8.11, Figure 8.10) as well as in dynamic graphics. In that way the plots give a good support for the interpretation of the results of the cluster analysis.

Späth (1985) recommended a scale-invariant adaptive cluster analysis that is based on the more general distances

$$d^2_{Q_k}(\mathbf{x}_i, \mathbf{x}_{i'}) = (\mathbf{x}_i - \mathbf{x}_{i'})^T \mathbf{Q}_k (\mathbf{x}_i - \mathbf{x}_{i'}) \tag{8.24}$$

between the two observations $\mathbf{x}_i$ and $\mathbf{x}_{i'}$, both belonging to the cluster k. Here, $\mathbf{Q}_k$ $(k = 1, 2, \ldots, K)$ are (different) positive definite matrices, for instance, the inverses of the within-cluster covariance matrices. However, there are some serious disadvantages, such as the lack of a geometrical presentation as well as a huge number of parameters that have to be estimated. Keep in mind that the number of parameters is K times a quadratic increase with the number of variables J.

8.5 The Hierarchical Cluster Analysis

All the methods of cluster analysis start from the assumption that distances between observations may be quantified numerically by data values. Before computing a hierarchy, a distance matrix $\mathbf{D} = (d_{ii'})$ has to be prepared. The distance matrix contains the pairwise distance values between the row points. A distance matrix $\mathbf{D}$ can be computed by typing

```
D = distance(X metric { w })
```

where $\mathbf{w}$ (column weights) is an optional input parameter. As usual, the data matrix is denoted by $\mathbf{X}$. The following types of metric are supported in XploRe: Euclid (Euclidean distance), L1 (absolute metric, Manhattan metric), Maximum (supremum norm), Cosine, Chisquare (χ^2-distance), Tanimoto, Jaccard, and Matching (simple matching coefficient). The corresponding formulas are described in detail by Mucha (1992). For example, the absolute metric d between two observations $\mathbf{x}_i$ and $\mathbf{x}_{i'}$ is given by

$$d_Q(\mathbf{x}_i, \mathbf{x}_{i'}) = \sum_{j=1}^{J} q_j \, |x_{ij} - x_{i'j}| \ ,$$

and the measure Cosine is computed as

$$d^2(\mathbf{x}_i, \mathbf{x}_{i'}) = 2 - 2 \, \frac{\sum_{j=1}^{J} x_{ij} x_{i'j}}{\sqrt{\sum_{j=1}^{J} x_{ij}^2 \, \sum_{j=1}^{J} x_{i'j}^2}} \ .$$

The last three coefficients, Tanimoto, Jaccard, and Matching, are based on binary data (0-1-data); since there is an internal temporary dichotomization, the input matrix $\mathbf{X}$ may contain arbitrary data values (if $x_{ij} < 1$, then $x_{ij} = 0$; if $x_{ij} \geq 1$, then $x_{ij} = 1$).

With the help of XploRe, further distance measures can be computed (see below). Otherwise one can handle mixed data in a simple but nevertheless flexible way by adding two (or more than two) distance matrices to a new one. For example, $\mathbf{D}_B$ contains the *simple matching* distances (which are computed on the basis of the binary variables only) and $\mathbf{D}_L$ contains the L1 distances based on metric variables. By transposing the matrix $\mathbf{X}$ before typing the command `distance`, the distances between the column points are computed. Afterward, a hierarchical cluster analysis of the variables can be computed.

About eight hierarchical ascending clustering methods are available. The methods are called by typing the command

```
(p t u) = agglom(D method K { f { m } } )
```

where the scalar f as well as the vector $\mathbf{m}$ (row weights, masses of observations) are optional input parameters. Usually, the distance matrix (denoted by $\mathbf{D}$) is the result of the command `distance`. Otherwise, an external distance matrix $\mathbf{D}$ can be read in, or $\mathbf{D}$ is the result of the matrix language of XploRe. K is the number of clusters, at most, that correspond to the output parameter, the partition $\mathbf{p}$. The clustering methods that can be chosen by the user are the following: Single (single linkage, or nearest neighbor), Complete (complete linkage, or furthest neighbor), Mean (simple average linkage), Median (median method), Average (average linkage), Centroid, Ward (Ward's minimum variance method), and Lance (flexible method of Lance and Williams). For instance, Ward's minimum variance method is based on the squared weighted Euclidean distance (8.1). Here, these two objects (clusters, observations, or a cluster and an observation) $\overline{\mathbf{x}}_k$ and $\overline{\mathbf{x}}_{k'}$ with the masses n_k and $n_{k'}$, respectively, create a new cluster by fusion if

$$d_Q(\overline{\mathbf{x}}_k, \overline{\mathbf{x}}_{k'}) = \frac{n_k n_{k'}}{n_k + n_{k'}} \sum_{j=1}^{J} q_j \left(\overline{x}_{kj} - \overline{x}_{k'j}\right)^2$$

is a minimum. As usual, the weights of the variables are denoted by q_j. In that way, the sum of within-cluster variances becomes minimal for a hierarchy of clusters. For example, Gordon (1981) gives a detailed description of both distance measures and clustering algorithms. The output parameters are the partition $\mathbf{p}$, the dendrogram $\mathbf{u}$, and the "cluster-dendrogram" $\mathbf{t}$ which corresponds to the chosen number of clusters K. For example, a dendrogram (or in a similar way, a cluster-dendrogram) can be prepared by

```
f = tree (u 0.0 center xaxis)
show(f s2d),
```

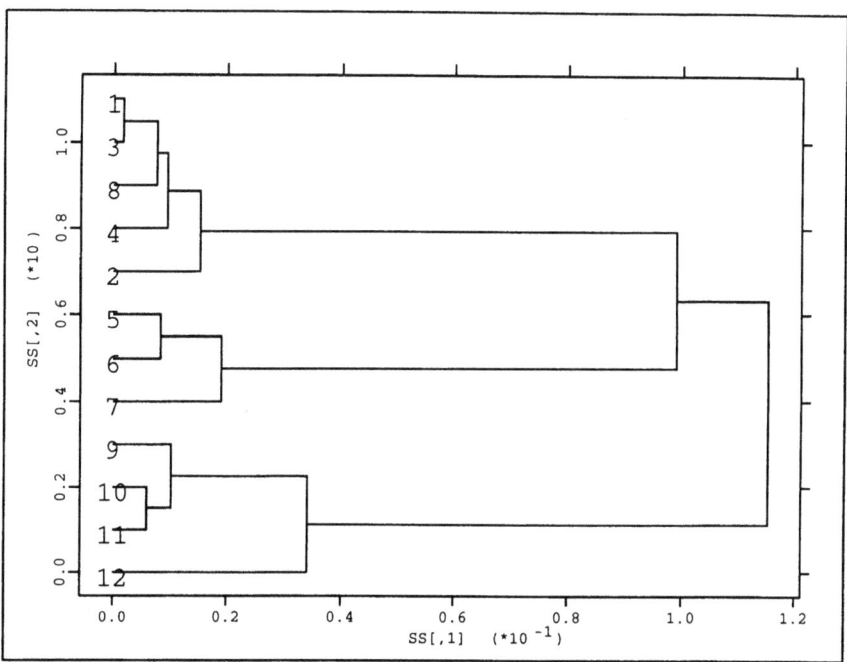

FIGURE 8.13. Dendrogram of the rows of Table 8.1. The Ward method based on the chi-square distance is applied.

where the input vector **u** is the result of the command `agglom` described above. Figure 8.13 shows the dendrogram of the Ward's clustering of the rows of the contingency Table 8.1. The macro `wardcont` is applied which also performs correspondence analysis plots as well as the hierarchical cluster analysis of the column points (Figure 8.14).

How can one get a partition into R clusters from the hierarchy $(R < K)$? One may repeat the cluster analysis with the cluster number R as the input parameter of `agglom` or `wardcont`. Another way is to use the XploRe function `recode` that performs any chosen partition of the dendrogram. For instance, suppose that Figure 8.13 is obtained from clustering the 12 rows of Table 8.1 into 12 clusters by

```
(p c) = wardcont(x 12 9)
```

If one is interested in a partition **b** of the $I = 12$ row points of into three clusters one can type:

```
func(recode)
b = recode(#(5 8 12) p)
```

Note that the clusters of a dendrogram are numbered by $1,2,...,K$ from the top to the bottom. Figure 8.13 shows the names of the rows, not the names of the clusters.

15	27	10	10	0	2	0	1	0
198	656	38	11	5	37	57	1	0
159	644	160	61	21	57	14	0	0
113	404	71	3	26	155	57	2	1
260	1303	470	103	105	855	545	14	17
72	738	366	93	48	489	490	6	29
8	246	88	21	116	209	188	5	5
41	435	145	69	5	118	10	0	1
39	349	136	168	33	35	8	6	2
33	262	128	112	83	96	17	37	10
4	132	74	114	48	23	7	15	7
24	246	122	126	77	169	139	78	51

TABLE 8.1. Contingency table on a bird study from Mucha (1992, p. 179).

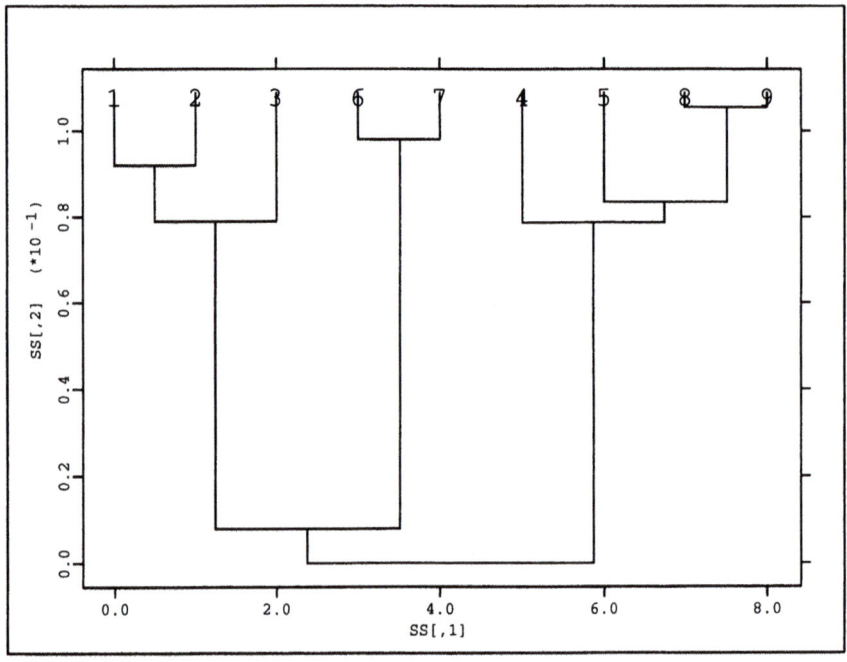

FIGURE 8.14. Dendrogram of the columns of Table 8.1. The Ward method based on the chi-square distance is applied.

8.6 Classification and Regression Tree (CART)

Several CART methods are described by Breiman et al. (1984). The algorithms work similar to the monothetic hierarchical descending cluster

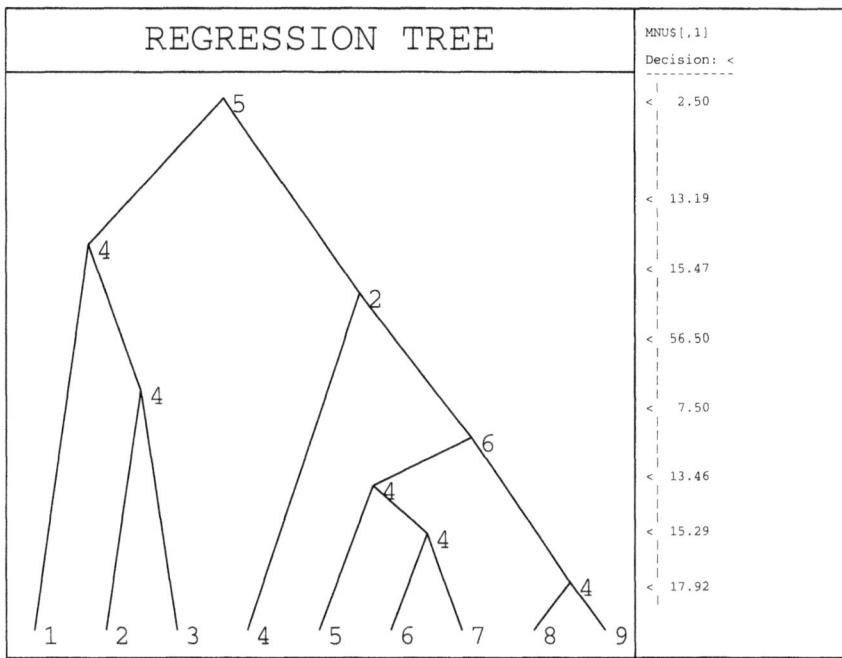

FIGURE 8.15. A regression tree on stage C prostata cancer patients. The nonterminal nodes are marked by the split variable, whereas the small window on the right contains the corresponding split value of the split variable.

analysis (which considers one variable at a step in order to divide the set of points into two subsets), but with the aim of an optimum prediction of a given independent variable. The independent variable may be a metric or categorical one. In the former case, a regression tree is formed, whereas in the latter case, a classification tree is built up. Generally, the results of the CART techniques are presented in dendrograms. Figure 8.15 shows a regression tree computed with XploRe. Basically, the regression surface is approximated by a linear combination of step functions

$$m(x) = \sum_{k=1}^{K} c_k I \{x \in N_k\}.$$

Here, the c_k are constants and the N_k are disjoint hyperrectangles with sides parallel to the coordinate axes (I denotes the indicator function). The estimate of $m(x)$ is the average response (that is, the mean value of cluster N_k; Breiman et al., 1984). In order to minimize the residual sum of squares of the independent variable, a recursive partitioning with $K = 2$ is applied.

8.7   The Investigation of the Stability of Adaptive Weights

As mentioned above, the true cluster structure is unknown in practice. For this reason, we determine the stability of clustering results in simulation studies. The stability of results of cluster analysis can be measured by the Rand's value (Rand, 1971; Hubert and Arabie, 1985). In Mucha (1992), as an example, adaptive K-means clustering of the well-known Iris data was carried out. A stable partition $P(I, 3)$ is reached after four cycles. At the end, just six misclassifications are counted (see Figure 8.11). It was shown by the author that this solution is very stable against random selection of observations (called internal stability) as well as against perturbation of the data values by random generated errors (called external stability), and against random weighting the observations.

How can the internal stability be investigated? One approach is that each simulation l starts with a random selection of $I/2$ observations. Afterward, the adaptive K-means clustering method computes a partition $R^{(l)}(I/2, 3)$ as well as the adaptive weights of variables in the manner described above. The partition $P(I, 3)$ of the whole data set is compared with the partition $R^{(l)}(I/2, 3)$. The stability of results of cluster analysis can be assessed by several measures of correspondence between partitions (Hubert and Arabie, 1985). The macro **measure** computes the well-known Rand's measure as well as other important measures of correspondence between two partitions. For instance, the median values of the Rand's values in the case of the Iris data are: 1.000 (for 2 clusters), 0.996 (3 clusters), 0.967 (4 clusters), 0.916 (5 clusters), 0.907 (6 clusters), and 0.902 (7 clusters); see Mucha and Klinke (1993) for details. There is a very high stability associated with two as well as three clusters. Otherwise, the median values of Rand's measure are very low in the case of random generated data (normal distribution without any cluster structure): 0.520 (for 2 clusters), 0.678 (3 clusters), 0.733 (4 clusters), 0.782 (5 clusters), 0.808 (6 clusters), and 0.838 (7 clusters). Further details as well as the tables of the simulation results are described by Mucha and Klinke (1993). With XploRe it is quite easy to draw the median or mean Rand's values of correspondence as curves along the number of clusters 2,3, etc. In that way two or more "stability curves" can be compared.

XClust is an XploRe library containing a collection of macros for cluster analysis and classification. The macros are based mainly on functions like **kmeans, distance** (computation of a matrix of distances such as (8.5), χ^2, Jaccard,...), **agglom** (hierarchical cluster analysis), **discrim** (linear discriminant analysis), and **cart** (classification and regression trees).

REFERENCES

Andrews, D.F. and Herzberg, A.M. (1985). *Data: A Collection of Problems from Many Fields for the Student and Research Worker*, Springer Series in Statistics, Springer-Verlag, New York.

Breiman, L., Friedman, J.H., Olshen, R. and Stone, C.J. (1984). *Classification and Regression Trees*, Wadsworth, Belmont.

Dillon, W.R. and Goldstein, M. (1984). *Multivariate Analysis*, John Wiley & Sons, New York.

Gordon, A.D. (1981). *Classification*, Vol. 16 of *Monographs on Statistics and Applied Probability*, Chapman and Hall, London.

Greenacre, M.J. (1988). Clustering the rows and columns of a contingency table, *Journal of Classification* **5**: 39–52.

Hubert, L.J. and Arabie, P. (1985). Comparing partitions, *Journal of Classification* **2**: 193–218.

Jones, M.C. and Sibson, R. (1987). What is projection pursuit? (with discussion), *Journal of the Royal Statistical Society, Series A* **150**(1): 1–36.

Lubischew, A.A. (1962). On the use of discriminant functions in taxonomy, *Biometrics* **18**: 455–477.

Mardia, K.V., Kent, J.T. and Bibby, J.M. (1979). *Multivariate Analysis*, Academic Press, Duluth, London.

Mucha, H.J. (1992). *Clusteranalyse mit Mikrocomputern*, Akademie Verlag, Berlin.

Mucha, H.J. and Klinke, S. (1993). Clustering techniques in the interactive statistical computing environment XploRe, *Discussion Paper 9318*, Institut de Statistique, Université Catholique de Louvain, Louvain-la-Neuve, Belgium.

Nishisato, S. (1980). *Analysis of Categorical Data: Dual Scaling and Its Applications*, University of Toronto Press, Toronto.

Rand, W.R. (1971). Objective criteria for the evaluation of clustering methods, *Journal of the American Statistical Association* **66**(336): 846–850.

Silverman, B.W. (1986). *Density Estimation for Statistics and Data Analysis*, Vol. 26 of *Monographs on Statistics and Applied Probability*, Chapman and Hall, London.

Späth, H. (1985). *Cluster Dissection and Analysis. Theory, FORTRAN Programs, Examples*, Ellis Horwood Limited, Chichester.

Ward, J.H. (1963). Hierarchical grouping methods to optimize an objective function, *Journal of the American Statistical Association* **58**(301): 236–244.

9
Exploratory Projection Pursuit

Sigbert Klinke[1] and Jörg Polzehl[2]

9.1 Introduction

"Projection Pursuit" (PP) stands for a class of exploratory projection techniques. This class contains methods designed for analyzing high dimensional data using low-dimensional projections. The main idea is to describe "interesting" projections by maximizing an objective function or projection pursuit index.

The idea of Exploratory Projection Pursuit (EPP) as it was introduced by Kruskal (1969; 1972) looks for nonlinear structure contained in the data. The approach was successfully implemented for exploratory purposes by Friedman and Tukey (1974). Alternative projection indices have been proposed by Huber (1985), Jones and Sibson (1987), Friedman (1987), Hall (1989a), and Cook, Buja and Cabrera (1993), among others. The idea has been applied to regression analysis (Friedman and Stuetzle, 1981b), density estimation (Friedman, Stuetzle and Schroeder, 1984), classification (Friedman and Stuetzle, 1981a), and discriminant analysis (Polzehl, 1993) . For projection pursuit regression the approximation of the regression function is characterized in Donoho and Johnstone (1989), and convergence rates are obtained in Hall (1989b). Good references about PP are given in Jones and Sibson (1987) and Huber (1985).

In Exploratory Projection Pursuit, we try to find "interesting" low-dimensional projections of the data. For this purpose, an index-function $I(\alpha, \beta, \ldots)$ dependent on the projection vectors $\alpha, \beta, \ldots$ is used. The function will be defined such that "interesting" views are the local and global maxima of the function.

This approach naturally accompanies the usual exploratory techniques in multivariate analysis such as principal component analysis (PCA) of a random vector X. In PCA, the index function is just the variance of a linear combination $\Gamma^T X$ subject to the normalizing constraint $\Gamma^T \Gamma = I$, or

[1]Institut de Statistique, Université Catholique de Louvain, B-1348 Louvain la-Neuve, Belgium; and Institut für Statistik und Ökonometrie, Humboldt-Universität zu Berlin, D-10178 Berlin, Germany.
[2]Konrad-Zuse-Institut für Informationstechnik Berlin, D-10711 Berlin, Germany.

other standard multivariate techniques. We will concentrate on the classical Projection Pursuit purpose of finding nonnormal projections of the data, searching for information not revealed by the covariance structure.

We will consider five different index-functions for two-dimensional projections: three based on orthogonal polynomials (Hermite, Legendre, Natural Hermite) and two based on kernel density estimates (Friedman–Tukey, Entropy).

The aim is to have a closer look at the behavior of several two-dimensional indices. In the following sections we explain the definition of the indices, and the result of the application for two artificial and one real datasets. We try to investigate their behavior for different bandwidths with respect to numerical aspects and for practical purposes. In the case of the real dataset, we will try to detect structure known from background information. In the last section, we will look at the computational aspects of the indices.

9.2   Projection Pursuit Indices

9.2.1   The Kernel-Based Indices

The indices suggested by Friedman and Tukey (1974) and Huber (1985) are based on kernel density estimates.

The index of Friedman and Tukey (1974) was designed to capture a cluster structure contained in the data. It has the form

$$I_{FT}(\alpha) = s(\alpha) * d(\alpha),$$

where $s(\alpha)$ only depends on the covariance structure and $d(\alpha)$ describes properties of a "local" density of the projected data. The first term can be avoided by sphering.

We will assume that the p-dimensional random variable X to be sphered and centered, that is, $E(X) = 0$ and $Var(X) = I_p$. This will remove the effect of location, scale, and correlation structure.

Using kernel density estimates, the "empirical" form of the Friedman–Tukey index can be expressed as in Jones and Sibson (1987):

$$\hat{I}_{FT,h}(\alpha) = \frac{1}{n^2 h} \sum_{i=1}^{n} \sum_{j=1}^{n} K\left(\frac{\alpha^T(x_i - x_j)}{h}\right),$$

or for two dimensions as:

$$\hat{I}_{FT,h}(\alpha, \beta) = \frac{1}{n^2 h^2} \sum_{i=1}^{n} \sum_{j=1}^{n} K\left(\frac{\alpha^T(x_i - x_j)}{h}, \frac{\beta^T(x_i - x_j)}{h}\right).$$

Normally we have to take different bandwidths, but after sphering it is reasonable to use the same bandwidth h in both directions. This turns out to be an estimate of

$K(r) = I(r < 1)$	$\frac{1}{\pi}$	Uniform
$K(r) = I(r < 1)$	$\frac{3}{\pi}(1 - r)$	Triangle
$K(r) = I(r < 1)$	$\frac{2}{\pi}(1 - r^2)$	Epanechnikov
$K(r) = I(r < 1)$	$\frac{3}{\pi}(1 - r^2)^2$	Quartic
$K(r) = I(r < 1)$	$\frac{4}{\pi}(1 - r^2)^3$	Triweight
$K(r) = I(r < 1)$	$\frac{1}{4(1-2/\pi)}\cos(\frac{\pi r}{2})$	Cosine
$K(r) =$	$\frac{1}{2\pi}\exp(\frac{-r^2}{2})$	Gaussian

TABLE 9.1. Common kernels.

$$I_{FT}(\alpha, \beta) = \int_{I\!\!R^2} p_Y^2(y) dy,$$

with $Y = (\alpha^T X, \beta^T X)$ and p_Y is the marginal density of Y. A high value of I_{FT} corresponds to a large departure from a parabolic density. Huber (1985) proposed the entropy index as a measure of dissimilarity to the normal distribution

$$E_Y(\log(p_Y(y))) = I_E(\alpha, \beta) = \int_{I\!\!R^2} p_Y(y) \log(p_Y(y)) dy.$$

which is uniquely minimized for $E(X) = 0$ and $Var(X) = I_p$ by the standard normal density, that is, an "uninteresting" view is identified with a "normal" marginal distribution.

$I_E(\alpha, \beta)$ can be estimated by

$$\hat{I}_{E,h}(\alpha, \beta) = \frac{1}{n} \sum_{i=1}^{n} \log \left(\frac{1}{nh^2} \sum_{j=1}^{n} K\left(\frac{\alpha^T(x_i - x_j)}{h}, \frac{\beta^T(x_i - x_j)}{h} \right) \right),$$

using a kernel density estimate of $p_Y(y)$.

The kernels in Table 9.1 are bivariate extensions of some common univariate kernels described in Härdle (1991).

9.2.2   The Polynomial-Based Indices

The main idea of polynomial-based indices is to approximate a criterion measuring the departure from a reference distribution (in our case, from the normal) using expansions based on orthogonal polynomials. Following Cook et al. (1993), the three indices considered here can be written in the form

$$\int_{I\!\!R^2} (p_Y(y) - \phi(y))^2 g(y) dy, \tag{9.1}$$

	a	b	$\omega(x)$	h_n
Legendre polynomials P_n	-1	1	1	$\dfrac{2}{2n+1}$
Hermite polynomials H_n	$-\infty$	∞	e^{-x^2}	$\sqrt{\pi}2^n n!$
Hermite polynomials H_{e_n}	$-\infty$	∞	$e^{-x^2/2}$	$\sqrt{2\pi}n!$

TABLE 9.2. Creating values a, b, and h_n for orthonormal polynomials.

where $\phi(y)$ is the bivariate standard normal density and $g(y)$ is a weight function. The orthogonal polynomials $f_n(x)$ applied in the approximation of the indices are defined by

$$\int_a^b \omega(x) f_n(x) f_m(x) dx = \begin{cases} 0 & n \neq m \\ h_n & n = m \end{cases} \text{ with } n, m \geq 0.$$

See Table 9.2 for a, b, and h_n.

The Legendre index based on Legendre polynomials P_j was introduced by Friedman (1987) with the idea of upweighting distances in the center of the distribution rather than in the tails. In practice, it was noticed that this index is attracted by skewed distributions. Recognizing theoretical problems with heavy-tailed distributions, Hall (1989a) introduced an index using Hermite polynomials. Cook et al. (1993) proposed an index, called Natural Hermite, which has a different weight function. The intention of the Natural Hermite index is to come back to Friedman's original purpose.

All of these polynomial-based projection indices (Legendre, Hermite, and Natural Hermite) try to measure nonnormality of the data. In Cook et al. (1993), additional indices are given which are attracted by special (bivariate) distributions (Central mass, Central hole).

The idea of the Legendre index is to transform the components of the projected data $Y_1 = \alpha^T X$ and $Y_2 = \beta^T X$ with the standard normal cdf Φ:

$$Z = (2\Phi(\alpha^T X) - 1, 2\Phi(\beta^T X) - 1).$$

If the projected data Y are normally distributed, then Z will be uniformly distributed on the square $I^2 = [-1, 1] \times [-1, 1]$. The integral-squared-distance between the probability density $p_Z(z)$ of Z and the uniform density $p_U(z) = 0.5$,

$$I_L(\alpha, \beta) = \int_{I^2} (p_Z(z) - p_U(z))^2 dz = \int_{I^2} p_Z^2(z) dz - 0.25$$

is used to measure the departure from normality. $p_Z(z)$ is now expanded in terms of a product Legendre expansion:[3]

$$\int_{I^2} p_Z^2(z)dz - 0.25 = \int_{I^2} \left(\sum_{i,j=0}^{\infty} a_{ij} P_i(z_1) P_j(z_2) \right)^2 dz - 0.25.$$

The coefficients a_{ij} are given by

$$\sum_{k,l=0}^{\infty} a_{kl} P_k(z_1) P_l(z_2) = p_Z(z),$$

and integration on the square I^2 gives

$$\int_{I^2} \sum_{k,l=0}^{\infty} a_{kl} P_k(z_1) P_l(z_2) P_i(z_1) P_j(z_2) dz = \int_{I^2} p_Z(z) P_i(z_1) P_j(z_2) dz.$$

Using the property of the Legendre polynomials,

$$\sum_{k,l=0}^{\infty} a_{kl} \int_I P_k(z_1) P_i(z_1) dz_1 \int_I P_l(z_2) P_j(z_2) dz_2 = E(P_j(z_1) P_j(z_2)),$$

we get finally

$$a_{ij} = (i+0.5)(j+0.5) E(P_j(z_1) P_j(z_2)),$$

which leads to replacing the expectations by the corresponding sample means to

$$\begin{aligned}
I_L(\alpha, \beta) &= \int_{I^2} \left(\sum_{i,j=0}^{\infty} a_{ij} P_i(z_1) P_j(z_2) \right)^2 dz - 0.25 \\
&= \sum_{i,j,k,l=0}^{\infty} a_{ij} a_{kl} \int_{I^2} P_i(z_1) P_j(z_2) P_k(z_1) P_l(z_2) dz - 0.25 \\
&= \sum_{i,j=0}^{\infty} a_{ij}^2 (i+0.5)^{-1}(j+0.5)^{-1} - 0.25 \\
&= \sum_{i,j=0}^{\infty} (i+0.5)(j+0.5)(E(P_i(z_1) P_j(z_2)))^2 - 0.25.
\end{aligned}$$

Truncating the sum at order J and replacing the expectation by the corresponding sample means, we get

$$\hat{I}_L(\alpha, \beta) = \sum_{i=0}^{J} \sum_{j=0}^{J-i} (i+0.5)(j+0.5) \left(\frac{1}{n} \sum_{k=1}^{n} P_i(Z_{1,k}) P_j(Z_{2,k}) \right)^2 - 0.25.$$

---

[3] Here and in the following formulas $\sum_{i,j=0}^{\infty}$ is to be interpreted as $\sum_{i=0}^{\infty} \sum_{j=0}^{\infty}$.

Following Cook et al. (1993), I_L can be written as in (9.1) with

$$g(y) = \frac{1}{2\phi(y)}. \tag{9.2}$$

The weight function shows that the tails of the distribution are upweighted in the case of the Legendre index. Because of this, Hall (1989a) proposed the Hermite index, where Hermite polynomials H_j are used instead of Legendre polynomials to expand the density of Y:

$$I_H(\alpha, \beta) = \int_{\mathbb{R}^2} (p_Y(y) - \phi(y))^2 dy,$$

that is, $g(y) = 1$ is used as the weight function. Defining

$$h_n(x) = \sqrt{\frac{\sqrt{\pi}}{2^{n-1} n!}} H_n(x)\phi(x) \qquad \text{with} \qquad \int_{\mathbb{R}} h_n(x) h_m(x) = \delta_{n,m}$$

leads to

$$
\begin{aligned}
I_H(\alpha, \beta) &= \int_{\mathbb{R}^2} (p_Y(y) - \phi(y))^2 dy \\
&= \int_{\mathbb{R}^2} \left(\sum_{i,j=0}^{\infty} a_{i,j} h_i(y_1) h_j(y_2) - \frac{1}{2\sqrt{\pi}} h_0(y_1) h_0(y_2) \right)^2 dy \\
&= \sum_{i,j,k,l=0}^{\infty} a_{i,j} a_{k,l} \int_{\mathbb{R}^2} h_i(y_1) h_k(y_1) h_j(y_2) h_l(y_2) dy \\
&\quad - \frac{1}{\pi} \sum_{i,j=0}^{\infty} a_{i,j} \int_{\mathbb{R}^2} h_i(y_1) h_0(y_2) h_j(y_1) h_0(y_2) dy \\
&\quad + \frac{1}{4\pi^2} \int_{\mathbb{R}^2} h_0^2(y_1) h_0^2(y_2) dy \\
&= \sum_{i,j=0}^{\infty} a_{i,j}^2 - \frac{1}{\pi} a_{0,0} + \frac{1}{4\pi^2}.
\end{aligned}
$$

The coefficients $a_{i,j}$ can be calculated by

$$\sum_{k,l=0}^{\infty} a_{k,l} h_k(y_1) h_l(y_2) = f(y_1, y_2),$$

and integration over $\mathbb{R}^2$ leads to

$$\sum_{k,l=0}^{\infty} a_{k,l} \int_{\mathbb{R}^2} h_i(y_1) h_k(y_1) h_j(y_2) h_l(y_2) \, dy = \int_{\mathbb{R}^2} f(y_1, y_2) h_i(y_1) h_j(y_2) \, dy,$$

and finally to

$$a_{i,j} = E_Y(h_i(y_1)h_j(y_2)).$$

This results in the following estimates:

$$\hat{a}_{i,j} = \frac{1}{n}\sum_{k=1}^{n} h_i(\alpha^T X_k)h_j(\beta^T X_k),$$

$$\hat{I}_{H,J}(\alpha,\beta) = \sum_{i=0}^{J}\sum_{j=0}^{J-i} \hat{a}_{i,j}^2 - \frac{1}{\pi}\hat{a}_{0,0} + \frac{1}{4\pi^2}.$$

The third polynomial-based index developed by Cook et al. (1993) is the so called "Natural Hermite" index:

$$I_N(\alpha,\beta) = \int_{\mathbb{R}^2} (p_Y(y) - \phi(y))^2 \phi(y)dy,$$

(that is, $g(y) = \phi(y)$) which can be estimated by

$$\hat{I}_{N,J}(\alpha) = \sum_{i=0}^{J}\sum_{j=0}^{J-i} (\hat{a}_{i,j} - b_{i,j})^2,$$

with

$$\hat{a}_{i,j} = \frac{1}{n}\sum_{k=1}^{n} \frac{1}{\sqrt{i!j!}} He_i(\alpha^T X_k)He_j(\beta^T X_k)\phi(\alpha^T X_k, \beta^T X_k),$$

$$b_{i,j} = b_i b_j,$$

$$b_{2i} = \frac{(-1)^i\sqrt{(2i)!}}{\sqrt{\pi}i! \, 2^{2i+1}},$$

$$b_{2i+1} = 0.$$

Here the $a_{i,j}$ represent the coefficients in the expansion for $p_Y(y)$ and $b_{i,j}$ represent the coefficients in the expansion of $\phi(y)$. This index upweights the differences in the center of the distribution.

9.2.3 Other Indices

A great variety of indices are possible. Many indices based on functionals which are minimized by the Gaussian distribution are possible. The entropy index as described before is one example. Jee (1985) proposed the use of the Fisher information.

Jones and Sibson (1987) suggested the use of an approximation of the Entropy index based on mixed central moments:

$$I_C(\alpha,\beta) = \frac{1}{12}(\kappa_{30}^2 + 3\kappa_{21}^2 + 3\kappa_{12}^2 + \kappa_{03}^2) + \frac{1}{48}(\kappa_{40}^2 + 4\kappa_{31}^2 + \kappa_{22}^2 + \kappa_{13}^2 + \kappa_{04}^2).$$

The restriction on the third and fourth moments makes the index only sensitive to certain deviations from normality, namely skewness and kurtosis.

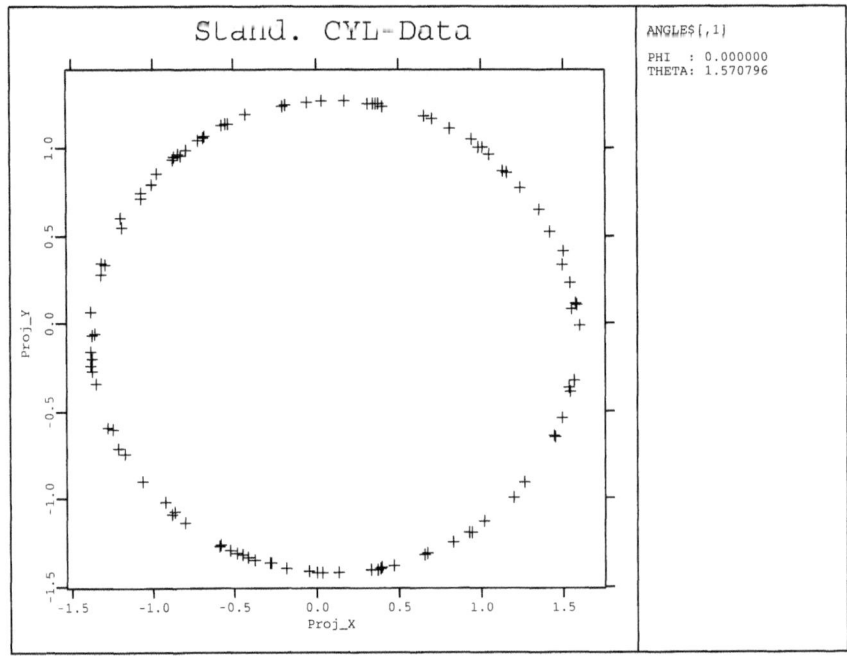

FIGURE 9.1. Optimal projection of CYL data.

9.3   The Index Functions in Practice

9.3.1   The Datasets

For our explanations we will use two artificial datasets. The first dataset (CYL data) contains 100 random points located on the surface of a cylinder of length 1 and radius 1. In case of the CYL data we have a situation with one striking two-dimensional projection (see Figure 9.1).

The second dataset (TET data) consists of 250 points located on the supporting hyperplanes of a tetrahedron given by its facets. Three hyperplanes contain 50 normally distributed random points each, while the fourth hyperplane accommodates 100 normally distributed random points. In the case of the TET data, there are six interesting views, determined by the pairs of the supporting hyperplanes. The lower row of Figure 9.2 contains the projection determined by the intersection of the basis plane with one of the others. The upper row contains projections determined by pairs of the first three planes, that is, the structure displayed contains 100 points in the case of the upper row and 150 in the case of the lower.

Three-dimensional datasets are used to enable us to plot the index functions. For every two-dimensional projection of a three-dimensional data set, the projection plane can be uniquely identified by the normal vector, which can be expressed in terms of two angles, φ and ϑ.

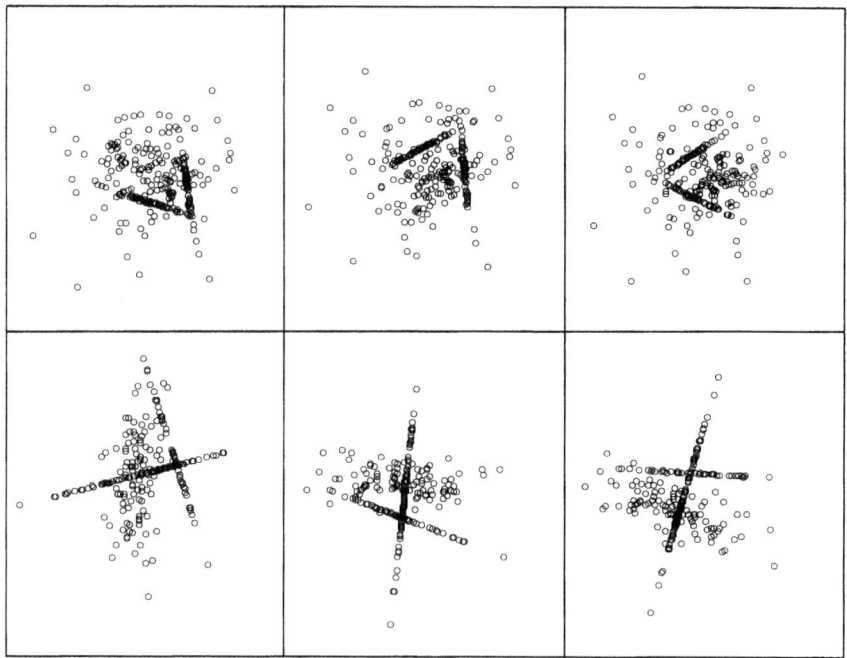

FIGURE 9.2. Optimal projection of TET data.

The location of the projections can be seen in Figure 9.4. The projections of the upper row are the three projections indicated by the vertices of the triangle in the upper part of the picture, and the projections of the lower row are on the band in the middle of the picture. The TET dataset will be used to demonstrate interesting effects concerning the choice of bandwidths and polynomial orders.

9.3.2   The Kernel-Based Projection Indices

The uniform kernel leads to discontinuous piecewise constant indices, while Triangle and Epanechnikov kernels provide continuity but no differentiability. Because of the difficulties with maximization of the index function involved in the Projection Pursuit approach, the easy-to-compute uniform kernel is not helpful. We restricted ourself to the Epanechnikov kernel for easy computing and continuity. Of course, the Quartic and Triweight kernels secure the existence of derivatives of the indices and are much easier to compute than the alternative Gaussian kernel. But with the Epanechnikov kernel, it is possible to get speed improvements for larger sample sizes (Klinke, 1993).

The essential aim in Exploratory Projection Pursuit seems to be to select an appropriate bandwidth. For a small bandwidth, we would expect the index function to have numerous local maxima. Most of these maxima will correspond to observations grouped randomly in bivariate projections. If the bandwidth grows, the number of maxima should decrease; the local maxima will melt together. If the bandwidth becomes large, the index function should have only a few (maybe only 1) global maxima.

Finding a global maximum is a difficult task if there are numerous local maxima. In Exploratory Projection Pursuit, the aim is usually to find a set of distinct meaningful local maxima rather than to find a unique global one because projections corresponding to local maxima may also provide "interesting" views of the data.

Silverman (1986) gives the following formulas for a bandwidth that minimizes the asymptotic mean integrated squared error. If the kernel is a radially symmetric probability density function and the unknown density has bounded and continuous second derivatives, AMISE for a given h results in

$$AMISE = \frac{\int_{\mathbb{R}^d} K^2(t)dt}{nh^d} + \frac{1}{4}h^4 \left(\int_{\mathbb{R}^d} t_1^2 K(t)dt\right)^2 \left(\int_{\mathbb{R}^d} tr(\nabla^2 f)^2(t)dt\right),$$

which leads to

$$h_{opt}^{d+4} = \frac{d}{n} \left(\int_{\mathbb{R}^d} K^2(t)dt\right) \left(\int_{\mathbb{R}^d} t_1^2 K(t)dt\right)^{-2} \left(\int_{\mathbb{R}^d} tr(\nabla^2 f)^2(t)dt\right)^{-1},$$

(see Scott, 1992, 6.48) where d is the dimensionality of the projected data. As a rule of thumb, we plug in the multivariate normal density instead of f, which gives

$$\int_{\mathbb{R}^d} tr(\nabla^2 \phi)^2(t)dt = (4\pi)^{d/2}(d/2 + d^2/4).$$

Using the data from our examples, the normal reference bandwidth becomes:

$$h_{NR}^{d+4} = \frac{d}{n} R(K) \left(\int_{\mathbb{R}^d} t_1^2 K(t)dt\right)^{-2} (4\pi)^{-d/2}\frac{4}{d(d+2)}, \tag{9.3}$$

which leads to a normal reference bandwidth $h_{NR} \approx .38$ in the case of the CYL data, and to $h_{NR} \approx .33$ in the case of the tetrahedron data.

9.3.3   The Polynomial-Based Indices

The polynomial-based indices under consideration measure differences between the underlying density of the projected data and a standard normal density. In order to detect differences that are not caused by location and covariance structure, the data have to be standardized in a first step.

Even light deviations from a standardized situation lead to marked changes in the index functions. The effect can be thought of as a mapping of information by location and covariance effects.

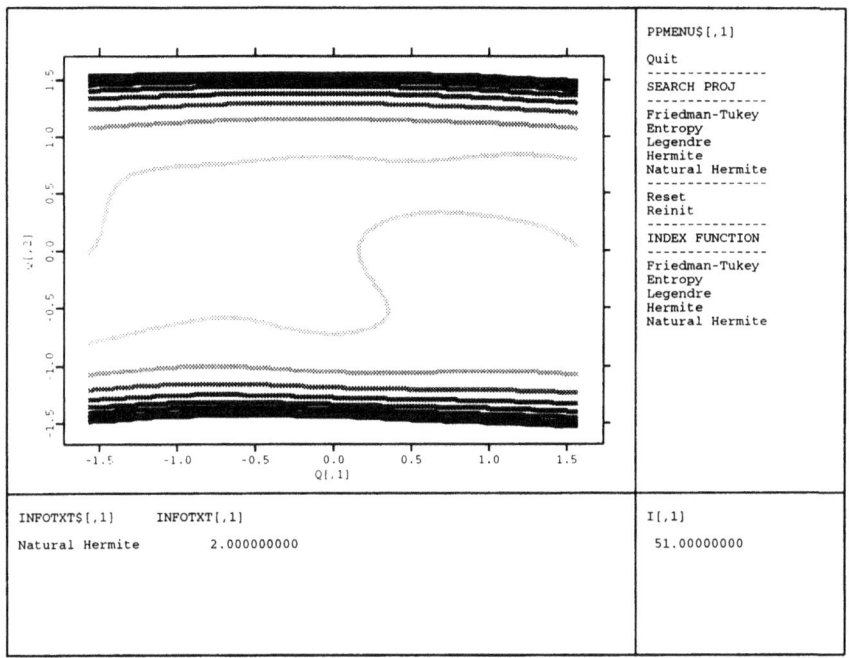

FIGURE 9.3. Index function of cylinder data with Natural Hermite index (J=2).

The problem of bandwidth selection for kernel indices translates into a problem of specification of an appropriate order J of the orthogonal series approximations in the case of the polynomial indices. Using a small order usually corresponds to a crude approximation of the marginal density of the projected data, while a high order will allow a more detailed modeling. The situation is similar to the case of kernel indices in the sense that a high order J will cause numerous local maxima of the index function, while a small J may hide interesting maxima by looking for simple structures.

An unpleasant problem of the Legendre index is that it is not invariant with respect to rotation inside the projection plane. This effect is avoided by the use of rotation symmetric kernels in the case of kernel-based indices.

9.3.4   The Behavior of the Indices

Figures 9.3 to 9.8 illustrate the behavior of the index functions for the data. The figures show contour plots of the index functions computed on a net of 50×50 points for (φ, ϑ). The levels used correspond to the 0.10, 0.46, 0.68, 0.81, 0.88, 0.93, 0.96, and 0.97 quantiles (blue to red, light gray to black) of index values on the net.

For the cylinder data, bandwidths in the magnitude of the normal refer-

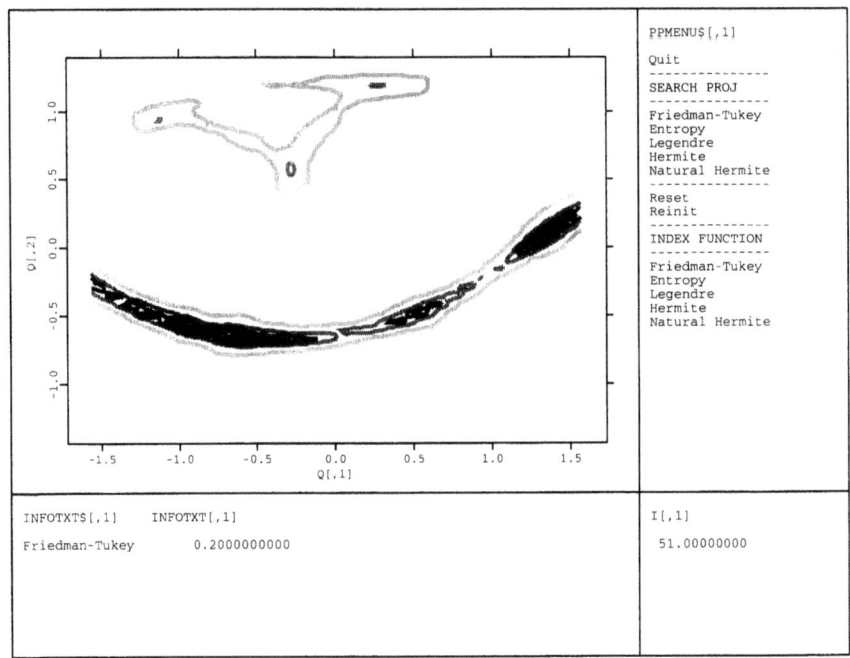

FIGURE 9.4. Index function of tetrahedron data with Friedman–Tukey index (h=0.2) (see color insert).

ence bandwidth behave well in this example, while large bandwidths lead to oversmoothing effects, blurring the structure; for more detailed pictures see Klinke and Polzehl (1994).

The Legendre index, and to some extent the Hermite index, show evident problems in handling even this clear structure in the case of a small order J. The Natural Hermite index behaves much better with these data, as is noted in Cook et al. (1993), and as seen in Figure 9.3. Even for larger values of J, the estimated criteria are sufficiently smooth, providing only a small number of local minima.

The situation is much more complicated for the second dataset. Figures 9.4 to 9.6 show contour plots of the Friedman–Tukey index. In the case of the smallest bandwidth, $h = 0.2$, we get local maxima corresponding to all interesting projections, although for this bandwidth there are 141 local maxima on the net in the case of the Friedman–Tukey index (see Figure 9.4). Increasing the bandwidth leads to smearing effects, still presenting interesting projections of the basis plane but melting together some maxima corresponding to the projections searched for. Compare the change in functions from Figure 9.4 to Figure 9.5. In the top left is a triangular region corresponding to the three interesting projections of the first row

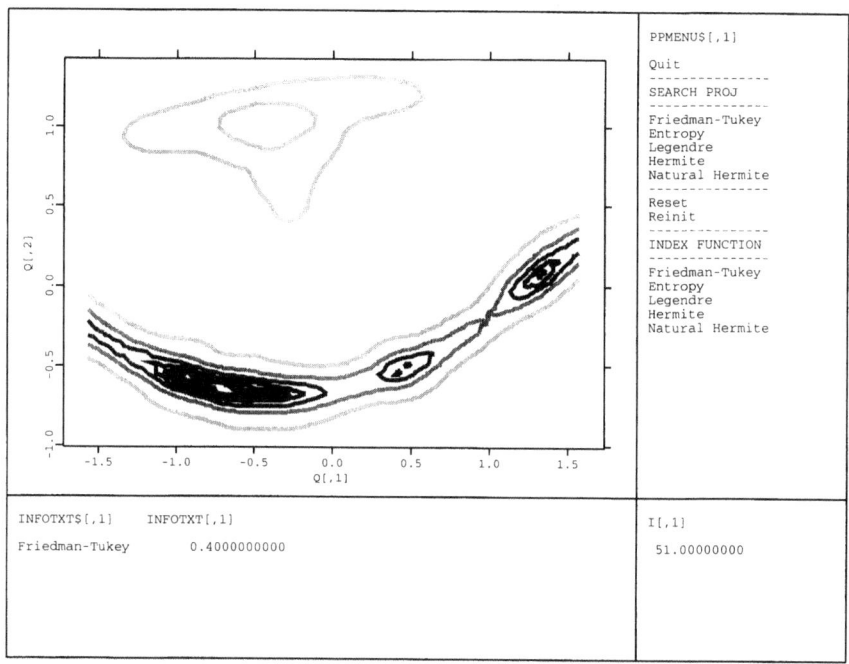

FIGURE 9.5. Index function of tetrahedron data with Friedman–Tukey index (h=0.4) (see color insert).

of Figure 9.2. In Figure 9.4 all three projections can be detected, but in Figure 9.5, the three projections are not distinguishable from their local surroundings. The normal reference bandwidth is clearly too large to show the underlying structure of a tetrahedron. The maxima corresponding to the projections in the upper row of Figure 9.5 are smoothed away, resulting in a local maximum without information while the basis plane can still be identified for $h = 0.8$. The largest bandwidth, $h = 1.6$, leads to strong oversmoothing effects, hiding all structure (see Figure 9.6).

In the case of small order $J \leq 3$, all polynomial indices fail completely to find the structure (see Figure 9.7). For $J > 4$, the basis plane is identified. In the case of the Legendre index with order $J = 10$, the information about the structure can be identified, while the Hermite (see Figure 9.8) and Natural Hermite indices are likely to require much higher values of J. It can be observed that for a medium J, a sufficient number of local maxima exists but global information considered in the polynomial approximation leads to smoothing, which is sufficient to hide the structure. A substantial increase of the order J of the approximation leads to a substantial increase in computing time.

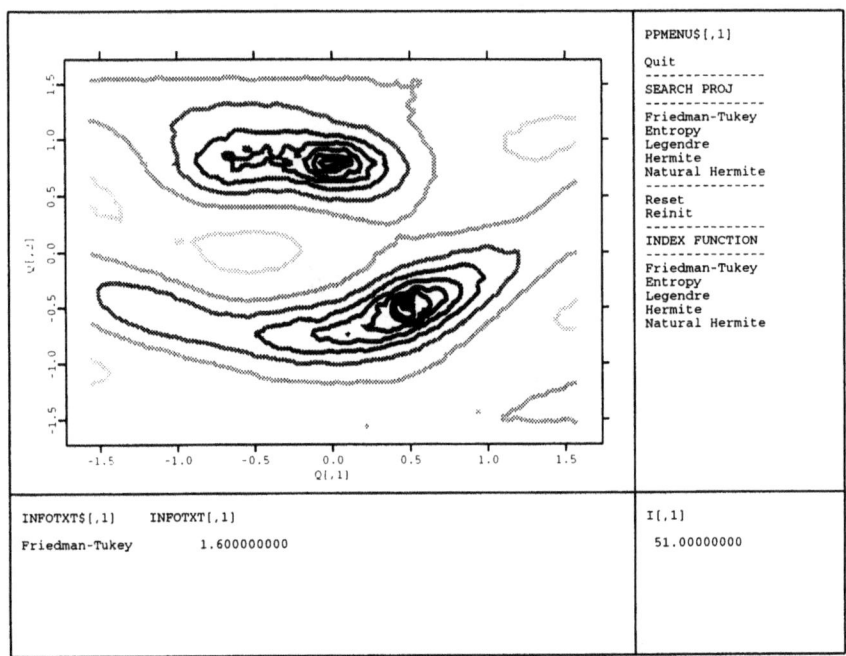

FIGURE 9.6. Index function of tetrahedron data with Friedman–Tukey index (h=1.6).

9.4   What Will Be Found on Real Data?

The dataset we use here is the Swiss Banknote dataset found in Flury and Riedwyl (1981). The dataset consists of 200 observations: 100 genuine and 100 forged banknotes. The six variables are measurements of the size of the banknotes (X_1, width of banknote; X_2, height on left side; X_3, height on right side; X_4, lower margin; X_5, upper margin; and X_6, diagonal of a inner box). Figure 9.9 shows a principal component plot of the first two principal components of the dataset. We can distinguish clearly the forged and genuine banknotes. If we use the Hermite index with order 7, we get the picture in Figure 9.10. This picture clearly shows three clusters: the upper one consists mostly of forged bank notes and one genuine bank note (observation 70).

The other clusters contain forged or genuine banknotes. An easy interpretation would be that at least two gangs of forgers were falsifying Swiss banknotes. It seems that one of the genuine banknotes is wrongly classified.

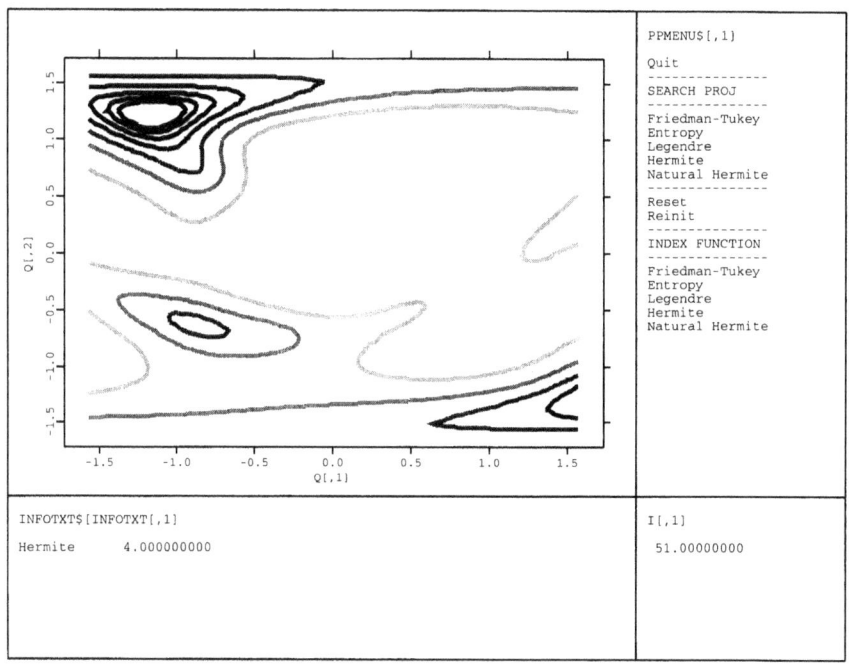

FIGURE 9.7. Index function of tetrahedron data with Hermite index
(J=4) (see color insert).

The projection and loading vectors from the EPP step are

$$
p = \begin{pmatrix}
0.896562 & 0.381411 \\
-0.087805 & 0.632784 \\
-0.248246 & 0.423262 \\
-0.349619 & 0.510308 \\
-0.0441331 & 0.113526 \\
0.0515641 & 0.04071
\end{pmatrix}
\quad
l = \begin{pmatrix}
-0.1643288 & 0.0958421 \\
-0.4423175 & 0.3462294 \\
0.0023849 & 0.6045508 \\
0.7167979 & -0.6467759 \\
0.6894302 & -0.5717335 \\
0.2225492 & -1.2337853
\end{pmatrix}.
$$

The projection vectors p could be interpreted as factor loadings with
respect to the standardized variables Y; and the loading vectors l as factor loadings with respect to the original variables X. The first projection
turns out to depend mainly on the first component of Y while the second
reflects a mean of the second, third, and fourth components of Y. An in-
terpretation of $\alpha^T \Sigma^{-1/2}$, $\Sigma = Cov(X) = (\sigma_{ij})$, and $\alpha^T \Sigma^{-1/2} diag(\sigma_{ii})^{-1/2}$
as factor loadings with respect to the unsphered variables as in PCA based
on covariance or correlation structure may be of interest.

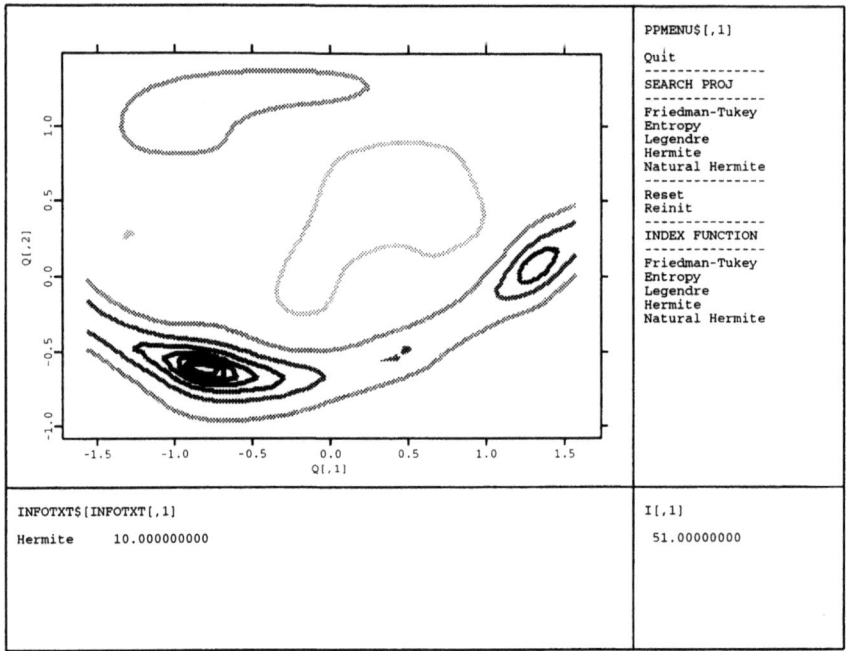

FIGURE 9.8. Index function of tetrahedron data with Hermite index (J=10).

EPP is a technique that allows us to use interactive graphics for interpretation. The macro

`ppinter (x p)`

links the projected dataset of the EPP with each of the variables (strips of data in lower plot). By simple linking, we can visualize the relationship between the projection and the original variables. In the upper plot, we see the projection for the Swiss banknotes found with `PPEXPL`. The lower plot shows jittered dotplots of each of the six variables. The variables are rescaled on $[0,1]$ via $xr_j = (x_j - \min_j(x_j))/(\max_j(x_j) - \min_j(x_j))$. The text window on the right shows the projection vector for X and Y found by EPP.

We mark the three clusters and they appear in univariate projections into the coordinate (variable) axes to assist interpretation based on differences of banknote sizes. Figure 9.11 shows clearly that all the points in the first cluster have low values in the variable 6 (inner box diagonal measurement of bank note). The second cluster seems to be linked to the high values of the sixth variable and the low variables of the fourth variable. The third cluster is linked to the low values of the sixth and the high values of the fourth variable.

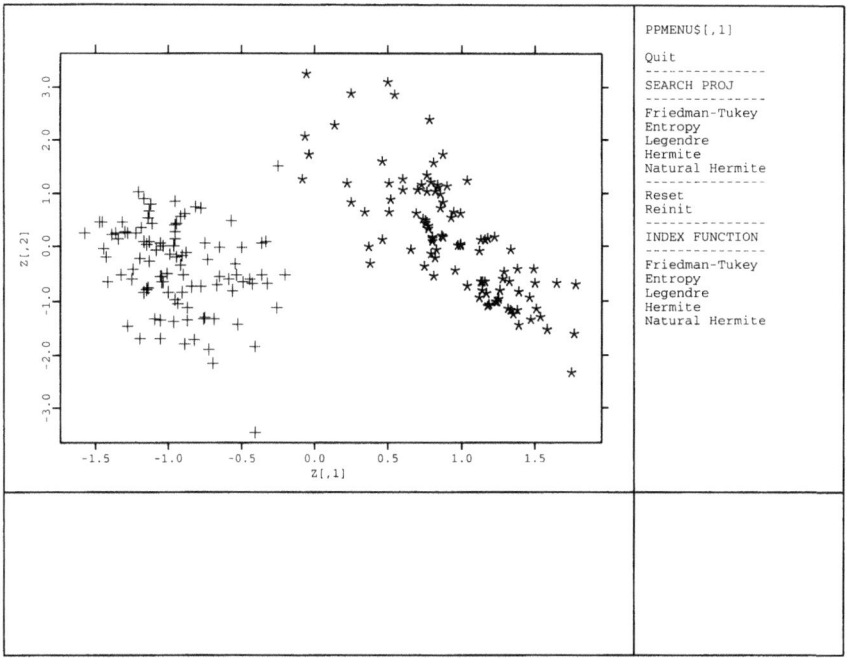

FIGURE 9.9. Basic screen of the **PPEXPL** macro. Swiss banknote data projected on the first two principal components. Red crosses indicate genuine banknotes and blue stars indicate the forged bank notes (see color insert).

A possible next step would be to mark each variable to see how it behaves in the projection. However, we will not find anything interesting with this dataset.

9.5   Computational Aspects

9.5.1   The Polynomial-Based Indices

The main savings in terms of computational speed for the polynomial indices are obtained using recursive relationships of the orthogonal polynomials $p_Y(y)$ $(P_0(x) = H_0(x) = H_{e_0}(x) = 1, P_1(x) = H_1(x) = H_{e_1}(x) = x)$:

- Legendre polynomial $P_j(x) = \dfrac{(2j+1)P_{j-1}(x) - jP_{j-2}(x)}{j+1}$;

- Hermite polynomial $H_j(x) = 2xH_{j-1}(x) - 2(j-1)H_{j-2}(x)$; and

- Hermite polynomial $H_{e_j}(x) = xH_{e_{j-1}}(x) - (j-1)H_{e_{j-2}}(x)$.

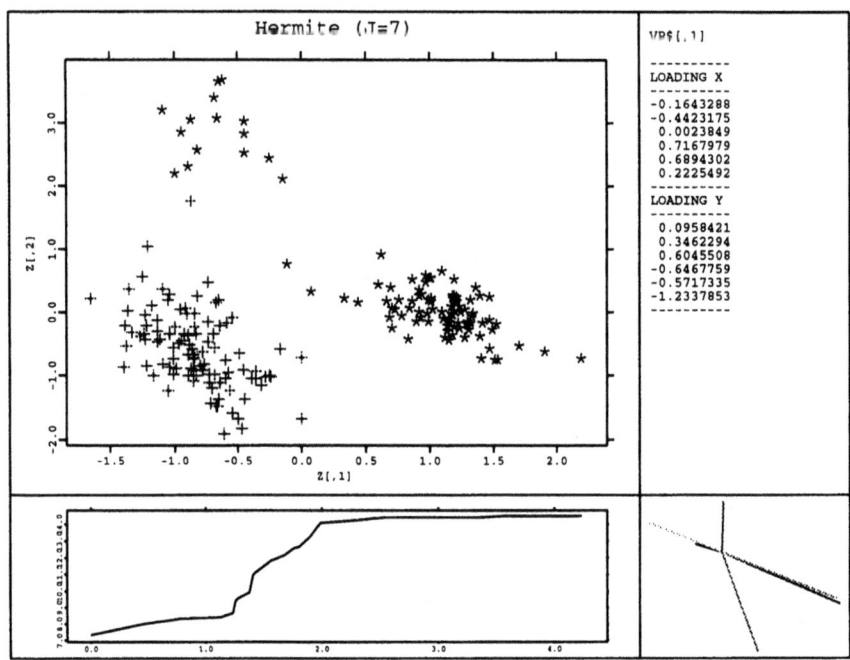

FIGURE 9.10. Projection with Hermite index (J=7). The right upper window shows the loadings with respect to X; the right lower window shows a graphical representation of the loadings. The left lower window shows how the index function has increased (see color insert).

The missing invariance of the Legendre index requires optimization with respect to an additional variable describing rotations inside the plane. Many applications will require a large order J to avoid the hiding effects described in 9.3.3. This, of course, increases the computational effort.

9.5.2   The Kernel-Based Indices

We recall that we are mainly interested in finding maxima. Every transformation of the index function that preserves the maxima of this function and minimizes the number of operations will speed up the calculations.

Thus the formula for the Friedman–Tukey index

$$\hat{I}_{FT,h}(\alpha, \beta) = \frac{1}{n^2 h^2} \sum_{i=1}^{n} \sum_{j=1}^{n} K\left(\frac{\alpha^T (x_i - x_j)}{h}, \frac{\beta^T (x_i - x_j)}{h}\right)$$

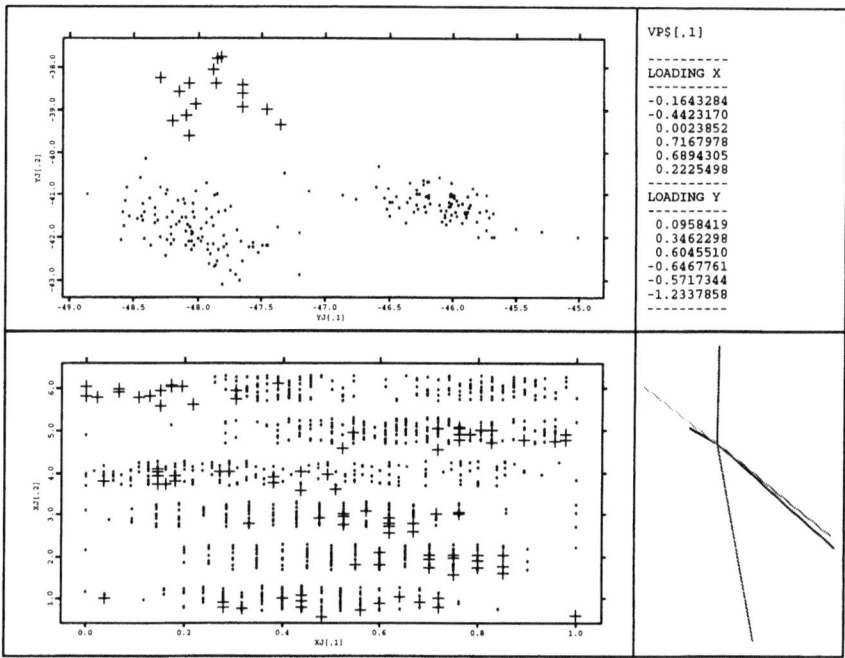

FIGURE 9.11. First cluster (forged banknotes) linked to original variables.

can be rewritten as

$$\hat{I}_{FT,h}(\alpha, \beta) = \frac{1}{n^2 h^2} \sum_{i=1}^{n} \left(K(0,0) + 2 \sum_{j=1}^{i-1} K\left(\frac{\alpha^T (x_i - x_j)}{h}, \frac{\beta^T (x_i - x_j)}{h} \right) \right).$$

Of course, the positive constant multiplicative term $1/(n^2 h^2)$, the additive constant term $nK(0,0)$, and the 2 can be neglected, leading to

$$\tilde{I}_{FT,h}(\alpha, \beta) = \sum_{i=1}^{n} \sum_{j=1}^{i-1} K\left(\frac{\alpha^T (x_i - x_j)}{h}, \frac{\beta^T (x_i - x_j)}{h} \right).$$

Because of the included logarithm, the Entropy index cannot be transformed in such a way. Thus, the estimate of the Entropy index results in

$$\tilde{I}_{E,h}(\alpha, \beta) = \sum_{i=1}^{n} \log \left(\sum_{\substack{j=1 \\ i \neq j}}^{n} K_h\left(\frac{\alpha^T (x_j - x_i)}{h}, \frac{\beta^T (x_j - x_i)}{h} \right) \right). \quad (9.4)$$

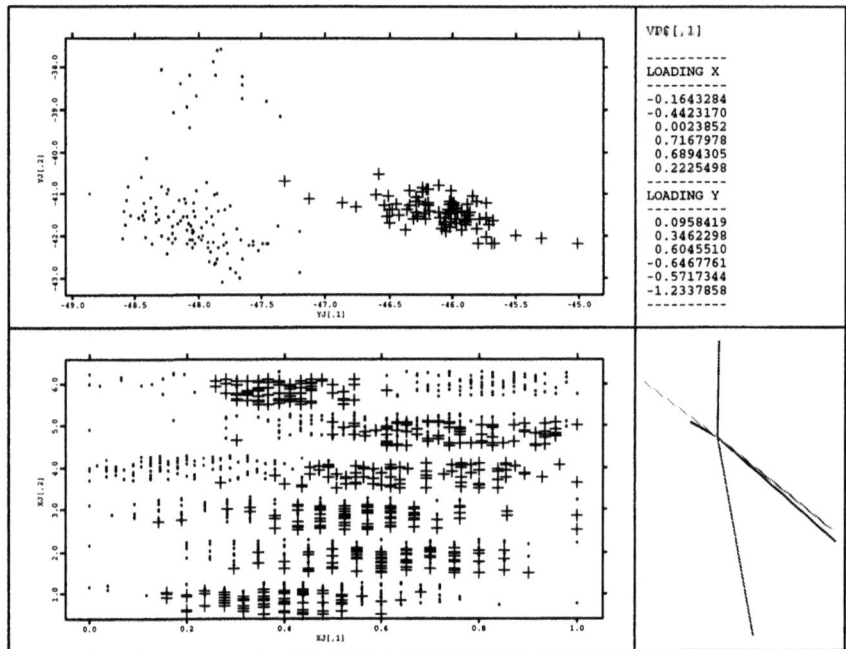

FIGURE 9.12. Second cluster (forged banknotes) linked to original variables.

Kernel	Ratio
Uniform	1.00
Epanechnikov	1.16
Quartic	1.35
Triweight	1.41
Triangle	1.77
Gaussian	5.17
Cosine	5.78

TABLE 9.3. Ratio of evaluation time of kernels compared to Uniform kernel.

The transformed index functions heavily depend on the number of evaluations of the kernel function. Let us now look at the speed of kernel evaluations. They differ considerably,[4] as can be seen in Table 9.3. So we see that, for example, the Epanechnikov kernel needs 16% more time for evaluation of the kernel values compared to the Uniform kernel. On average, the polynomial kernels (Uniform, Quartic, Epanechnikov, Triangle, and Triweight)

---

[4]Bivariate kernels programmed in C with the Zortech C++ 3.0 - compiler on a PC 486 with 50 Mhz.

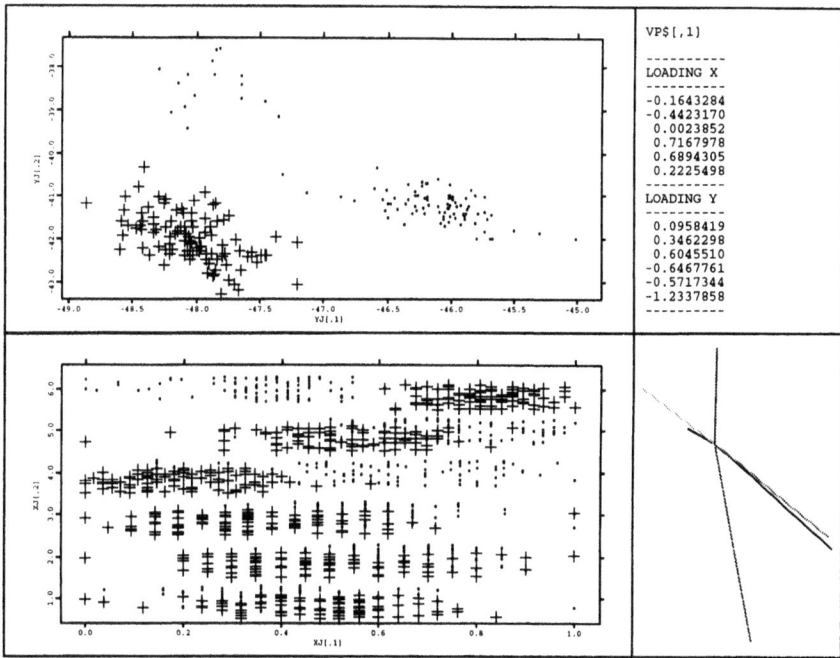

FIGURE 9.13. Third cluster (genuine banknotes) linked to original variables.

are five to seven times faster to calculate than the transcendental kernels (Cosine, Gaussian). The limited support can be used explicitly in the case of the polynomial kernels, which leads to additional advantages compared to the Gaussian in the case of explicit computation of the convolutions. Of course, there are other ways to use the Gaussian kernel efficiently, for example, the Fast Fourier Transform (Wand, 1994).

If we sort the data through the first component, we have only to run from a datapoint with index idx_{low} to an index idx_{high}.

Binning

The main advantage of binning is that it reduces the computational costs of density- (and regression-) estimates by a slight loss of accuracy.

The binning or WARPing first makes a "regular" tessellation of the real space $I\!R^p$, if p is the dimension of the data X. Every part of the tessellation, also called bin or cell, should be connected and disjoint. Additionally, the bins should exhaust the $I\!R^p$; then one point, the bin center, is taken to represent all the datapoints that fall in such a bin. Usually the gravity center of a bin is taken. The set of bin centers S should have the property that for every bin center the direct neighbors have the same distance. In the one-dimensional case such tessellation are the equally spaced intervals.

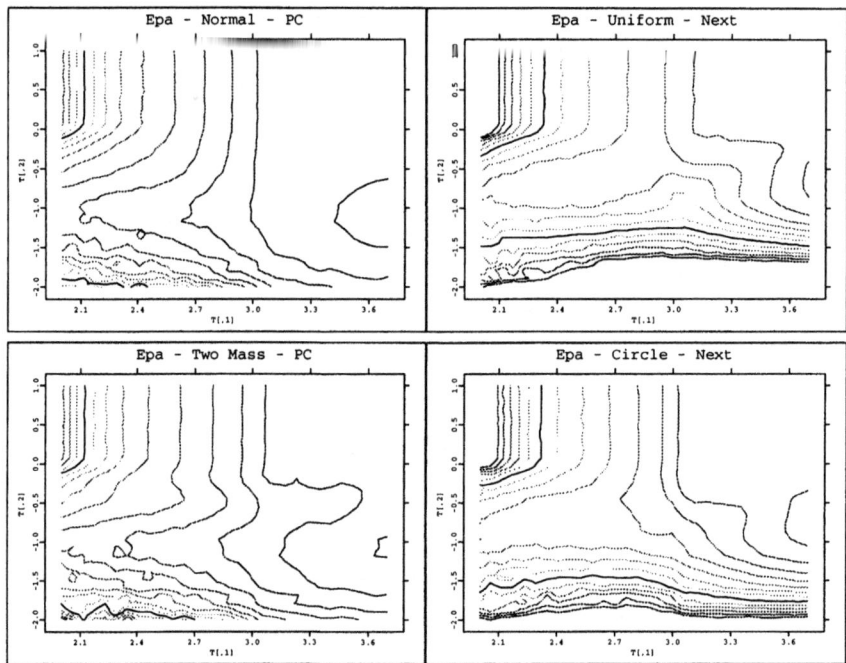

FIGURE 9.14. Ratio of computational time for the density estimation with binned and unbinned data for different data and computers (see color insert).

We gain in the one-dimensional case for density estimation mainly from a decrease of datapoints. But we obtain an additional advantage: the distance between bin centers can be expressed as multiples of the binwidth $\delta = \min_{p_1, p_2 \in S; p_1 \neq p_2} | p_2 - p_1 |$. So we have to evaluate the kernel only at the points $i\,\delta$ with $i = 0, 1, 2$, etc.

In the two-dimensional case, the advantage of reducing the sample size will be lost. If we take a small grid with 50 bins in each variable, we will get $50 \times 50 = 2500$ bins for a squared grid in the two-dimensional case. If we have a data set with 1000 observations, we will get, on the average, 20 observations per bin in the one-dimensional case. In the two-dimensional case, this number reduces to 0.4 observations per bin.

Figure 9.14 indicates how fast the binning method in two dimensions is for different bandwidths and different sample sizes (2500 bins). On the horizontal axis, we have the common logarithm of the sample size; on the vertical axis, we have the common logarithm of the bandwidth. The data are in the unit square $[0, 1]^2$. The four datasets consists of a normal distribution (NORMAL), a distribution on a ring (CIRCLE), a mixture of two normal distributions (TWO MASS), and a uniform distribution. The Epanechnikov kernel is used.

The thick black line indicates here that the ratio of computational time of the binned version and the unbinned version is 1. The blue thin lines indicate ratios of 0.9, 0.8, ..., which means that the binned program needs only 90%, 80%, ... of the time to calculate the same density. The yellow and red lines indicate ratios of 1.1, 1.2, ...

This has to be compared with Figure 3 in Fan and Marron (1994), which shows a speed improvement of factor 10 and greater for binned density estimation over direct density estimation in the one-dimensional case. That would mean that the ratio becomes less than 0.1, which is in fact the case for the most right line. Nevertheless, in the interesting area between −1.5 and 0.0 (the common logarithm of the bandwidth), the ratio is less than one, and thus, we gain some advantage using binning.

Because we replace the (continuous) kernel by a discrete step function on a square, the value of the index function will vary slightly if we rotate the data. This effect could be diminished by using linear binning, that is, replacing the kernel by a piecewise linear function. A computationally efficient algorithm will distribute the mass 1 of each observation in 4 bins, which leads in the worst case to an increase of the number of bins by a factor of 4. Figure 9.14 shows that we could then use the direct estimate.

9.6   The PPEXPL Macro

In XploRe, the macros for Exploratory Projection Pursuit are part of the ADDMOD library. The following sequence in XploRe

```
x = read ("bank2")
library ("addmod")
p = ppexpl (x 0)
```

will load the Swiss banknote dataset and call the macro for Exploratory Projection Pursuit. The data will automatically be centered and sphered. The screen is divided into four windows. The dataset is projected along the first two principal components. On the right side a menu appears.

At first you will be questioned whether the output should be colored or in gray-scale. This is only important if you want to plot index functions. The next menu consist of two parts. The SEARCH PROJ part tries to find an optimal projection according to the selected index function. The actual best projection is displayed in the 2-D window. The left lower text-window gives information about the selected index function, the smoothing parameter for the density estimation, the value of the index function for the actual projection, and a step parameter. The Reset selection allows us to reset the projection to the initial projection based on principal component analysis. The Reinit selection selects a random projection as initial projection. The INDEX FUNCTION makes a contour plot of the index function.

The additional parameter 0 tells the macro that the whole process is interactive. If you put in a float vector instead of a 0, the macro can run completely noninteractive. The float vector ch should have the following entries:

$$ch[1,1] = \begin{cases} 3 & \text{Use color} \\ 4 & \text{Use gray scale} \end{cases}$$

$$ch[i,1] = \begin{cases} 5 & \text{Find best projection with Friedman–Tukey index} \\ 6 & \text{Find best projection with Entropy index} \\ 7 & \text{Find best projection with Legendre index} \\ 8 & \text{Find best projection with Hermite index} \\ 9 & \text{Find best projection with Natural Hermite index} \end{cases}$$

$$ch[i+1,1] = \text{Bandwidth or order}$$

$$ch[i,1] = \begin{cases} 16 & \text{Index Function with Friedman–Tukey index} \\ 17 & \text{Index Function with Entropy index} \\ 18 & \text{Index Function with Legendre index} \\ 19 & \text{Index Function with Hermite index} \\ 20 & \text{Index Function with Natural Hermite index} \end{cases}$$

$$ch[i+1,1] = \text{Bandwidth or order}$$
$$ch[i+2,1] = \text{Number of Gridpoints}$$

$$ch[i,1] = \begin{cases} 0 & \text{Quit} \\ 1 & \text{Quit} \\ 11 & \text{Reset} \\ 12 & \text{Reinit} \end{cases}$$

The numbers presented here may change, since the macro is still under development. The correct numbers can be found by counting on which line of the menu the appropriate entry appears.

Any other value of ch[i,1] will have no effect. In the noninteractive mode, the program tries to print the result to a specified device.

For example, the following program

```
proc()=main()
  x = read ("bank2")
  x = x[,4:6]
  library ("addmod")
  ppexpl (x #(3 20 5 50 0))
endp
```

will draw colored contour lines of the index function for the Natural Hermite index with order 5 for the last three variables of the Swiss Banknote dataset based on a grid with $50 \times 50 = 2500$ projections. The 0 quits the PPEXPL macro.

REFERENCES

Cook, D., Buja, A. and Cabrera, J. (1993). Projection pursuit indexes based on orthonormal function expansions, *Journal of Computational and Graphical Statistics* **2**(3): 225–250.

Donoho, D.L. and Johnstone, I.M. (1989). Projection based approximation and a duality with kernel methods, *Annals of Statistics* **17**(1): 58–106.

Fan, J. and Marron, J.S. (1994). Fast implementations of nonparametric curve estimators, *Journal of Computational and Graphical Statistics* **3**(1): 35–56.

Flury, B. and Riedwyl, H. (1981). Graphical representation of multivariate data by means of asymmetrical faces, *Journal of the American Statistical Association* **76**(376): 757–765.

Friedman, J.H. (1987). Exploratory projection pursuit, *Journal of the American Statistical Association* **82**(397): 249–266.

Friedman, J.H. and Stuetzle, W. (1981a). Projection pursuit classification. unpublished manuscript.

Friedman, J.H. and Stuetzle, W. (1981b). Projection pursuit regression, *Journal of the American Statistical Association* **76**(376): 817–823.

Friedman, J.H., Stuetzle, W. and Schroeder, A. (1984). Projection pursuit density estimation, *Journal of the American Statistical Association* **79**(387): 599–607.

Friedman, J.H. and Tukey, J.W. (1974). A projection pursuit algorithm for exploratory data analysis, *IEEE Transactions on Computers* **C-23**(9): 881–890.

Hall, P. (1989a). On polynomial-based projection indices for exploratory projection pursuit, *Annals of Statistics* **17**(2): 589–605.

Hall, P. (1989b). On projection pursuit regression, *Annals of Statistics* **17**(2): 573–588.

Härdle, W. (1991). *Smoothing Techniques, With Implementations in S*, Springer-Verlag, New York.

Huber, P. (1985). Projection pursuit, *Annals of Statistics* **13**(2): 435–475.

Jee, J.R. (1985). Exploratory projection pursuit using nonparametric density estimation, *Proceedings of the Statistical Computing Section*, American Statistical Association, Statistical Computing Section, Washington, D.C., pp. 335–339.

Jones, M.C. and Sibson, R. (1987). What is projection pursuit? (with discussion), *Journal of the Royal Statistical Society, Series A* **150**(1): 1–36.

Klinke, S. (1993). A fast implementation of projection pursuit indices, *Discussion Paper 9320*, Institut für Statistik und Ökonometrie, Humboldt-Universität zu Berlin, Berlin, Germany.

Klinke, S. and Polzehl, J. (1994). Experiences with bivariate projection pursuit indices, *Discussion paper 9410*, Institut de Statistique, Université Catholique de Louvain, Louvain-la-Neuve, Belgium.

Kruskal, J.B. (1969). Toward a practical method which helps uncover the structure of a set of observations by finding the line tranformation which optimizes a new "index of condensation", *in* R.C. Milton and J.A. Nelder (eds), *Statistical Computation*, Academic Press, New York, pp. 427–440.

Kruskal, J.B. (1972). Linear transformation of multivariate data to reveal clustering, *in* R.N. Shepard, A.K. Romney and S.B. Nerlove (eds), *Multidimensional Scaling: Theory and Applications in the Behavioural Sciences*, Vol. 1, Seminar Press, London, pp. 179–191.

Polzehl, J. (1993). Projection pursuit discriminant analysis, *Discussion Paper 9320*, CORE, Université Catholique de Louvain, Louvain-la-Neuve, Belgium.

Scott, D.W. (1992). *Multivariate Density Estimation: Theory, Practice, and Visualization*, John Wiley & Sons, New York, Chichester.

Silverman, B.W. (1986). *Density Estimation for Statistics and Data Analysis*, Vol. 26 of *Monographs on Statistics and Applied Probability*, Chapman and Hall, London.

Wand, M.P. (1994). Fast computation of multivariate kernel estimators, *Journal of Computational and Graphical Statistics*. in press.

10
Generalized Linear Models

Joseph Hilbe[1] and Berwin A. Turlach[2]

10.1 Introduction

In this chapter we shall discuss a class of statistical models that generalize the well-understood normal linear model. A normal or Gaussian model assumes that the response Y is equal to the sum of a linear combination $X^T\beta$ of the d-dimensional predictor X and a Gaussian distributed error term. It is well known that the least-squares estimator $\hat{\beta}$ of β performs well under these assumptions. Moreover, extensive diagnostic tools have been developed for models of this type.

The assumptions imposed by a linear model are simply too strict for many practical data situations; for example, from the assumption that the error term is Gaussian distributed, it follows that Y has a continuous distribution. Thus the theory of the standard or normal linear model fails when dealing with count data, that is, when Y takes only positive integer values. The standard model assumptions also fail when dealing with proportional data, or continuous data that is constrained to positive-only values. However, the assumption of Gaussian distributed errors may be relaxed in the linear model. In this case we do not know the exact distributions of the estimated parameters, which we need for making viable inferences, but rather work with approximate distributions. Extensive research has been done to relax different assumptions of the normal model in order to adapt it to other situations. Nelder and Wedderburn (1972) showed that all of these approaches, which tried to generalize the normal linear model, can be reduced into a single framework, which they termed *Generalized Linear Models*.

Nelder and Wedderburn argued that an essential part of the normal linear model is the assumption that the expected value μ of Y is directly dependent on a linear combination $\eta = X^T\beta$ of X. They constructed the expected value μ to depend, via a given transformation function G, on

[1]Department of Sociology and Graduate College, Committee on Statistics, Arizona State University, Tempe, AZ 857210, USA.

[2]CORE and Institut de Statistique, Université Catholique de Louvain, B-1348 Louvain-la-Neuve, Belgium; and Statistics, Centre for Mathematics and Its Applications, The Australian National University, Canberra ACT 0200, Australia.

the linear combination η (called the *linear predictor*). By imposing at the same time some distributional assumption on Y, they stayed within the framework of maximum-likelihood estimation. This technique made the estimation of these models tractable and enables us to expand the range of linear models.

In the next section, we shall present some of the basic theory of generalized linear models (henceforward referred to as GLM). For a more complete account, the reader is referred to the research monograph of McCullagh and Nelder (1989) or to the introductory book of Dobson (1990) (also see Hilbe (1994a) for theory and comparative software implementation). Section 10.3 will show how to implement a GLM in XploRe and will describe the extensive XploRe GLM library. In Section 10.4 we shall demonstrate the use of XploRe on an example using the Gamma distribution. In Section 10.5 we shall discuss the geometric and negative binomial models and how the latter may be used with overdispersed count data. Finally, we shall show how GLM methodology may be extended to model types of survival data.

10.2   Some Theory

In a GLM we assume that the variable Y has a distribution that belongs to the exponential family, that is, the density function $f_Y(y)$ (if Y is continuous) or the probability mass function $f_Y(y) = \Pr[Y = y]$ (if Y is discrete), has the distributional form:

$$f_Y(y; \theta, \phi) = \exp\left\{ \frac{y\theta - b(\theta)}{a(\phi)} + c(y, \phi) \right\}. \tag{10.1}$$

If ϕ is known in (10.1) (for example, the Poisson distribution), we have a one-parameter exponential family. Otherwise, we usually have a two-parameter family (for example, the normal distribution) and declare the parameter θ as the parameter of interest and regard ϕ as a nuisance parameter. Written in the form (10.1), θ is called the *canonical parameter*.

Under suitable conditions (which are fulfilled by the distributions herein discussed), the following relations hold:

$$\mathrm{E}\left[\frac{\partial}{\partial \theta} \log f_Y(y; \theta, \phi) \right] = 0,$$

$$\mathrm{E}\left[\frac{\partial^2}{\partial^2 \theta} \log f_Y(y; \theta, \phi) \right] = -\mathrm{E}\left[\left(\frac{\partial}{\partial \theta} \log f_Y(y; \theta, \phi) \right)^2 \right]. \tag{10.2}$$

Hence,

$$\mathrm{E}[Y] = b'(\theta), \tag{10.3}$$

$$\mathrm{Var}[Y] = b''(\theta)a(\phi). \tag{10.4}$$

Thus, the expectation of Y depends only on θ and the variance splits into two factors, one depending only on θ and the other on ϕ. Since there is a direct relation between μ and θ, the factor in the variance of Y depending on θ can also be written as a function of μ. The resulting function is referred to as the *variance function* $V(\mu)$.

It is usually assumed that $a(\phi)$ is constant across observations and has the form $a(\phi) = \phi/w$, where w are known prior weights. Specified in this manner, ϕ is referred to as the *dispersion parameter*. The dispersion parameter is also denoted by σ^2. Again, it is assumed that this parameter is the same for all observations.

But how are we to interpret the predictor(s) X? As mentioned in the introduction, an essential characteristic of a GLM is that the observation Y does not depend directly on a linear combination $X^T\beta$ of X. Rather, the mean μ of Y depends on $X^T\beta$ via a (fixed) link function[3] $\check{G}$, that is,

$$\check{G}(\mu) = \eta \iff \mu = G(\eta),$$

where $\eta = X^T\beta$ is called the *linear predictor*.

The connection between the mean μ and the linear predictor η immediately yields a relation between the canonical parameter θ and the predictor X, namely,

$$b'(\theta) = G(X^T\beta) \iff X^T\beta = \check{G}(b'(\theta)).$$

If $X^T\beta = \eta = \theta$ holds, we say that we use the *canonical link function*. This is an important special link function, since for this link function a sufficient statistic exists. Moreover, certain theoretical and practical problems are easier to solve with the canonical link than with a noncanonical link.

In Table 10.1 we have summarized some commonly used distributions and their characteristics. Note that the negative binomial distribution only fits into the framework described above if we assume that the parameter k is known. We shall elaborate on how to deal with this point later.

Now that we have made the connection between the predictor variable X and the response Y, we could estimate β, the (vector of) parameters in which we are actually interested. However, to do this we need a criterion that tells us how well our model fits the data for a given β. Such a criterion would allow us to discriminate between different values of β and to choose β such that our model best fits the data.

Once we have specified a distribution for Y, a natural candidate for β would be the maximum likelihood estimator. We have specified the dis-

---

[3]Note that we deviate here from the usual notation in generalized linear models. Normally, the function that we denote with $\check{G}$ is denoted with G and what we call G is the inverse link function. We have chosen this notation to be consistent with the notation in other chapters of this book, where the regression function is working on the regressors (especially Chapters 5 and 11).

	Notation	Range of y	$b(\theta)$	$\mu(\theta) = E[Y|\theta]$	Canonical link: $\theta(\mu)$	Variance function: $V(\mu)$	$a(\phi)$
Normal	$N(\mu,\sigma^2)$	$(-\infty,\infty)$	$\theta^2/2$	θ	identity	1	σ^2
Poisson	$P(\mu)$	$[0,\infty)$ y integer	$\exp(\theta)$	$\exp(\theta)$	log	μ	1
Binomial	$B(m,\pi)/m$	$[0,m]/m$ my integer	$\log(1+e^\theta)$	$e^\theta/(1+e^\theta)$	logit	$\pi(1-\pi)$	$1/m$
Gamma	$G(\mu,\nu)$	$(0,\infty)$	$-\log(-\theta)$	$-1/\theta$	reciprocal	μ^2	$1/\nu$
Inverse Gaussian	$IG(\mu,\sigma^2)$	$(0,\infty)$	$-(-2\theta)^{1/2}$	$-(-2\theta)^{-1/2}$	squared reciprocal	μ^3	σ^2
Geometric	$GE(\mu)$	$[0,\infty)$ y integer	$-\log(1-e^\theta)$	$e^\theta/(1-e^\theta)$	$\log(\mu/(1+\mu))$	$\mu+\mu^2$	1
Negative Binomial	$NB(\mu,k)$	$[0,\infty)$ y integer	$-\log(1-e^\theta)/k$	$e^\theta/(k(1-e^\theta))$	$\log(k\mu/(1+k\mu))$	$\mu+k\mu^2$	1

TABLE 10.1. Characteristics of some univariate distributions in the exponential family.

In this table μ (respectively π for the binomial distribution) denotes the mean value as given in (10.3). θ denotes the canonical parameter of the distribution as defined in (10.1).
The parameterization of the gamma distribution is such that its variance is μ^2/ν.
The parameterization of the negative binomial distribution is such that its variance is $\mu + k\mu^2$.

Family	Deviance function
Normal	$\sum(y - \hat{\mu})^2$
Poisson	$2\sum\left(y\log\left(\frac{y}{\hat{\mu}}\right) - (y - \hat{\mu})\right)$
Binomial ($\hat{\mu} = m\hat{\pi}$)	$2\sum\left(y\log\left(\frac{y}{\hat{\mu}}\right) + (m - y)\log\left(\frac{m - y}{m - \hat{\mu}}\right)\right)$
Gamma	$2\sum\left(-\log\left(\frac{y}{\hat{\mu}}\right) + \frac{y - \hat{\mu}}{\hat{\mu}}\right)$
Inverse Gaussian	$\sum\frac{(y - \hat{\mu})^2}{\hat{\mu}^2 y}$
Geometric	$2\sum\left(y\log\left(\frac{y}{\hat{\mu}}\right) - (1 + y)\log\left(\frac{1 + y}{1 + \hat{\mu}}\right)\right)$
Negative Binomial	$2\sum y\log\left(\frac{y}{\hat{\mu}}\right) - \frac{1 + ky}{k}\log\left(\frac{1 + ky}{1 + k\hat{\mu}}\right)$

TABLE 10.2. Deviance functions for the distributions in Table 10.1.

tribution in (10.1) by (θ, ϕ) (respectively (μ, ϕ)). Thus, if we have a vector $\boldsymbol{y} = (y_1, \ldots, y_n)$ of observations (together with the corresponding x_i) and write the vector of mean values as $\boldsymbol{\mu} = (\mu_1, \ldots, \mu_n)$, then the log likelihood is

$$l(\boldsymbol{\mu}, \phi; \boldsymbol{y}) = \sum_{i=1}^{n} \log f_Y(y_i; \theta(\mu_i), \phi). \tag{10.5}$$

The maximum likelihood estimate $\hat{\beta}$ for β is now the vector such that the corresponding $\hat{\boldsymbol{\mu}}$ maximizes $l(\bullet, \phi; \boldsymbol{y})$.

The maximum achievable log likelihood is $l(\boldsymbol{y}, \phi; \boldsymbol{y})$, arising from a *full model* with n parameters. The discrepancy of two fits is measured by twice the difference of the log likelihoods achieved by the two models, assuming that they are nested. Thus we can measure the discrepancy between a fit with $\hat{\beta}$ ($\hat{\boldsymbol{\mu}}$) and the full model by

$$\begin{aligned} 2\{l(\boldsymbol{y}, \phi; \boldsymbol{y}) - l(\hat{\boldsymbol{\mu}}, \phi; \boldsymbol{y})\} &= 2\sum_{i=1}^{n} w_i\{y_i(\tilde{\theta}_i - \hat{\theta}_i) - b(\tilde{\theta}_i) + b(\hat{\theta}_i)\}/\phi \\ &= D(\boldsymbol{y}; \hat{\boldsymbol{\mu}})/\phi, \end{aligned}$$

where $\hat{\theta} = \theta(\hat{\mu})$, $\tilde{\theta} = \theta(y)$ and it is assumed that $a_i(\phi)$ (see (10.4)) has the form $a_i(\phi) = \phi/w_i$. Typically, β is estimated by minimizing the function $D(\boldsymbol{y}; \hat{\boldsymbol{\mu}})$, known as the *deviance function*. Table 10.2 provides the deviance functions for the distributions considered in Table 10.1.

Another important measure of discrepancy is the generalized Pearson χ^2 statistic, which takes the form

$$\chi^2 = \sum_{i=1}^{n}(y_i - \hat{\mu})^2/V(\hat{\mu}),$$

where $V(\bullet)$ is the variance function of the chosen distribution (given, for example, in Table 10.1). Note that for the normal (linear) model these two discrepancy measures are identical, but for the other models they are not. Moreover, for the normal (linear) model, they have exact χ^2-distributions whereas for the other models (distributions), only asymptotic results are available. In practical situations involving finite samples, it is difficult to say which of these two measures performs better.

Mention should also be made of the role residuals play in GLM. A real data analysis must incorporate an assessment of whether the chosen model is in fact appropriate to the data. An important tool for this task is residual analysis. For standard GLMs, three types of residuals exist. First there is the *Pearson residual* r_P, which measures, for each observation, its contribution to the Pearson χ^2 statistic

$$r_{P,i} = \frac{y_i - \hat{\mu}_i}{\sqrt{V(\hat{\mu}_i)}}.$$

Second, there are *deviance residuals* r_D, which do the same for the deviance function; that is, if $D(\boldsymbol{y}; \hat{\boldsymbol{\mu}}) = \sum_{i=1}^{n} d_i$ then the deviance residual for the ith observation is

$$r_{D,i} = \text{sign}(y_i - \hat{\mu}_i)\sqrt{d_i}.$$

These two types of residuals are identical for the Gaussian (linear) model in which the residuals have a normal distribution. However, for the other GLM the residuals are not necessarily normally distributed. Anscombe (1953) proposed defining a residual by using a function $A(y)$ to make the distribution of $A(y)$ "as normal as possible". We mention these different types of residuals at this point since we shall show an example of a residual analysis in Section 10.4. However, a more detailed account would be beyond the scope of this chapter; the interested reader is referred to Anscombe (1953) or Cox and Snell (1968).

10.3   Implementation

As discussed in the previous section, a GLM specifies a distribution for the response variable Y. Thus, for any link function $\check{G}$, the log likelihood function, given by (10.5), is completely specified as a function of the observations y and x and the vector of parameter β (ignoring ϕ for the moment). Thus we can find the maximum likelihood estimate $\hat{\beta}$ by numerical maximization of (10.5), say by a Newton–Raphson algorithm.

On the other hand, assume that the observations y_i are close to their mean μ_i. Then a one term Taylor-expansion yields

$$\check{G}(y_i) \approx \check{G}(\mu_i) + (y_i - \mu_i)\check{G}'(\mu_i), \tag{10.6}$$

where $\check{G}'(\mu_i)$ denotes the function $\frac{\partial}{\partial\mu}\check{G}(\mu)$ evaluated at the point μ_i.

Assume that we have a current estimate $\hat{\beta}$ (and thus, $\hat{\eta} = X^T\hat{\beta}$ and $\hat{\mu} = G(\hat{\eta})$) and denote $\check{G}(y_i)$ by z_i. Then equation (10.6) reads

$$z_i \approx \hat{\eta}_i + (y_i - \hat{\mu}_i)\check{G}'(\mu_i), \tag{10.7}$$

where z_i is called the *adjusted dependent variable* and has (assuming for the moment that $\hat{\eta}_i$ and $\hat{\mu}_i$ are fixed) a variance of

$$\mathrm{Var}[z_i] = W_i^{-1} = V(\mu_i)\check{G}'(\hat{\mu}_i)^2. \tag{10.8}$$

Hence, another way of finding the estimate $\hat{\beta}$ is the following:

> Given a current estimate $\hat{\beta}^{(i)}$, calculate the adjusted dependent variable. Regress this variable with weights $W = (W_1, \ldots, W_n)$ on X to obtain a new estimate $\hat{\beta}^{(i+1)}$. Iterate this process until convergence is reached.

If you write down the details of this algorithm, you will see that it is nearly the same as the Newton–Raphson algorithm. The difference is that the matrix of second derivatives (the *Hessian* matrix) which appears in the Newton–Raphson algorithm is replaced by its expectation (use equation (10.2) to see this). Hence, this alternative algorithm is a modification of the Newton–Raphson algorithm and is known as *Fisher scoring* or *iteratively reweighted least squares*. Because of its ease of implementation, practically all implementations of GLM fitting procedures use Fisher scoring instead of a Newton–Raphson method.

A skeleton GLM algorithm based on Fisher scoring is presented below. When implemented in XploRe, we have utilized several shortcuts to enhance efficiency. This example outlines the code for Bernoulli models; for others, the initialization is slightly different, and for binomial models consideration must be made of the binomial denominator:

```
mu  = (y+0.5)/2             /* initialization              */
eta = G(mu)                 /* link                        */
dev = 0
DO {
    u = (y-mu)G'            /* derivative                  */
    w = 1/(V*G'^2)          /* weight                      */
    z = (eta + u)           /* adjusted dependent variable */
    B = (X'WX)^(-1) X'WZ    /* weighted least squares      */
    eta = sum(B*X)          /* linear predictor            */
    mu = G(eta)             /* inverse link                */
```

```
    odev = dev                 /* save last deviance value    */
    dev = binomial deviance
    deltaD = dev - odev        /* convergence check           */
} UNTIL (abs(deltaD)<tolerance)
```

where B is $\hat{\beta}$, V is the variance, and G' is the derivative of the link function with respect to $\hat{\mu}$.

In order to implement the fitting of a GLM in XploRe, type at the command line edit(example1) (or any other name) to start the XploRe editor and enter the following code:

```
PROC(beta)=MAIN()
  dat  = read(sim)
```

This code reads a data set which comes with the XploRe package. It can be used to fit a Bernoulli (binomial) model since the third column of this data set contains a 0-1 response. In the current example we will use the canonical logit link.

```
x    = matrix(rows(dat))~dat[,1:2]
y    = dat[,3]
mu   = (y+0.5)/2
eta  = ln(mu./(1-mu))
dev  = 2*( SUM(XLNXDY(y mu)) + SUM(XLNXDY((1-y) (1-mu))) )
```

As this point we have read the data and constructed the design matrix x and the vector y with the response variable. We then initialized our first estimate for β and calculated the corresponding η, μ, and deviance. We use the built-in function XLNXDY which allows us to calculate terms of the form $x \log(x/y)$ without the numerical problems caused when $x = 0$. If this function is not used, code must be written for the deviance to allow for this possibility. Now we will start our iteration.

```
DO
  w    = mu.*(1-mu)
  z    = eta + (y-mu)./w
  xw   = x.*w
  beta = INV(XTY(xw x))*XTY(xw z)
  eta  = x*beta
  mu   = 1./(1+exp(-eta))
  odev = dev
  dev  = 2*( SUM(XLNXDY(y mu)) + SUM(XLNXDY((1-y) (1-mu))) )
UNTIL( ABS(dev-odev)/dev .< 0.0001 )
beta
ENDP
```

In this example, we see one of the simplifications that occurs if the canonical link function is used. We have $\check{G}(\mu) = V(\mu)^{-1}$ and thus by (10.8), the weights used to regress the adjusted dependent variable are just $V(\hat{\mu}_i)$. The first line in the loop calculates the weights, which are then used in the second line to calculate the adjusted dependent variable as given by (10.7). The next line calculates, using pointwise multiplication, the matrix XW, where X is the design matrix and W is the diagonal matrix with the weights W_i on the diagonal (note that this matrix is clearly symmetric). This allows us to efficiently calculate in the next step (that is, without actually transposing any of the involved matrices, a memory-intensive operation in XploRe) the update of our estimate β by a weighted least square step. We then update the values for μ and η and calculate the new deviance. Finally we check whether the algorithm has converged. In this case, the iterations cease and XploRe displays the values of the last estimate $\hat{\beta}$ for β.

After typing this code into the editor, press <F4> to run it. You will see the algorithm converging after a few iterations, displaying the last value of beta thereafter.

The above example is a fairly simple implementation of a GLM fitting procedure. There are more sophisticated ways of checking for convergence. Generally one also would like to be able to use prior weights or an offset in the linear predictor η. In addition, the calculation of standard errors for the parameter vector β and of the different types of residuals should be desirable outputs.

The essential difference between the full Newton–Raphson algorithm and its Fisher scoring modification is that the former employs the observed rather than the expected information matrix. Both yield identical results when dealing with canonical linked GLMs, for example, logistic regression. However, the differences in information become apparent when dealing with noncanonical linked models; for example, probit, cloglog, and log-linked negative binomial. Since estimates for the standard errors are derived from the information matrix, Fisher scoring results in slightly different standard errors than do models based on the full Newton–Raphson algorithm. Fortunately the differences are usually negligible unless employed on very small data situations. However, the Newton–Raphson algorithm based on observed values of the information matrix is a more accurate estimation method.

The above described algorithms based on Fisher scoring can be adjusted so that the observed rather than the expected values of the information matrix are used in the fitting process (Aitkin, Anderson, Francis and Hinde, 1989). Of course, this is only necessary for noncanonical links. One only needs to modify the weights and adjusted dependent variable. This is accomplished by adding a new line to the above algorithm and amending another. Following the calculation of weights, w, we add a calculation for observed weights, w_o. We also amend the adjusted dependent variable as

```
w_o = w + (y-mu) ( V*G'' + V'*G')/(V^2*G'^3)
z   = eta + (w*u)/w_o
```

with the least squares regression weighted by w_o. All major noncanonical links have been adjusted in this manner and are offered as options to the user. However, be warned that convergence may take substantially longer in the case of probit and cloglog regressions due to the extra coding involved.

Start	Fourth and Fifth Letter		Sixth to Eighth Letter	
GLM	NO	Normal	ID	identity
GLM	PO	Poisson	LN	logarithm
			ID	identity
GLM	BI	Binomial	LO	logit
			PR	probit
			CLL	complementary-log-log
GLM	GA	Gamma	IN	reciprocal
			LN	logarithm
			ID	identity
GLM	IG	Inverse Gaussian	IN2	squared reciprocal
			LN	logarithm
GLM	GE	Geometric	CL	canonical link
			LN	logarithm
GLM	NB	Negative Binomial	CL	canonical link
			LN	logarithm
GLM	With all of the above		POW	power link

TABLE 10.3. Naming conventions for XploRe macros. These macros use the Fisher scoring algorithm. The versions that use the Newton–Rhapson algorithm have 0 as the eighth letter.

Full-scale GLM implementations have already been placed into XploRe using the macrolanguage library GLM. You can load this library by simply typing the command library(glm). Issuing this command enables one to load functions for fitting generalized linear models.

The names of the functions that fit a GLM are built up by the structure shown in Table 10.3. Thus, if you want to fit a gamma model with a logarithmic link, you should use the routine GLMGALN. Instead of calling a specific function directly, however, you can also call the function DOGLM, which provides an interactive interface for the functions of the GLM library.

The functions for residual calculations are similarly named in the library. They start with GLMRES and have seventh and eighth letter identifications of the appropriate distribution (as given in Table 10.3). There is also an interactive interface called GLMRESID which may be called from within DOGLM. Hence, one can perform a complete fitting analysis for different models (for example, same distribution but different link functions) and appropriate residual analysis entirely within DOGLM.

In the next section, we shall demonstrate the use of the GLM-library for an example using the gamma distribution. A detailed description of all the functions and their implementation is too lengthy for this chapter. The reader is referred to the XploRe manual or to the on-line help, that is, type, for example, DOGLM, position the cursor on top of this "keyword", and press <F10>.

10.4   Example for a Gamma Model

Two data sets come with the XploRe GLM library to illustrate the Gamma model. These data sets are CARDAMAG.DAT and DROSOP.DAT; a short description of them is provided in CARDAMAG.TXT and DROSOP.TXT. Both data sets are from McCullagh and Nelder (1989, Chapter 8). In the following we will show how to analyze the first data set to obtain the results found in McCullagh and Nelder (1989).

10.4.1   Car Insurance Claims (CARDAMAG.DAT)

The data in the file CARDAMAG.DAT present the average claims for damage to an owner's car for both privately owned and comprehensively insured cars. The data originally comes from Baxter, Coutts and Ross (1980). The fourth column of the data set gives the average claim, which is stratified with respect to three explanatory variables. Those variables are given in the first three columns and represent the policyholder's age (PA, divided into eight levels identified by the numbers 1 through 8), the car group (CG, divided into four levels and coded by 1 through 4), and the age of the car (CA, divided and coded in four levels). The fifth column gives the number of claims for each combination of the explanatory variables.

Let Y_{ijk} and m_{ijk} denote the observed average claim and the observed number of claims if the variable PA has the value i ($= 1, \ldots, 8$), CG equals j ($= 1, \ldots, 4$) and CA equals k ($k = 1, \ldots, 4$). Since Y_{ijk} is an average of m_{ijk} claims, it is reasonable that the variance of Y_{ijk} should be inversely proportional to m_{ijk}, at least if we assume that for each combination of i, j, and k, the claims for that combination are independent, identical distributed random variables (i.i.d.). If we assume, as did McCullagh and Nelder (1989, Section 8.4.1), that the coefficient of variation is constant, that is, that the variance of Y_{ijk} is proportional to the squared expectation and that the explanatory variables have a linear influence on the response Y_{ijk} on a reciprocal scale, we arrive at the following model (considering only the main terms for the moment):

$$
\begin{aligned}
\mu_{ijk} = \mathrm{E}[Y_{ijk}] &= (\mu_0 + \alpha_i + \beta_j + \gamma_k)^{-1}, \\
\mathrm{Var}[Y_{ijk}] &= \sigma^2 \mu_{ijk}/m_{ijk}
\end{aligned}
\tag{10.9}
$$

where α_i, β_j, and γ_k are the parameters corresponding to the different levels for the explanatory variables PA, CG, and CA.

This model can be fitted by a GLM using the gamma distribution with canonical inverse link with m_{ijk} as prior weights. In the following, we shall illustrate how to do this with the GLM-library of XploRe. From the command line, we issue the following commands:

```
library(glm)
dat = read(cardamag)
i   = #(2,3,4,5,6,7,8)
x   = dat[,1].=i
i   = #(2,3,4)
x   = x~(dat[,2].=i)
x   = x~(dat[,3].=i)
x   = matrix(rows(x))~x
y   = dat[,4]
w   = dat[,5]
```

With these commands, we first load the GLM-library and read the data set. Then we start to create the design matrix by creating the dummy vectors for the different levels of the first variable using a "corner-point" constraint. That is, if we were to try to estimate all the α_i $(i = 1,\ldots,8)$ in our model (10.9), this would lead to a singular design matrix and an ill-posed problem. Thus, we arbitrarily set α_1 to zero; in this way α_i $(i = 2,\ldots,8)$ measures the contrast between PA having the value 1 and PA having the value i. The same is then done for the second and third factor variable. After this, we attach a constant term as the first column to our design matrix (which is actually not necessary since we will use DOGLM). Finally, we extract the response variable, which is stored in the fourth column, and use the values in the fifth column (the number of claims in that specific group) as prior weights.

We are now ready to call DOGLM for fitting the model. We have to assign the values that DOGLM returns to some variables and pass to this routine the design matrix x and the vectors y and w together with the response variable and prior weights. We do this by issuing the command:

```
(itres beta se t bvar stat) = DOGLM(x y w)
```

When DOGLM starts, the screen is split up into three windows. In the left window, you see a list of the possible distributions that you can choose. Use the arrow keys to scroll to the line with the gamma distribution and hit the <RETURN> key to select. After pressing <RETURN>, you will see a second menu appearing in the right window of the screen which offers you the possible link functions. Again, use the arrow keys to choose the inverse function as link function.

Following the selection of the distribution and link functions,[4] DOGLM

---

[4] In this example we have chosen the canonical link function (see Table 10.1). Thus the Fisher scoring and the Newton–Rhapson algorithms are equivalent. If you select a noncanonical link you are asked which of these algorithms you want to use (the default is the Newton–Rhapson algorithm).

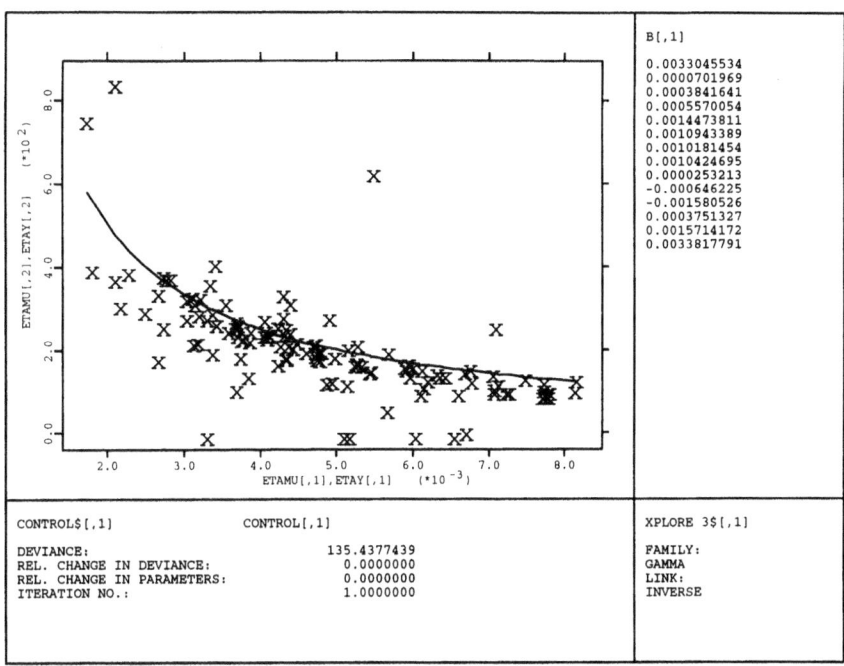

FIGURE 10.1. Initial display in GLMGAIN showing the initial fit and the data in the upper left. The upper right shows the initial estimates for $\hat{\beta}$, and the lower left shows the initial deviance. The lower right shows which model we are fitting.

asks us if we want to add an intercept term. This question provides for the case where we have not already put a constant term into our design matrix. Since we have included a constant term, we answer by pressing <N> and <RETURN>. Next, DOGLM asks if we want to see each step in the iterative fitting of the model. The default is not to display these intermediate steps. However, for educational purposes or if one notices that the iterative fitting procedure does not converge, it may be helpful to look at each step of the iteration. For this example we decide to look at each step and thus answer by pressing Y and <RETURN>. Finally, DOGLM queries us for two parameters that define the criterion when convergence is reached. Press <RETURN> both times to accept the default values.

DOGLM will now call the function GLMGAIN to do the actual fitting. Since we have chosen to look at each step of the iterative fitting procedure, GLMGAIN will stop immediately, showing a first picture. In this picture, the upper left shows the original data and the current fit based on the initial estimates $\hat{\beta}$. The initial estimates are shown in the upper right. The lower right contains the information about which model we are fitting and the lower left displays the initial deviance. During the iterative fitting of the model, this window

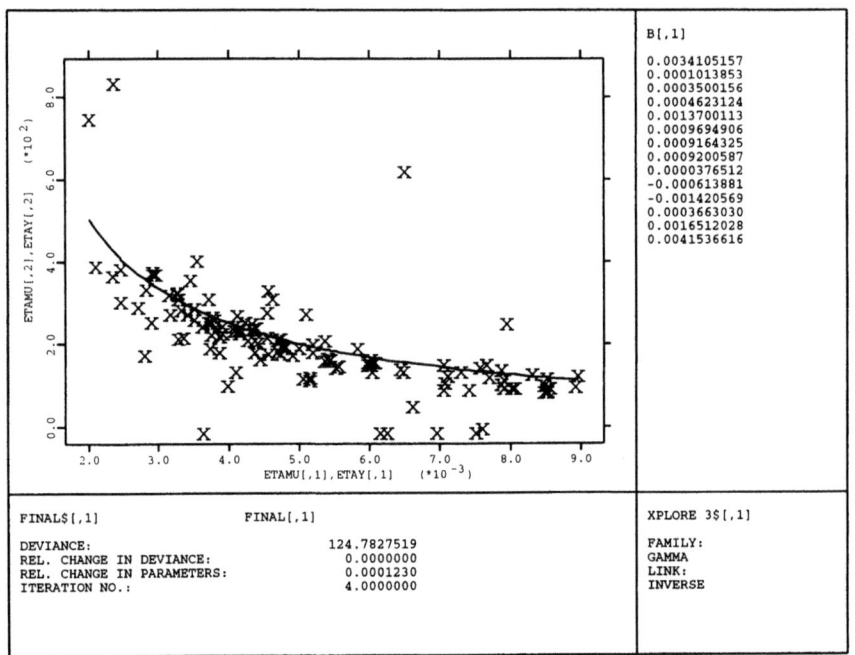

FIGURE 10.2. Final display in GLMGAIN showing the final fit and the data in the upper left. The upper right shows the final estimates for $\hat{\beta}$ and the lower left shows the final deviance.

will always be updated, showing the deviance for the current values of $\hat{\beta}$, the relative change in the deviance to the previous fit, and the relative change in the estimates $\hat{\beta}$. The current iteration number is also provided. These are the values that are used to determine whether convergence is reached. The last questions of DOGLM allow you to specify a threshold and a maximal number of iterations. The iteration will stop if either the maximal number of iterations is reached or one of the relative changes is below the given threshold. At this point, GLMGAIN stops to give the user the option of editing the 2-D picture with the data and the current fit as desired. During the iterative fitting, this picture will be updated and will always be in the style the user has chosen. After some editing, you should see a picture as Figure 10.1. Now press <ESC> to start the iterative fitting procedure.

Notice that GLMGAIN begins with an iterative fitting of the model, updating the information on the screen following each step. After four iterations, GLMGAIN will stop and display the screen shown in Figure 10.2. You can see that convergence was reached since the relative difference in the value of the deviance between the fit in the third and fourth iterations is below our threshold. The function GLMGAIN stops at this point to give the user another opportunity to change the picture and to make a printout of these results. If you now press <ESC>, you will return to the routine DOGLM.

```
BETA, S.E.[,1]BETA, S.E.[,2]BETA, S.E.[,3]                              BVAR[,1]

  0.0034105157      0.0004066686      8.386474499                     0.0000001654
  0.0001013853      0.0004245111      0.238828403                    -0.000000144
  0.0003500156      0.0004013436      0.872109582                    -0.000000144
  0.0004623124      0.0003995412      1.157108069                    -0.000000144
  0.0013700113      0.0004078890      3.358784698                    -0.000000144
  0.0009694906      0.0003937599      2.462136248                    -0.000000144
  0.0009164325      0.0003969693      2.308572941                    -0.000000145
  0.0009200587      0.0004045028      2.274542442                    -0.000000145
  0.0000376512      0.0001641221      0.229409713                    -0.000000019
 -0.000613881       0.0001654261      3.710909130                    -0.000000020
 -0.001420569       0.0001757171      8.084410993                    -0.000000020
  0.0003663030      0.0000981396      3.732467890                    -0.000000006
  0.0016512028      0.0002206729      7.482579351                    -0.000000006
  0.0041536616      0.0004303884      9.650961617                    -0.000000008
```

```
STAT[,1]        STAT$[,1]

 109.0000000       DEGREES OF FREEDOM
 109.0000000       N-P
 124.7827519       DEVIANCE
 131.7861908       PEARSON'S CHI^2
   1.1447959       SCALE
```

FIGURE 10.3. Numerical results of the fitting of the model as displayed by DOGLM.

After returning from GLMGAIN, which estimated the model, DOGLM will show us a plot of the observations Y and the fitted mean values $\hat{\mu}$ against the estimated linear predictor $\hat{\eta} = X^T \hat{\beta}$. This is always done by DOGLM to show the final fit in case the interactive option was not selected. Press <ESC> to leave this picture. DOGLM will then show a screen with the numerical results of the fitting procedure; this screen is shown in Figure 10.3. In the upper right, one sees the estimates $\hat{\beta}$ in the first column. The estimated standard errors and t-values for each estimate are displayed in the second and third columns, respectively. The upper right shows the estimated covariance matrix for $\hat{\beta}$. Unfortunately this window is so small that it can only display one column of the matrix, but you can scroll through the other columns by pressing simultaneously <CTRL> and the arrow keys. The lower left window provides summary statistics: model degrees of freedom (number of observations with nonzero prior weight minus the number of parameters), deviance statistic, Pearson χ^2 statistic, and an estimate for the model's scale parameter σ^2.

The above-mentioned values are partially summarized again in Table 10.4. You may want to compare them with the results of McCullagh and Nelder (1989, Table 8.3, page 299). If you do so, note that they base their scale estimate on the Pearson χ^2 statistic ($\hat{\sigma}^2$ is the Pearson χ^2 divided by the

Level	Age group (PA)		Car group (CG)		Car age (CA)	
			$\hat{\mu}_0 = 34105$ (4067)			
2	1014	(4245)	377	(1641)	3663	(981)
3	3500	(4013)	-6139	(1654)	16512	(2207)
4	4623	(3995)	-14206	(1757)	41537	(4304)
5	13700	(4079)				
6	9695	(3936)				
7	9164	(3970)				
8	9201	(4045)				

TABLE 10.4. Parameter estimates and standard errors ($\times 10^7$) for our model.

degrees of freedom), whereas XploRe (like most other packages) uses the deviance to estimate the scale parameter.

After pressing <ESC> to quit the display of numeric results, DOGLM will ask you if you want to perform a residual analysis. If answered with Y <RETURN>, the routine GLMRESID will be called by DOGLM to allow an interactive residual analysis. On entering GLMRESID, the screen is split up into two windows and you will see on the left a menu offering the different types of residuals. As usual, we can scroll with the cursor keys through the different options and select by pressing the <RETURN> key. For our example, we choose Pearson residuals and press <RETURN>. After this choice, an additional menu appears on the right side of the screen offering different kinds of standardization. To continue with this example, choose the standardization by the scale parameter. With these choices you should see, after some editing, the picture shown in Figure 10.4.

When looking at Figure 10.4, you may wish to know the values of the individual residuals. It is not difficult to access them given the interactive graphic possibilities of XploRe. Just press <F3>, and move the cursor to the residual whose value you wish to know. If you do this and move the corner to the big residual in the upper left region, you should see the picture shown in Figure 10.5. You can see that this residual belongs to observation 21 and has a value of approximately 3.524.

But what if you want to know the values for the observed X and Y variables for this observation? Press <F3> to quit the display of the label box (note, **do not** press <ESC>; this would make you escape completely from the picture). Press <F10> to edit the data that you are displaying. When you press <F7>, you will see in the upper left corner a menu that displays all of the different data sets that you could link to the data set currently in memory. On the command line, we read our original data into a matrix called **dat**. Thus, the matrix was created on the highest level; we should

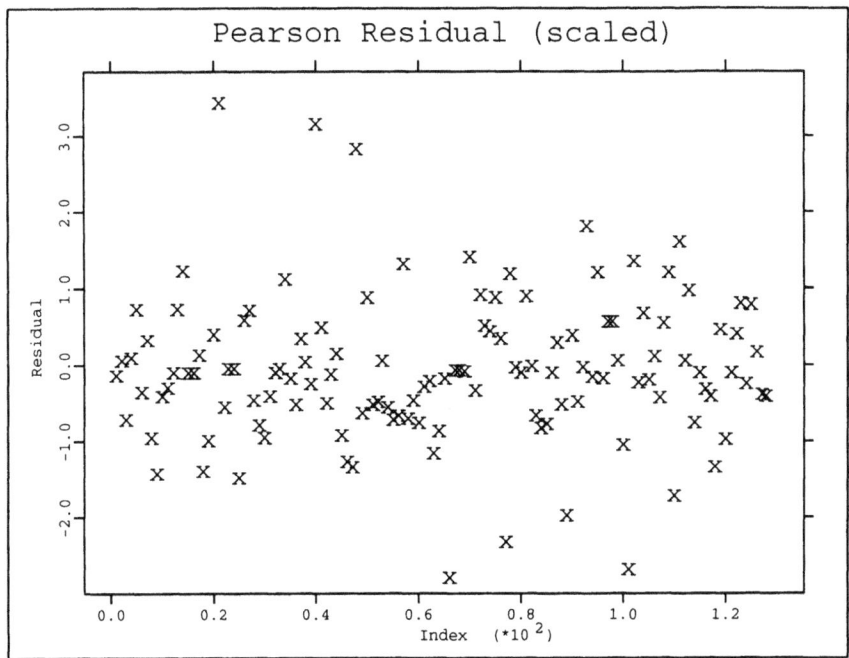

FIGURE 10.4. Initial display of the Pearson residuals.

look for it in the menu under "1 DAT". Place the cursor on this entry and press <RETURN> to select. Upon pressing <ESC> to quit this editing session and subsequently pressing <F3>, you should see the picture shown in Figure 10.6. You can see that this residual belongs to the 21st observation, for which the Policyholder's age was in category 2, the car group belonged to group 2, and the car age was in category 1. The average claim for this combination of explanatory variables is £420, for which there were 59 claims.

If you now press <ESC> to quit the picture with the display of the residuals, GLMRESID asks if you want to quit the residual analysis. If you answer with <Y> <RETURN>, you will exit the routine and come back to DOGLM. An answer of <N> <RETURN> brings you back to the menu where you may again choose which type of residuals you would like to investigate. If you do this and choose Anscombe residuals (with standardization by the estimated scale parameter), you should see on the screen the picture shown in Figure 10.7 (after some editing). Remember, Anscombe residuals are based on a transformation of the Y variable which should make the residuals look "as normal as possible". We shall leave it to the reader to decide whether Figure 10.7 depicts normally distributed residuals.

After leaving GLMRESID we come back to DOGLM and are asked whether we want to leave this routine. If we answer with <Y> <RETURN>, we will quit DOGLM and return to the command line from where we started. On answering <N> <RETURN>, we return to the first menu of DOGLM and can start

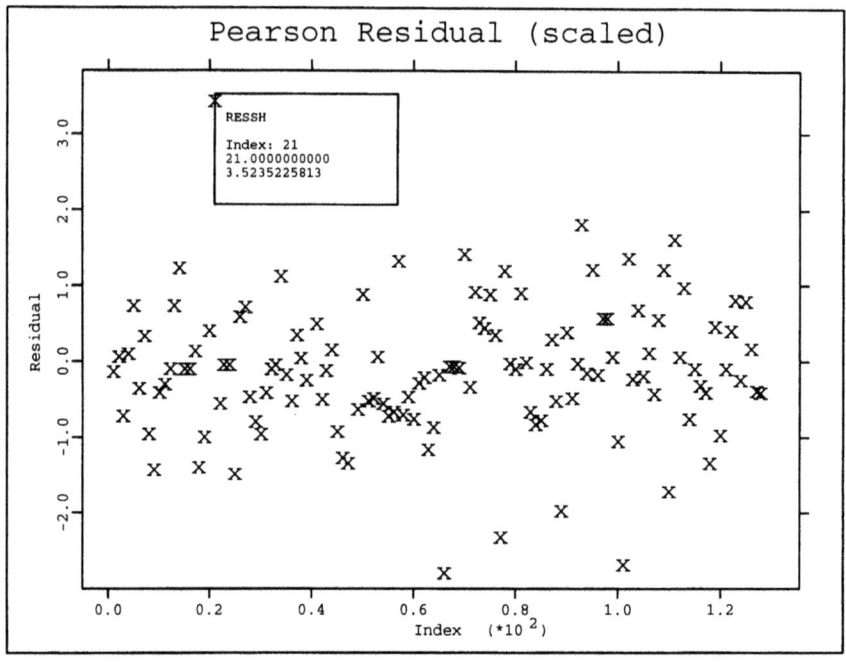

FIGURE 10.5. Display of the Pearson residuals and checking some values with <F3>.

another analysis. For example, we could now choose the normal distribution and thus fit a weighted normal linear model as was done in the original analysis by Baxter et al. (1980). This, also, we will leave as an exercise to the reader.

A final question that should be addressed for this example is whether one should incorporate interaction terms into the model (10.9). In generalized linear models, the difference of the deviance of two **nested** models has an approximate χ_p^2-distribution, where the degrees of freedom p is equal to the difference of the degrees of freedom of the two models. So we can think of starting with a simple model containing only the constant term, fitting it, and recording the deviance and degrees of freedom for that model. Then we add one variable and do the same again. After this, we add one by one the variables (main terms) and then the interaction terms, each time fitting the model and recording the deviance of each fit and the corresponding degrees of freedom. After fitting all of these models, we could test which terms to include in the model by testing whether the difference in deviance for the models with and without a specific term or set of terms is significant when compared to the critical value of the appropriate χ_p^2-distribution.

Clearly, such a procedure would be cumbersome to do in an interactive interface such as DOGLM, so we will write a small program that will call the

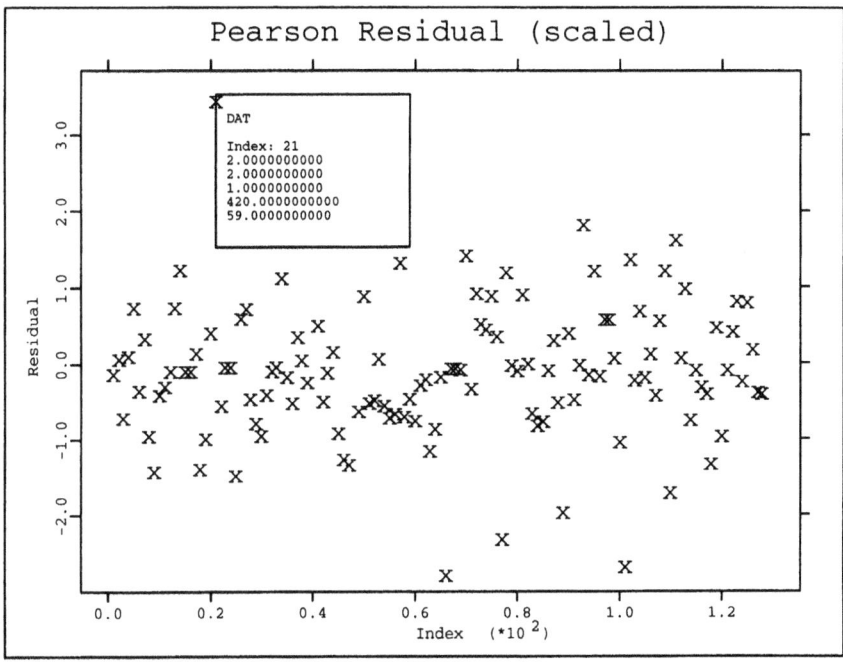

FIGURE 10.6. Display of the Pearson residuals and checking some values with <F3> after linking another data set.

GLMGAIN function directly and implement the whole procedure described above. Type in the command line `edit(example2)` and enter the code given below. This code implements the requested procedure by starting with the model containing only the constant term and then adding, one after the other, the main effect terms and the first-order interaction terms. After entering the code, you just press <F4> to run it.

```
proc(dev df)=main()
  library(glm)
  dat = read(cardamag)
  x   = matrix(rows(dat))
  y   = dat[,4]
  w   = dat[,5]
  (itres beta bvar stat)=glmgain(x y 1 w)
  dev = stat[3,1]
  df  = stat[1,1]
  writecon(27)
  dev~df
  ind = #(2,3,4,5,6,7,8)
  x   = x~(dat[,1].=ind)
  (itres beta bvar stat)=glmgain(x y 1 w)
  df  = df|stat[1,1]
```

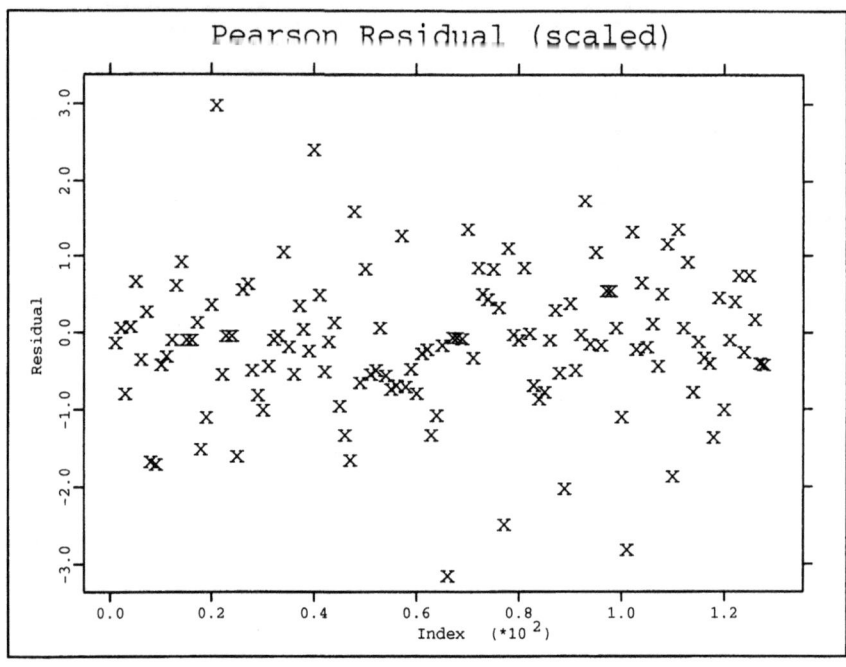

FIGURE 10.7. Display of Anscombe residuals.

```
dev = dev|stat[3,1]
writecon(27)
dev~df
ind = #(2,3,4)
x   = x~(dat[,2].=ind)
(itres beta bvar stat)=glmgain(x y 1 w)
df  = df|stat[1,1]
dev = dev|stat[3,1]
writecon(27)
dev~df
x   = x~(dat[,3].=ind)
(itres beta bvar stat)=glmgain(x y 1 w)
df  = df|stat[1,1]
dev = dev|stat[3,1]
writecon(27)
dev~df
x   = x~(x[,2:8].*x[,9])
x   = x~(x[,2:8].*x[,10])
x   = x~(x[,2:8].*x[,11])
(itres beta bvar stat)=glmgain(x y 1 w)
df  = df|stat[1,1]
dev = dev|stat[3,1]
```

```
writecon(27)
dev~df
x    = x~(x[,2:8].*x[,12])
x    = x~(x[,2:8].*x[,13])
x    = x~(x[,2:8].*x[,14])
(itres beta bvar stat)=glmgain(x y 1 w)
df   = df|stat[1,1]
dev = dev|stat[3,1]
writecon(27)
dev~df
x    = x~(x[,9:11].*x[,12])
x    = x~(x[,9:11].*x[,13])
x    = x~(x[,9:11].*x[,14])
(itres beta bvar stat)=glmgain(x y 1 w)
df   = df|stat[1,1]
dev = dev|stat[3,1]
ddev = -diff(dev)
ddf  = -diff(df)
createdisplay(maintmp, 1 2, text)
show(dev~df text1, ddev~ddf~(ddev./ddf)~(1-cdfc(ddev ddf)) text2)
endp
```

Since we are now calling GLMGAIN directly, the sequence in which we have to pass the parameters has changed. For DOGLM, we had to pass the design matrix x, then the vector with the response variable y, followed by the vector w of prior weights. If we call any of the functions of the GLM-library that fit a GLM directly, the third parameter that is passed contains the information regarding whether one wants to have the interactive display of the results after each fit and the parameters that control the iteration procedure. If we use DOGLM, we are asked interactively for this information and DOGLM puts all the parameters into the correct order before calling the routine that does the fitting. But here we have to take care ourselves. Note how we just pass a 1 as third parameter to tell GLMGAIN that we do not want to have the interactive display, but that we accept the default values of the parameters that control the fitting process (use the <F10> key to learn more about this third parameter).

After each fit, the deviance and degrees of freedom for the prior fitted model are extracted from the returned variables and displayed, together with the deviances and degrees of freedom of all the models fitted before. Note the use of writecon(27), which immediately quits the text window after displaying the values so that the program can continue without our interaction. After displaying the values, the program updates the design matrix x to incorporate a further main or interaction term. You should verify the manner in which the dummy vectors for fitting an interaction term are constructed from the dummy vectors used to fit the main term of the two variables involved in this interaction term. After fitting the last model, the program splits the screen into two horizontal text windows and displays in the upper text window one last time the deviance for each

```
CAT[,1]        CAT[,2]

   649.8713098      122.0000000
   567.6935215      115.0000000
   339.3848712      112.0000000
   124.7827519      109.0000000
    90.7487458       88.0000000
    70.9869112       67.0000000
    65.5846259       58.0000000
```

```
CAT[,1]        CAT[,2]       CAT[,3]        CAT[,4]

    82.1777883     7.00000000    11.73968404    0.0000000000
   228.3086503     3.00000000    76.10288343    0.0000000000
   214.6021194     3.00000000    71.53403978    0.0000000000
    34.0340061    21.00000000     1.62066696    0.0359398579
    19.7618346    21.00000000     0.94103974    0.5363483914
     5.4022854     9.00000000     0.60025393    0.7979113906
```

FIGURE 10.8. Results of incorporating interaction terms into model (10.9).

fitted model together with the corresponding degrees of freedom. In the lower text window, the first difference in the deviance, together with the corresponding degrees of freedom, the mean deviance (that is, the difference of the deviances divided by the degrees of freedom), and the p-value for these differences, is displayed. The resulting screen is shown in Figure 10.8 and you may want to compare it with the results of McCullagh and Nelder (1989, Table 8.2, p. 299). After analyzing these results, press <ESC> to return to the command line.

10.5 Negative Binomial Regression

The modeling of log-linear count data is a common statistical occurrence. For the most part, one models such data using Poisson regression. However, it is often the case that the assumptions upon which the Poisson model is based are violated. Perhaps the most common problem is having the variance of the response greater than its mean. When this is the case, the model is called Poisson overdispersed. Such overdispersion results in the standard errors being underestimated, hence giving an overly optimistic view of the variable's worth to the model. A typical adjustment for this type of overdispersion is to scale the standard errors by the square root of either the deviance or the χ^2-based dispersion. This will work quite satisfactorily for models with only a small amount of overdispersion, but the adjustment only acts upon the standard errors. For larger amounts of overdispersion, it would be expected that the estimates themselves should change. However, the model can no longer be Poisson.

The log linked negative binomial model (LNB) can be used effectively to model many types of otherwise Poisson-overdispersed data situations. The negative binomial distribution has a variance that can be parameterized as $\mu + \mu^2/\nu$. Since the Poisson variance is simply μ and the gamma distribution variance is μ^2/ν, the negative binomial may be considered as a Poisson-Gamma mixture. In effect, one may think of the counts entering cells or time frames as gamma distributed, with the shape of the distribution of events entering each cell determined by the value of ν. This is in contrast to the Poisson assumption that events enter cells in a random uniform manner. One may consider other mixture distributions as well, for example, the Poisson-Inverse Gaussian. However, these mixtures have no exponential family distribution as a basis. They may be modeled quite easily as quasi-likelihood models, but their interpretation and diagnostics may become difficult.

We have reparameterized the heterogeneity or shape factor for the negative binomial models to be $k = 1/\nu$ so that there is a directly proportional relationship between the value of μ and the amount of overdispersion present in the data. Hence, the negative binomial (NB) variance function becomes $\mu + k\mu^2$. The linear predictor and deviance functions have likewise been re-parameterized to reflect this relationship. When using the negative binomial model to accommodate Poisson-overdispersion, one must use the log-linked form of the model. When so structured, the prime difference between the Poisson model and the LNB model is the variance and deviance functions. A full mathematical discussion, together with simulation studies of the negative binomial models, can be found in Hilbe (1994b).

GLM methodology requires that k be entered into the model as a constant. Generally, k has values ranging from 0.01 to 2. When k is 1.0, the model is in fact a geometric regression, having properties somewhat similar to those of a log-gamma or exponential distribution. XploRe provides the capability to model data using the canonical link as well as other power links, including the log-link power of 0. The former program is called GLMNBCL, while the power-based negative binomial is called GLMNBPOW. These modules may be accessed directly or by using the DOGLM menu. We have also placed separate stand-alone programs for geometric and log-geometric models into the XploRe library—GLMGECL and GLMGELN, respectively. Geometric models have no heterogeneity factor k, so may be called directly, as, for example, GLMGELN(x y).

We have mentioned that the value of k represents the amount of overdispersion present in an otherwise Poisson model. But how does one decide on the proper value of k to enter into the model? In fact, this is quite simple. One adjusts k so that the χ^2 or deviance-based dispersion value approximates 1.0. XploRe provides the deviance-dispersion on output (in the parameter stat). The χ^2 dispersion may be calculated as the value χ^2/df, where df is the degrees of freedom. The Poisson model assumes that the dispersion or scale factor is 1.0. Greater values indicate overdispersion. Hence, one uses the LNB to adjust the estimates and standard errors by reducing the value of the dispersion to 1. Hilbe (1994b) has shown that optimal adjustment occurs by reducing the χ^2 dispersion to 1 rather than the deviance-based dispersion. However, for well-specified models, the results will be nearly identical.

LNB models are typically used after one has determined that a Poisson model is overdispersed. One may provide an initial estimate of k for a LNB model by setting k equal to the inverse of the Poisson χ^2 dispersion. One may also simply use the default of $k = 1$ (geometric). If the resultant χ^2 dispersion is under 1, then reset k to a lower value. For example, if after using the default $k = 1$ the χ^2 dispersion is 0.5, then reset $k = 0.5$. The new χ^2 value should be much closer to 1.0. Several adjustments may be required.

XploRe also includes a program called GLMLNB that iteratively searches for the optimal value of k and estimates parameters and standard errors. In essence, GLMLNB embeds the LNB GLM in a larger loop that reduces the value of the χ^2 dispersion to 1.00. It is based on the same logic as described in the previous paragraph. Details may be found in Hilbe (1994b), but we shall outline the algorithm's structure for the interested reader.

```
Poisson y=response on predictors
chi2  = sum{(y - mu)^2/mu}
disp  = chi2/df
alpha = 1/disp
DO {
   odisp = disp
   LNB y=response on predictors, k=alpha
   chi2 = sum{(y-mu)^2/(mu+k*mu^2)}
   disp = chi2/df
   alpha = disp * alpha
   deltad = disp - odisp
} UNTIL (abs(deltad)<tolerance)
```

When the LNB aspect of the above algorithm is structured such that it uses the observed information matrix, the estimation is equivalent to a full Newton–Raphson maximum likelihood model.

Negative binomial models may generally be accessed in XploRe through the DOGLM menu. The canonical link specification, $\ln(\mu/(1 + k\mu))$, calls GLMNBCL, while the log link option utilizes the capability of GLMNBPOW. A power of 0 sets the link as natural log. Again, when modeling Poisson-overdispersed data, one should use the log-linked negative binomial. If the data is overdispersed in such a manner that k equals 1.0, the model is geometric. Using GLMGECL and GLMGELN will produce identical results to the respective negative binomial models with k set to 1.0.

A caveat should be given with respect to overdispersion. Overdispersion can be the result of various factors. When data are clustered or correlated, the basic GLM model may be inappropriate. Recall the description of how events enter cells in Poisson and negative binomial models. In the latter case, events enter as gamma distributed, the manner of distribution being specified by the shape parameter. Implicit in this characterization is that the shape parameter is constant across cells. When data are clustered or correlated, we have what is called a random effects model. Graphically, we now have nonconstant cell shape parameters. One cell or group of cells may have events entering with a shape parameter 0.6, and another 0.8. Models like these cannot be directly modeled using GLM methodology. Random effect models have been created as extensions to GLM, notably Generalized Estimating Equations (Liang and Zeger, 1986) and more recently, Double GLMs (Lee and Nelder, 1994), but they are beyond the scope of this discussion.

Log negative binomial regression enhances the capability of modeling many types of count data situations that cannot easily be handled using more traditional methods. Unfortunately, until now these means were largely unavailable. However, once the logic of the extra heterogeneity parameter is understood, one should have no difficulty in modeling appropriately overdispersed count data.

10.6 GLM Extensions: Parametric Survival Models

Many types of survival data may be modeled within the GLM framework. There are two distinguishing features of survival data: (1) the response is time to failure, and (2) the response may have censored values. Time is typically measured as a discrete variable, but may also be considered as continuous. Censoring occurs when a case is lost to the study, for example, when a patient withdraws from followup studies while still alive. Important modeling information may be lost if we simply delete the case from the final model; after all, a patient with a particular set of covariates did survive until the date of withdrawal. Other types of censoring may affect the model as well. We shall look at how GLMs may be extended to allow modeling of parametric survival data. We shall not examine the semi-parametric proportional hazards model of Cox in this section; however, a Poisson model can in fact be used to perform Cox regression. The adjustments to the GLM algorithm for Cox regression are much more complex and less efficient than those required for parametric models; hence, they are rarely used in practice.

Exponential and Weibull regression are by far the two most widely used parametric survival models. Exponential regression may easily be modeled using the GLM selections in DOCLM. It is widely known that without censoring the log-linked gamma distribution with standard errors not scaled is identical to the exponential regression model. However, the gamma model cannot deal with censoring. For that, we must turn to the Poisson model. The trick is to use the censoring variable, defined in binary terms as 1=not censored and 0=censored, as the response. The log of the time variable is then entered into the model as an offset. Using the standard Poisson model results in a log hazard rate parameterization of the exponential regression model.

Weibull regression must be programmed using XploRe's macro language. Essentially, the procedure is to embed a Poisson model within a larger iterative loop which is used to define the shape and scale parameters. The product of the changing shape parameter and log of the time variable is entered into the Poisson model as an offset at each iteration. To show the schema of the algorithm, we shall define t as the time variable, c as the censor variable, nc as the number of noncensored cases, and alpha as the shape:

```
nc=sum(c)
lt=ln(t)
r = 2 * sum(c * lt)
alpha = 1 : oldD = 0
DO {
  Poisson c=response on predictors, offset=(lt*alpha)
  dev = r - 2 * sum{c*ln(alpha*mu)-mu}
  deltaD = dev - oldD
  OldD = dev
  alpha1 = 1/(sum((mu-c)*lt/nc))
  alpha  = (alpha+alpha1)/2
} UNTIL (abs(deltaD)<tolerance)
shape = alpha
scale = 1/alpha
```

The Weibull estimates are produced by the Poisson output at the final iteration. Again, this is the log hazard rate parameterization. To calculate the log expected time parameterization, one would need to add an additional step following the final iteration loop. This consists of a final Poisson regression with an adjusted value of z. For noncensored observations, z has the new value of

```
z = {-(-eta+(c-mu)/mu) + (lt*alpha)}/alpha
```

For censored cases z is

```
z = 1 - eta/ln(alpha)
```

The regression is weighted by the product of μ and $alpha$ squared.

Other survival models may be programmed using the basic GLM format as well, including the extreme value, normal, lognormal, logistic, and log-logistic regressions. All use the same underlying logic regarding the calculation of a shape parameter and its role as an offset in the embedded GLM. However, gamma models require the full maximum likelihood approach, so they are not amenable to the GLM approach.

GLM methodology is a powerful base upon which to model a variety of data situations. We have seen how it may be used effectively to model types of survival models. However, this is not the end. An entire range of quasi-likelihood models plus the previously mentioned generalized estimating equations and double GLMs are based on this method. And, of course, we have Generalized Additive Models, which are GLMs with an extra internal loop that smooths designated continuous predictors. GAMs, as they are called, are the subject of Chapter 11.2. But all in all, as new GLM-based applications are continually being developed, it has become increasingly apparent that this methodology has become one of the most important statistical tools available to the researcher.

REFERENCES

Aitkin, M., Anderson, D., Francis, B. and Hinde, J. (1989). *Statistical Modelling in GLIM*, Vol. 4 of *Oxford Statistical Science Series*, Oxford University Press, Oxford.

Anscombe, F.J. (1953). Contribution to the discussion of H. Hotelling's paper, *Journal of the Royal Statistical Society, Series B* **15**(2): 229–230.

Baxter, L.A., Coutts, S.M. and Ross, G.A.F. (1980). Applications of linear models in motor insurance, *Proceedings of the 21st International Congress of Actuaries*, Zürich, pp. 11–29.

Cox, D.R. and Snell, E.J. (1968). A general definition of residuals (with discussion), *Journal of the Royal Statistical Society, Series B* **30**(2): 248–275.

Dobson, A.J. (1990). *An Introduction to Generalized Linear Models*, Chapman and Hall, London.

Hilbe, J.M. (1994a). Generalized linear models, *The American Statistician*. To appear.

Hilbe, J.M. (1994b). Log negative binomial regression as a generalized linear model, *Technical Report 26*, Graduate College Committee on Statistics, Arizona State University, Tempe.

Lee, Y. and Nelder, J.A. (1994). Double generalized linear models. Manuscript.

Liang, K.Y. and Zeger, S.L. (1986). Longitudinal data analysis using generalized linear models, *Biometrika* **73**(1): 13–22.

McCullagh, P. and Nelder, J.A. (1989). *Generalized Linear Models*, Vol. 37 of *Monographs on Statistics and Applied Probability*, 2nd edn, Chapman and Hall, London.

Nelder, J.A. and Wedderburn, R.W.M. (1972). Generalized linear models, *Journal of the Royal Statistical Society, Series A* **135**(3): 370–384.

11
Additive Modeling

Thomas Kötter[1] and Berwin A. Turlach[2]

11.1 Introduction

In Chapter 10, on *Generalized Linear Models*, we saw how the standard linear model with normal assumptions can be generalized to incorporate a broader range of models. The essential idea was to connect the conditional mean of the response variable via a *link function* to a one-dimensional projection of the explanatory variables (a *single index*).

In this chapter we will discuss methods and their implementation in XploRe which further generalize this idea. Given a GLM, there are essentially two ways of further generalizations. One is to abandon the specific parametric link function and replace it with an unknown "smooth" function. The other is to allow for more flexible functionals of the explanatory variables to which the conditional mean of the response variable is linked.

The latter approach is taken by *Generalized Additive Models*, which allow the conditional mean of the response variable to depend via a fixed link function on a sum of univariate functions, each function having one component of the vector of explanatory variables as argument. These functions do not need to have a parametric form and are estimated nonparametrically. Models of this type are discussed in Section 11.2.

The other approach, in which the specific link function is replaced by an unknown function, can be treated in many different ways. Moreover, this approach can be further generalized to allow the conditional mean of the response variable to depend on several projections of the explanatory variables (*multiple index model*). Such models can be estimated in various ways, for example, *projection pursuit methods*. We will concentrate here on two other ways of estimating such models. One of the alternative ways to estimate such models is *sliced inverse regression*, which will be discussed in Section 11.3. Another method that can be used for single index models is *average derivative estimation*. Average derivative estimation does not fit

---

[1]Institut für Statistik und Ökonometrie, Humboldt-Universität zu Berlin, D-10178 Berlin, Germany.

[2]CORE and Institut de Statistique, Université Catholique de Louvain, B-1348 Louvain-la-Neuve, Belgium; and Statistics, Centre for Mathematics and Its Applications, The Australian National University, Canberra ACT 0200, Australia.

exactly into this framework, since it tries to estimate the mean slope of the conditional mean of the response variable. But as we will see in Section 11.4, in single index models, this average slope is proportional to the unknown projection and thus we can apply average derivative estimation here.

11.2   Generalized Additive Models

In this section we will discuss an extension of generalized linear models to a class of models known as *generalized additive models*. These models, as presented here, have been proposed by Hastie and Tibshirani (1986) (see also Stone (1985; 1986)). The basic idea is to allow for more flexibility in the form of the linear predictor η. Hastie and Tibshirani (1986) propose to replace the linear predictor η, which is a linear combination of the regressor variable X, by an *additive* predictor, that is, a sum of d univariate (unknown) smooth functions $g_j(\bullet)$, each one operating on one component of X. Thus, our model becomes

$$\mathrm{E}[Y|X = x] = G\left(\alpha + \sum_{j=1}^{d} g_j(x_j)\right), \tag{11.1}$$

where $G(\bullet)$ is a fixed link function and the distribution of Y is assumed to belong to an exponential family (see Table 10.1 on page 198). To make the functions g_j identifiable, one usually assumes that $\mathrm{E}[g_j(X_j)] = 0$.

11.2.1   The Algorithms

The fitting of a generalized additive model consists of two parts: estimating the additive predictor $(\alpha + \sum g_j)$ and linking it to the function $G(\bullet)$ in an iterative manner. The first part requires solving a system of normal equations. One way to do this is backfitting (see Schimek, Neubauer and Stettner, 1994). The functions $g_j(\bullet)$ are evaluated by nonparametric smoothing techniques. For the second part the local scoring algorithm is applied.

The *local scoring* algorithm is practically identical with the *Fisher scoring* algorithm, used in generalized linear models which are described in Chapter 10, except that the *least squares step* is replaced by the *backfitting* step. The least squares step is used to update the estimate $\hat{\beta}$ for the linear predictor $X^T\beta$ in the generalized linear model; now we make a backfitting step to update the estimates for α and the g_j's.

The idea for the backfitting algorithm goes back at least to Friedman and Stuetzle (1985), who used it for *projection pursuit regression*, and Breiman and Friedman (1985), who employed it in their *alternating conditional expectation* algorithm.

A simple motivation for the backfitting algorithm is obtained if we assume that the additive model, that is, (11.1) with $G(u) = u$ the identity, is correct. Then we have, for any k,

$$
\mathrm{E}\left[Y - \alpha - \sum_{\substack{j=1 \\ j \neq k}}^{d} g_j(X_j) \middle| X_k \right] = g_k(X_k). \qquad (11.2)
$$

This suggests an iterative algorithm to calculate the $g_j(\bullet)$. A nonparametric smoother is one method to estimate the above conditional expectation. Thus, given a set of current estimates $\{\hat{\alpha}, \hat{g}_j\}$, we can improve these iteratively (that is, looping over $k = 1, \ldots, d$) by calculating the *partial residuals* from the observations $\{x_i, y_i\}$,

$$
\hat{r}_i = y_i - \hat{\alpha} - \sum_{\substack{j=1 \\ j \neq k}}^{d} \hat{g}_j(x_{ij}), \qquad i = 1, \ldots, n,
$$

and then applying a nonparametric smoother to these residuals to update the estimate $\hat{g}_k$. Thus, we replace the expectation operator in the "population" world (11.2) by a smoothing operator in the "sample" world. Note that we always use the current fit in each coordinate to estimate the partial residuals. Thus, the backfitting algorithm implemented in XploRe is a Gauss–Seidel variant for solving the system of normal equations mentioned above.

A generalized additive model library similar to the generalized linear model library described in Chapter 10 is implemented in XploRe. We will demonstrate the use of this library with an example below. The GAM library allows one to fit (nearly) all the models that are in the GLM library. The names of the macros in the GAM library follow the same scheme as outlined in Table 10.3 except that GLM is replaced by GAM at the start of the macro names.

The local scoring algorithm was implemented in the XploRe macro language since the calculations done here are easy to formulate in a matrix-oriented language and the routine is sufficiently fast. The backfitting algorithm, which estimates the functions g_j using nonparametric smoothing techniques, is computationally more intensive. Thus, this algorithm was partially implemented in C (as command backfit), and partially in the XploRe macrolanguage (as macro GAMBACKF). For the backfitting algorithm in XploRe, you have a broad choice of possible smoothers. The implementation allows you to use either the local linear smoother (Fan, 1992, 1993), the s-knn smoother (as described in Hastie and Tibshirani, 1990; Härdle, 1990), or the supersmoother (Friedman, 1984). The local linear smoother is also discussed in Chapter 5.

11.2.2 Illustration

In order to illustrate the fitting of a generalized additive model with the GAM library of XploRe, we choose a logistic model with two explanatory variables. The first of these variables enters linearly into the additive predictor, while the second has a quadratic influence:

$$\Pr[Y = 1 | X = x] = \mu,$$

where

$$
\begin{aligned}
X = (X_1, X_2)^T &\sim N_2(0, I_2); \\
\eta &= g_1(x_1) + g_2(x_2), \\
g_1(u) &= u, \qquad g_2(u) = u^2 - 1; \\
\mu &= \frac{\exp \eta}{1 + \exp \eta} = \frac{1}{1 + \exp(-\eta)}.
\end{aligned}
$$

The functions were chosen such that

$$E[g_1(X_1)] = E[g_2(X)] = 0.$$

To simulate a data set of 200 observations according to this model, we use the following XploRe code:

```
x = normal(200 2)
g1  = x[,1]
g2  = x[,2].*x[,2]-1
eta = g1+g2
px  = 1./(1+exp(-eta))
y   = uniform(200) .< px
```

The simulated data set is shown in Figure 11.1. This multidisplay was created by issuing the commands

```
createdisplay(exgam, 2 2, s2d s2d s2d d3d)
pl11 = eta~y~mask(rows(eta) 1 x)
pl12 = sort(eta~px 1)
pl21 = sort(x[,1]~g1 1)
pl31 = sort(x[,2]~g2 1)
pl4  = x[,1]~x[,2]~eta
writecon(27)
show(pl11 pl12 s2d1, pl21 s2d2, pl31 s2d3, pl4 d3d1)
dum = update(pl12 2 s2d1 solid line)
dum = update(pl21 1 s2d2 solid line)
dum = update(pl31 1 s2d3 solid line)
dum = update(pl4  1 d3d1 xaxis "x_1" yaxis "x_2" zaxis "eta")
display(exgam)
```

To fit the generalized additive model with the help of (different) smoothing methods, we have to type the commands

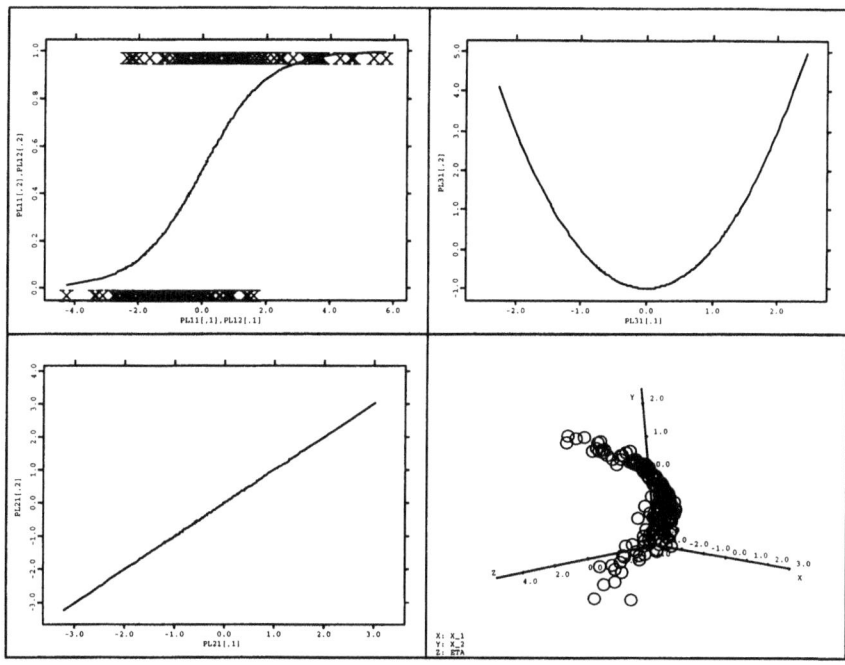

FIGURE 11.1. This picture shows the data set simulated by the code given in the text. The upper left shows the response y_i plotted against η_i together with $G(\eta_i)$. The lower left and upper right show the functions $g_1(x_1)$ and $g_2(x_2)$, respectively. A rotated view of the surface $\{x_1, x_2, \eta\}$ is given in the lower right.

Method	Deviance	$\hat{\alpha}$
Supersmoother	192.277	-0.097
Local linear smoother	196.9599	-0.076
s-knn smoother	204.869	-0.116

TABLE 11.1. Nongraphical results of the fits of the simulated generalized additive model shown in Figures 11.2 to 11.4. Given are the deviance upon convergence and the estimated value for α in (11.1). The theoretical value is $\alpha = 0$.

```
library(gam)
(itres beta alpha fx) = DOGAM(x y)
```

The results of these fits are shown in Figures 11.2 to 11.4. All plots in Figures 11.1 to 11.4 are on the same scale to make a direct comparison by visual inspection possible. The nongraphical results (deviance upon convergence and α) are summarized in Table 11.1.

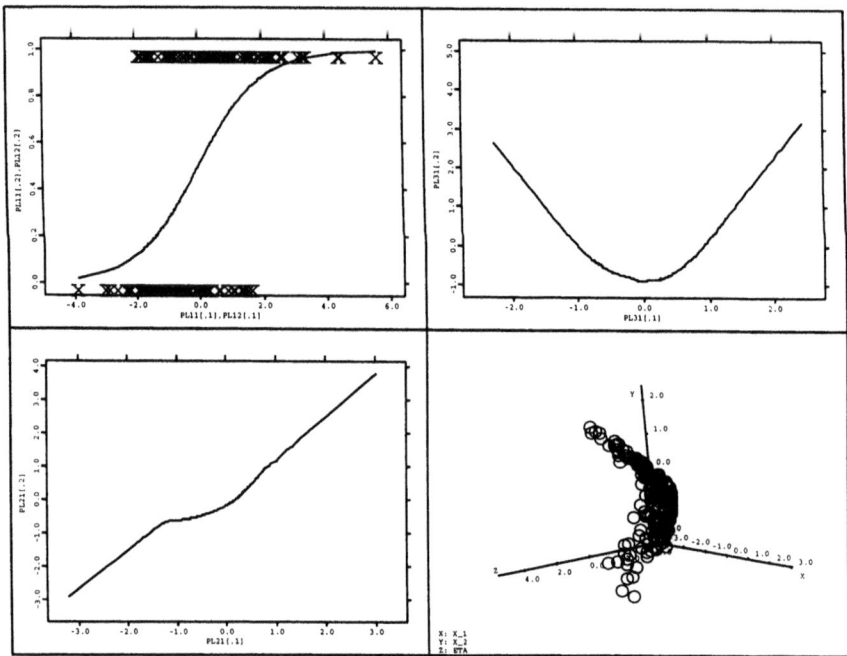

FIGURE 11.2. This figure shows the estimated model for the data set shown in Figure 11.1. The legend is the same as in Figure 11.1 with η, g_1, and g_2 replaced by $\hat{\eta}$, $\hat{g}_1$, and $\hat{g}_2$. The functions g_1 and g_2 were estimated with the supersmoother using a bass enhancement factor of 5.

Visually, the fit using the supersmoother is the best approximation to the true model. This fit also has the lowest deviance. But since the supersmoother is a nonlinear smoothing technique it seems to be impossible to obtain any rigorous distribution results regarding such a fit. Note also that we are using a bass enhancement factor of 5 for the (implicit) cross-validation within the supersmoother and thus we bias this cross-validation strongly towards over-smoothing.

The fits using the local linear smoother or the s-knn smoother offer no problems in identifying the quadratic influence of X_2. Both have problems, however, with the estimation of $g_1(\bullet)$; the local linear smoother in particular shows problems at the boundary. Given the results about the behavior of this smoother in the univariate case, this is somewhat surprising.

Finally, note that for our simulated model, we have $\alpha = 0$ in the notation of (11.1). If we look at the fitted values for α in Table 11.1 we see that they are "small." But again, there is no distribution theory that would allow us to test formally if this parameter is zero.

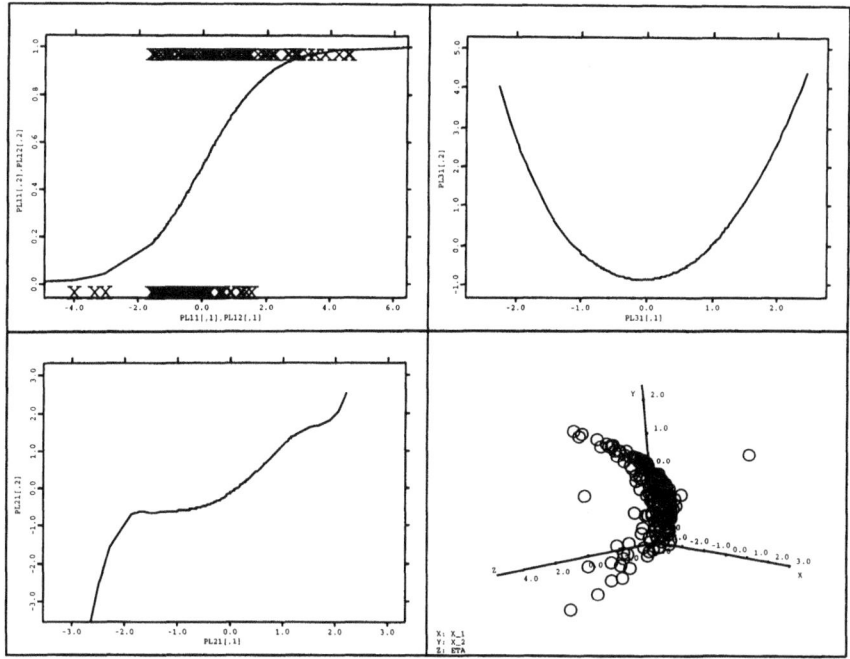

FIGURE 11.3. This figure shows the estimated model for the data set shown in Figure 11.1. The legend is the same as in Figure 11.1 with η, g_1, and g_2 replaced by $\hat{\eta}$, $\hat{g}_1$, and $\hat{g}_2$. The functions g_1 and g_2 were estimated with the local linear smoother with a bandwidth of 0.5. The lower left picture is truncated because of boundary problems. The fitted value for the smallest x_1 observation is -9.5, and the biggest x_1 observation has a fitted value of 7.3. These points are visible as outliers in the lower right.

11.3   Sliced Inverse Regression

Sliced inverse regression is a dimension reduction method proposed by Duan and Li (1991) and discussed in detail by Li (1991). The main goal of SIR is to reduce dimensionality for a given regression problem, not to find a smooth link function.

Given a response variable Y and a (random) vector $X \in \mathbb{R}^d$ of explanatory variables, SIR is based on the model:

$$Y = m(\beta_1^T X, \ldots, \beta_K^T X, \epsilon), \tag{11.3}$$

where $\beta_1, \ldots, \beta_K$ are unknown projection vectors, K is unknown and assumed to be less than d, $m : \mathbb{R}^{K+1} \to \mathbb{R}$ is an unknown function, and ϵ is the noise random variable with $\mathrm{E}[\epsilon \mid X] = 0$. We suppose that $\Sigma = \mathrm{Cov}[X]$ exists.

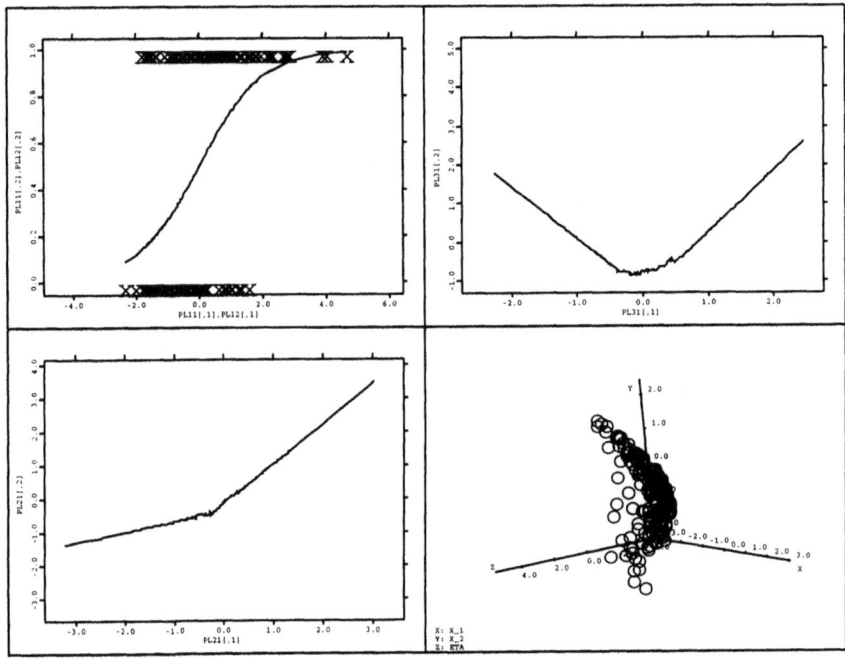

FIGURE 11.4. This figure shows the estimated model for the data set shown in Figure 11.1. The legend is the same as in Figure 11.1 with η, g_1, and g_2 replaced by $\hat{\eta}$, $\hat{g}_1$, and $\hat{g}_2$. The functions g_1 and g_2 were estimated with the s-knn smoother using a span of 0.6.

Model (11.3) describes the situation where the response variable Y depends on the d-dimensional variable X only through a K-dimensional subspace. The unknown β_i's, which span this space, are called *effective dimension reduction directions* (EDR-directions). The span is denoted as *effective dimension reduction space* (EDR-space) and because this K-dimensional space is built by K vectors, the β_i's have to be linearly independent. Otherwise the representation of model (11.3) would be redundant. So the aim is to estimate the base vectors of this space. The EDR-directions themselves are neither identifiable for length nor for direction. Only the space in which they lie is identifiable.

SIR tries to find exactly this K-dimensional subspace of the $\mathbb{R}^d$ which still carries the information between X and Y. If K is small, nonparametric methods could be applied for the estimation of m. A direct application of nonparametric smoothing is in high dimensions generally not possible due to the huge volume of such spaces; this fact is well known as the *curse of dimensionality* (see, for example, Huber, 1985).

The idea of SIR is to observe the inverse regression (IR) curve. That means instead of looking for $E[Y \mid x]$, we investigate $E[X \mid y]$, which consists actually of d one-dimensional regressions. One crucial advantage of SIR can be seen here: by doing only one-dimensional regressions, we avoid the above-mentioned problem of dimensionality.

The remaining problem is to find a connection between the IR and model (11.3). Assuming an additional property for the distribution of X, we get the following theorem (Li, 1991):

Given the model (11.3) and the assumption

$$\forall b \in I\!R^d : \ E\left[b^T X \,|\, \beta_1^T X = \beta_1^T x, \ldots, \beta_K^T X = \beta_K^T x\right] \ = \ c_0 + \sum_{i=1}^{K} c_i \beta_i^T x$$

$$(11.4)$$

then the centered inverse regression curve $E[X\,|\,Y=y] - E[X]$ lies in the linear subspace spanned by the vectors $\Sigma_X \beta_i$, $i = 1, \ldots, K$.

Assumption (11.4) is equivalent to the fact that X has an elliptically symmetric distribution (Cook and Weisberg, 1991), but it can be considerably weakened. Hall and Li (1993) showed that assumption (11.4) needs only to hold for the EDR-directions.

It is easy to see that for the standardized variable $Z = \Sigma^{-1/2}(X - E[X])$ the IR $E[Z \,|\, y]$ lies in $\text{span}(\eta_1, \ldots, \eta_K)$, where $\eta_i = \Sigma^{1/2}\beta_i$. That means that the conditional expectation $E[Z \,|\, y]$ is moving in $\text{span}(\eta_1, \ldots, \eta_K)$ depending on y. With b orthogonal to $\text{span}(\eta_1, \ldots, \eta_K)$, it follows that

$$b^T E[Z\,|\,y] \ = \ 0,$$

and further,

$$E[Z\,|\,y]E[Z^T\,|\,y] \times b \ = \ \text{Cov}[E[Z\,|\,y]] \times b \ = \ 0.$$

Here, the crucial consequence is that $\text{Cov}[E[Z \,|\, y]]$ is degenerated in each direction orthogonal to all EDR-directions η_i of Z.

This suggests the following algorithm. First, estimate $\text{Cov}[E[Z\,|\,y]]$ and then calculate the orthogonal directions of this matrix (for example, with eigenvalue/eigenvector decomposition). In general, the estimated covariance matrix will have full rank because of random variability, estimation errors and numerical imprecision. Therefore, we investigate the eigenvalues of the estimate and ignore eigenvectors having small eigenvalues. These eigenvectors $\hat{\eta}_i$ are estimates for the EDR-direction η_i of Z. We can easily rescale them to estimates $\hat{\beta}_i$ for the EDR-directions of X by multiplying with $\hat{\Sigma}^{-1/2}$, but then they are not necessarily orthogonal.

SIR is strongly related to PCA. If all data fall into a single slice, which means that $\widetilde{\text{Cov}}[E[Z \,|\, y]]$ is equal to $\widetilde{\text{Cov}}[Z]$, SIR coincides with PCA. Obviously, in this case any information about y is ignored.

For finding the EDR-directions it is sufficient to center the variable: $Z^* := X - E[X]$. Calculating the eigenvalues/eigenvectors of $Cov[E[Z^*\,|\,y]]$ leads also to the EDR-space.

Let us introduce some notation before illustrating the implementation of the SIR algorithm in XploRe. X, Y and Z are data matrices, not random vectors. The observations are stored in rows and single observations will be identified by small letters. The sample of size n is written as $\{(x_i, y_i)\}_{i=1}^{n}$. Thus,

$$
\begin{aligned}
\mathsf{X} &= (x_1, \ldots, x_n)^T, \quad x_i = (x_{i1}, \ldots, x_{id})^T, \\
\mathsf{Y} &= (y_1, \ldots, y_n)^T, \\
\bar{x} &:= \frac{1}{n} 1_n^T \mathsf{X}, \quad 1_n := \underbrace{(1, \ldots, 1)^T}_{n}, \\
\bar{\mathsf{X}} &:= \frac{1}{n} 1_n 1_n^T \mathsf{X} = (\bar{x}, \ldots, \bar{x})^T, \\
\hat{\Sigma} &:= \frac{1}{n-1} (\mathsf{X}^T \mathsf{X} - n\bar{x}\bar{x}^T).
\end{aligned}
$$

11.3.1   The SIR Algorithm

The algorithm to estimate the EDR-directions via SIR is as follows:

1. Standardize x:

$$
z_i := \hat{\Sigma}^{-1/2}(x_i - \bar{x}) \quad \text{or} \quad \mathsf{Z} := (\mathsf{X} - \bar{\mathsf{X}})\hat{\Sigma}^{-1/2}.
$$

2. Divide the range of y_i in S non-overlapping intervals *(slices)* H_s, $s = 1, \ldots, S$. n_s denotes the number of observations within slice H_s, and I_{H_s} the indicator function for this slice:

$$
n_s := \sum_{i=1}^{n} I_{H_s}(y_i).
$$

3. Compute the mean of z_i over all slices. This is a crude estimate for the *inverse regression* $\mathrm{E}[Z \,|\, Y \in H_s]$:

$$
\bar{z}_s := \frac{1}{n_s} \sum_{i=1}^{n} z_i \, I_{H_s}(y_i).
$$

4. Calculate the estimate for $\mathrm{Cov}[E[Z \,|\, y]]$:

$$
\hat{V} := n^{-1} \sum_{s=1}^{S} n_s \bar{z}_s \bar{z}_s^T.
$$

5. Identify the eigenvalues $\hat{\lambda}_i$ and eigenvectors $\hat{\eta}_i$ of $\hat{V}$.

Costs	Cause
nd	Mean $\bar{x}$
nd^2	Covariance Σ
d^3	$\Sigma^{-1/2}$
nd^2	Standardize X to Z
Sn	Compute n_s and $\bar{z}_s$
Sd^2	$\hat{V}$
d^3	Eigendecomposition of $\hat{V}$
d^3	Computation of β_i

TABLE 11.2. Costs of computation. In the column of costs, the terms are the order of the O function.

6. Transform the standardized EDR-directions $\hat{\eta}_i$ back to the original scale. Now the estimates for the EDR-directions are given by

$$\hat{\beta}_i := \hat{\Sigma}^{-1/2}\hat{\eta}_i.$$

The computational costs for the single steps can be found in Table 11.2. The overall complexity of the algorithm is $O(d^3 + nd^2 + Sn)$. As d is usually fixed, the leading term is $O(Sn)$. Unfortunately, the investigation of the statistical properties shows that choosing $S = O(n)$ guarantees for $\sqrt{n}$-consistency of the estimates for the eigenvalues and EDR-directions. In this case, we get a complexity of $O(n^2)$. But we can improve the algorithm if we sort the data before slicing with respect to Y. In this case, we need only $O(n)$ operations to compute n_s and $\bar{z}_s, s = 1, \ldots, S$. The advantage of sorting with respect to Y is that the slicing indicator depends only on Y (that is, conditioning on Y). That means that the matrix X are now already packed into slices. In order to perform the slicing, it is only necessary to find the start and end of each slice, which can be done by running exactly once over the length of the data, that is, $O(n)$ operations. Sorting itself can be done in $O(n \log(n))$ operations (see Sedgewick, 1990). The resulting complexity is $O(n \log(n))$, which is rather good computational complexity for nonparametric algorithms.

11.3.2 Implementation of the Algorithm

In general, it is convenient to avoid loops within matrix-oriented languages such as XploRe. We will use a technique based on indicator vectors, which can often be applied in order to avoid loops.

Let us first consider the interface. SIR computes EDR-directions and eigenvalues, which should be returned by two variables. The input is the data, which should be delivered in two separate variables, x and y. Since we would also like to control the method of slicing, we choose the following interface:

```
proc (edrdir eval) = SIR (x y h)
```

There are at least three different possibilities for slicing:

1. Define the number of slices (with constant width).

2. Define the number of elements within each slice.

3. Define a constant width for the slices.

If we want to slice the data by method 1, we interpret h as number of slices where $h \geq 2$. To distinguish method 1 from method 2 in the program we pass a negative value for h, if we want to choose the second method, that is, the number of elements per slice is given as negative number $h \leq -2$ and the number of elements is equal to abs(h). Finally, if $0 < h < 1$, we interpret h as the ratio of the width of a slice and the range of the response variable Y ($h = $ width/range) and choose method 3 for slicing the data.

Before we start with computations, the data are checked:

```
n = rows (x)          ;  n is number of x-observations
d = cols (x)          ;  d is dimension of the regressor space
error (d < 2 "Dimensionreduction with less than 2 dimensions !")
error (cols(y) <> 1 "y has to be a vector !")
error (rows(y) <> n "x and y have different length !")
```

We first standardize the data by computing $\hat{\Sigma}^{-1/2}$. One way to calculate this matrix is to perform a Jacobi decomposition: $\Sigma = \Gamma\Lambda\Gamma^T$ with $\Lambda = \text{diag}(l_1, \ldots, l_d)$, $l_i > 0$, $i = 1, \ldots, d$. Then $\Sigma^{-1/2} = \Gamma\Lambda^{-1/2}\Gamma^T$ where $\Lambda^{-1/2} = \text{diag}(1/\sqrt{l_1}, \ldots, 1/\sqrt{l_d})$. Note that the use of pointwise matrix multiplication (.* by computation of si2) within the program avoids the actual construction of the diagonal matrix:

```
xm  = mean(x)                        ; the overall mean vector of x
s   = (x'*x - n.*xm*xm') ./ (n-1)    ; unbiased covariance estimate
(eval evec) = eigsm (s)              ; eigenpair decomposition
si2 = evec*(sqrt(1./eval).*(evec')); Sigma ^ -1/2
z   = (x-xm') * si2                  ; standardized regressor
```

The difference x-xm' is computed where xm' is n times repeated (in rows). Such matching rules are automatically executed in XploRe. Sorting with respect to y is simply done by

```
data    = sort (y~z)        ; sort matrix regarding first column
zsorted = data[,2:cols(data)] ; get sorted z ...
ysorted = data[,1]          ; ... and sorted y
```

Now we have to iterate over the number of slices. To reduce the execution time of the macro, we will cut out the x-observations for each slice with the help of indicator vectors. For a given slice, we need a vector, say cutout, whose entries are true ($= 1$) for those x_i's lying within that slice, and false ($= 0$) otherwise. Then we get the matrix consisting of the slice elements simply by the command zslice = paf (zsorted cutout) (paf cuts out submatrices regarding an indicator vector).

11.3.3 Indicator Vector Instead of Loops

Assume that we have an indicator ind which is true for all x not sliced yet. Additionally, we know the upper end of the current slice, say uend. Then cutout can be calculated as the intersect of the indicator ind and the indicator ysorted <= uend, that is, (ysorted <= uend) .* ind, since intersections of indicator vectors can be easily computed by the pointwise product of the involved indicator vectors. This is the way to compute the slice indicator slind later on.

Let us consider the implementation of method 2, that is, each slice has abs(h) values.

```
h      = abs(h)              ; number of elements per slice
ns     = floor (n/h)         ; initial number of slices
condit = aseq (1 n 1)        ; conditional indicator is index
slends = aseq (1 ns 1) .* h  ; slice ends regarding condition
```

In the second line, the number of slices is calculated. By method 2, we are slicing on the index of y. As this variable is already sorted, the index is $(1, \ldots, n)$. The slice ends are the values of the vector $(h, 2h, \ldots, n_s h) = (1, \ldots, n_s) \times h$. Of course, if n % h ($n \bmod h$) is not zero there are some values left. The remaining data will be put into the first and last slice of the data.

```
if (h*ns <> n)                 ; are there observations left ?
  hk = floor ((n-h*ns)/2)      ; additional elements for 1. slice
  slends = slends + hk          ; shift slice ends
  slends[ns,1] = n              ; last slice goes always up to n
endif
```

Note that it is not necessary to check whether hs*ns <> n. If they are equal, hk is zero and slends is not changed by the lines of the if block. Step 4 of the SIR algorithm is mainly implemented along the following lines:

```
ind = matrix (n)               ; n x 1 matrix with ones
v   = matrix (d d 0)           ; d x d matrix with zeros
j   = 1                        ; loop counter over slices
while (j .<= ns)
  slind = (condit .<= slends[j,1]) .* ind
```

```
p      = sum (slind)                ; are there obs within slice j ?
if (p <> 0)
   ind    = ind - slind            ; update indicator ind
   zslice = paf (zsorted slind)    ; take slice observations
   zmean  = mean (zslice)          ; inverse regression
   v      = v + zmean*zmean' * rows (zslice) ; add up
endif
j = j+1                            ; increase loop counter
endo
```

Updating means that the just-sliced observations are removed from the index ind. Instead of mere averaging, other methods for (inverse) regression could be applied. Zhu and Ng (1993) considered weighted inverse regression based on kernel estimates.

Finally, the eigenvectors and eigenvalues of $\hat{V}$ have to be computed and the rescaling to the EDR-directions of x has to be done. Actually, the matrix v is an estimate for $n\hat{V}$. Generally, the factor n can be ignored as one is only interested in the ratio of the eigenvalues, not in the absolute amount. After standardizing these EDR-directions, the macro is finished.

```
(eval b) = eigsm (v ./ n)          ; eigenvecs/vals of v
b        = si2 * b                 ; rescale to EDR-directions
edrdir   = b ./ sqrt (sum (b.*b))  ; standardize to length one
endp
```

It is possible to establish asymptotic normality for the eigenvalues and for the statistic

$$\hat{\Psi}_K := \frac{\sum_{i=1}^{K} \hat{\lambda}_i}{\sum_{i=1}^{d} \hat{\lambda}_i},$$

where the eigenvalues are sorted $(\lambda_1 \geq \cdots \geq \lambda_d)$. $\hat{\Psi}_K$ can be interpreted as the variance explained by the first K EDR-vectors. The computation of the asymptotic variance of $\hat{\Psi}_K$ is quite complicated and time consuming. For details, see Kötter (1995). With that, confidence intervals for K for a predefined level α can be computed. A special kind of slicing (two elements in each slice) was investigated by Hsing and Carroll (1992). They also derived asymptotic normality and $\sqrt{n}$-consistency for their estimates. There are other possibilities to find K. Li (1991) gives an asymptotic result for $n \sum_{K+1}^{d} \hat{\lambda}_i$, which is asymptotically χ^2-distributed, if X is normal. Schott (1994) extended this result for X following distributions, which fulfill (11.4).

11.3.4   Remark

The number of different eigenvalues unequal to zero depends on the number
of slices. The rank of $\hat{V}$ cannot be greater than the *number of slices* $- 1$
(the z_i sum up to zero). This is a problem for categorical response variables,
especially for a binary response—only one direction can be found.

11.3.5   SIR II

In the previous sections we learned that it is valuable to consider the IR,
that is, $E[X \,|\, y]$. Now the question is, can we gain anything by considering
the conditional covariance $\text{Cov}[X \,|\, y]$? The answer is yes, as the following
example of Li shows:

$(X_1, X_2)^T \sim N(0, I_2)$, $Y = X_1^2$. Then $E[X_2 \mid y] = 0$ *because of the*
independency and $E[X_1 \,|\, y] = 0$ *because of the symmetry.*

This example demonstrates that the IR does not find the appropriate direction. By looking at the conditional variance

$$\text{Var}[X_1 \,|\, y] = E[X_1^2 \,|\, y] = y,$$

we would find β. This variance is varying with y while $\text{Var}[X_2 \,|\, y]$ is con-
stant.

The principle of SIR II is the same as before: investigation of the inverse
problem (here, instead of the conditional expectation, the conditional covariance). On the other hand, the theory of SIR II is more complicated.
The assumption of the elliptical symmetrical distribution of X has to be
sharpened. We must suppose normality for X.

Given this assumption, one can show that the vectors with the largest
distance to $\text{Cov}[Z \,|\, y] = E[\text{Cov}[Z \,|\, y]]$ for all y are the most interesting for
the EDR-space. An appropriated measure for the overall mean distance is
(Li, 1992)

$$E\left[\|\, (\text{Cov}[Z \,|\, y] - E[\text{Cov}[Z \,|\, y]]) \times b \|^2 \right] = b^T E[\ldots] b. \qquad (11.5)$$

With that, we conduct again an eigenpair decomposition, this time for
the above expectation $E\left[\|\text{Cov}[Z \,|\, y] - E[\text{Cov}[Z \,|\, y]]\|^2 \right]$; then we take the
rescaled eigenvectors with the largest eigenvalues as estimates for the un-
known EDR-directions.

11.3.6   The SIR II Algorithm

The algorithm of SIR II is very similar to the one for SIR; it differs in
only two points. Standardization and slicing is the same. Instead of merely

computing the mean, the covariance of each slice has to be computed. The estimate for the above expectation (11.5) is calculated after computing all slice covariances. Finally, decomposition and rescaling are conducted, as before.

1. Do steps 1 to 3 of the SIR algorithm.

2. Compute the slice covariance matrix $\hat{V}_s$:

$$\hat{V}_s = \frac{1}{n_s - 1} \sum_{i=1}^{n} I_{H_s}(y_i) z_i z_i^T - n_s \bar{z}_s \bar{z}_s^T.$$

3. Calculate the mean over all slice covariances:

$$\bar{V} = \frac{1}{n} \sum_{s=1}^{S} n_s \hat{V}_s.$$

4. Compute an estimate for $E[||.||^2]$:

$$\hat{V} = \frac{1}{n} \sum_{s=1}^{S} n_s \left(\hat{V}_s - \bar{V} \right)^2 = \frac{1}{n} \sum_{s=1}^{S} n_s \hat{V}_s^2 - \bar{V}^2.$$

5. Identify the eigenvectors and eigenvalues of $\hat{V}$ and scale back the eigenvectors. This gives estimates for the SIR II EDR-directions:

$$\hat{\beta}_i = \hat{\Sigma}^{-1/2} \hat{\eta}_i.$$

The computational costs are essentially the same as for SIR. The only difference is that computing $\hat{V}_s$ needs d times more operations than the computation of $\bar{z}_s$, but this is not crucial as d is usually constant. With that, the leading term is $O(Snd)$ and it can again be reduced to $O(n \log(n))$ operations by sorting with respect to Y if d is constant.

The implementation of this procedure does not differ much from that of SIR, so we do not present the complete XploRe code here. Only the few lines needed to compute $\hat{V}_s$ and $\hat{V}$ are shown. First, two variables, sig2a and sig2b, are initialized, which we need for computing the estimates for $\bar{V}$ and for $n^{-1} \sum_{s=1}^{S} n_s \hat{V}_s^2$.

Assume that the slicing is done as before. It is convenient to compute the number $n^{-1} \sum_{s=1}^{S} n_s \hat{V}_s^2$ regarding step 4 of the algorithm:

```
sig2a = matrix (d d 0)    ; initialize matrices
sig2b = matrix (d d 0)    ; to estimate E[Cov(x|y)]
ind   = matrix (n)        ; n x 1 matrix containing 1
nh    = 0                 ; number of slices obs
j     = 1
```

Because we have to compute an estimate for the covariance in each slice, we need at least two observations per slice (p > 1). In order to avoid that the last slice consists of only one observation, we check whether all slices before contain exactly (nh+p <> n-1) observations:

```
while (j .<= ns)
  slind = (condit .<= slends[j,1]) .* ind
  p     = sum (slind)                ; are there obs within slice j?
  if ((p > 1) && (nh+p <> n-1))
     nh        = nh + p
     ind       = ind - slind          ; update indicator ind
     zslice    = paf (zsorted slind)  ; take slice observations
     zmean     = mean(zslice)
     covslice = (zslice'*zslice - p.*zmean*zmean')/(p-1)
     sig2a     = sig2a + p.*covslice*covslice
     sig2b     = sig2b + p.*covslice
  endif
  j = j+1
endo
v = sig2a./n - sig2b*sig2b./(n.*n) ; estimate E[Cov(x|y)]
```

The eigendecomposition of $\hat{V}$ and the rescaling of the $\hat{\eta}_i$'s remain the same as for SIR.

11.3.7 Example

Now let us consider how SIR and SIR II work. The results are visualized by a small program (showsir) which displays the EDR-directions, the eigenvalues, and the data x, y. It produces four plots: the left two show the response variable versus the first respectively second direction. The upper right plot consists of a three-dimensional plot of the first two directions and the response. The last picture shows $\hat{\Psi}_K$, the ratio of the sum of the first K eigenvalues and the sum of all eigenvalues, similar to principal component analysis.

Let us generate data according to the following model:

$$y_i = \beta_1^T x_i + (\beta_1^T x_i)^3 + 4 \left(\beta_2^T x_i\right)^2 + \epsilon_i,$$

where the x_i's follow a three-dimensional normal distribution with zero mean and the identity matrix as covariance, $\beta_2 = (1, -1, -1)^T$, and $\beta_1 = (1, 1, 1)^T$. ϵ_i is standard normally distributed, $n = 300$. Corresponding to model (11.3), $m(u, v, \epsilon) = u + u^3 + v^2 + \epsilon$. The situation is depicted in Figure 11.5.

Both algorithms were conducted regarding slicing method 2 with 20 elements in each slice. The aim is to find β_1, respectively, β_2 with SIR. The data are designed such that SIR will detect β_1 because of the monotonic shape of $(\beta_1^T x_i + (\beta_1^T x_i)^3)$, while SIR II will search for β_2, as in this direction the conditional variance on y is varying.

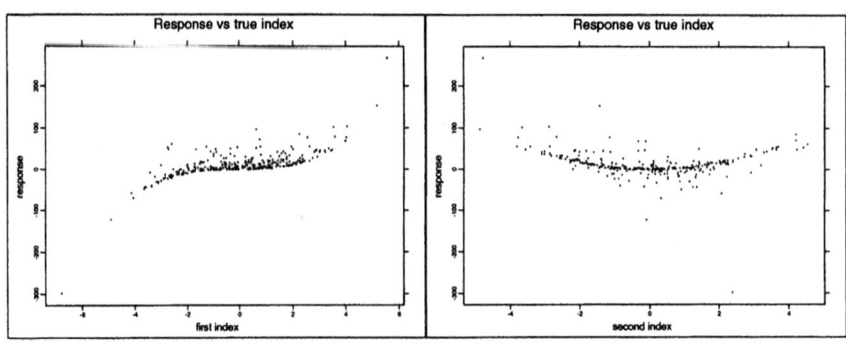

FIGURE 11.5. Plots of the true response versus the true indices. The monotonic and the convex shapes can be clearly seen.

$\hat{\beta}_1$	$\hat{\beta}_2$	$\hat{\beta}_3$
0.687	−0.720	0.109
0.534	0.482	−0.716
0.492	0.500	0.689

TABLE 11.3. SIR: EDR-directions for simulated data.

If we normalize the eigenvalues for the EDR-directions in Table 11.3 such that they sum up to one, the resulting vector is $(0.852, 0.077, 0.071)$. As can be seen in the upper left plot of Figure 11.6, there is a functional relationship found between the first index $\hat{\beta}_1^T x$ and the response. Actually, β_1 and $\hat{\beta}_1$ are nearly parallel. The normalized inner product $\hat{\beta}_1^T \beta_1 / \|\hat{\beta}_1\| \|\beta_1\| = 0.9894$ is very close to one, that is, the vectors are nearly parallel.

The second direction along β_2 is probably found due to the good approximation, but SIR does not see it clearly, because it is "blind" for the change of variance, as the second eigenvalue indicates.

For SIR II, the normalized eigenvalues are $(0.691, 0.191, 0.118)$, that is, about 69% of the variance is explained by the first EDR-direction (Table 11.4). Here, the normalized inner product of β_2 and $\hat{\beta}_1$ is 0.9992. Please note, that the estimator $\hat{\beta}_1$ estimates β_2 of the simulated model. In this case, SIR II found the direction where the variance varies regarding $\beta_2^T x$, similar to the example (Figure 11.7).

In summary, SIR has found the direction, which shows a strong relation regarding the conditional expectation between $\beta_1^T x$ and y, and SIR II has found the direction where the conditional variance is varying, namely, $\beta_2^T x$.

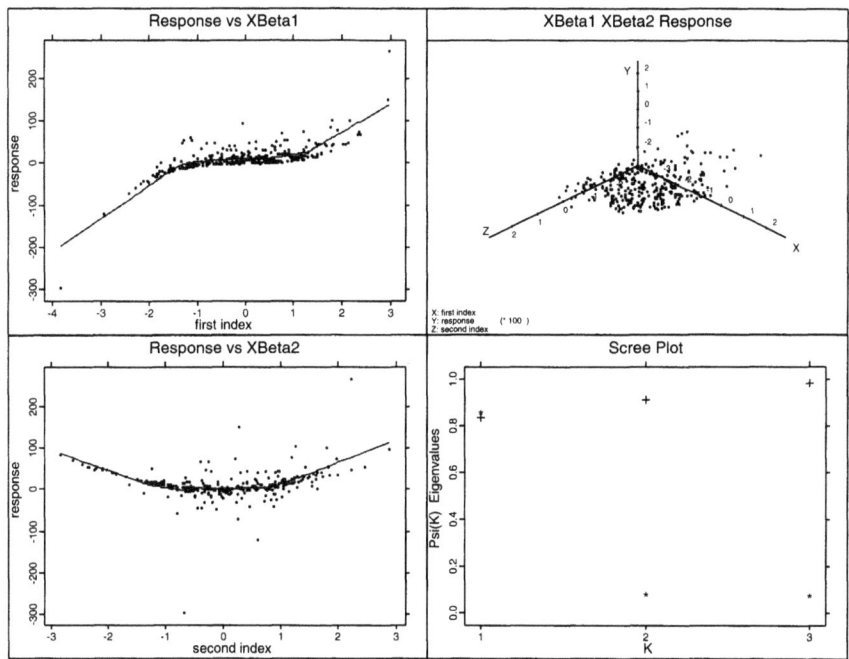

FIGURE 11.6. SIR: The left plots show the response versus the estimated EDR-directions. The upper right plot is a three-dimensional plot of the first two directions and the response. The lower right plot shows $\hat{\Psi}_i$ (*) and $\hat{\lambda}_i$ (+).

$\hat{\beta}_1$	$\hat{\beta}_2$	$\hat{\beta}_3$
0.608	0.251	0.768
−0.568	−0.602	0.600
−0.554	0.758	0.223

TABLE 11.4. SIR II: EDR-directions for simulated data.

The behavior of the two SIR algorithms is as expected. In addition, we learn that it is worthwhile to apply both versions of SIR.

Moreover, it is possible to combine SIR and SIR II (Cook and Weisberg, 1991; Li, 1991; Schott, 1994) directly, or to investigate even higher conditional moments. For the latter it seems to be difficult to obtain theoretical results.

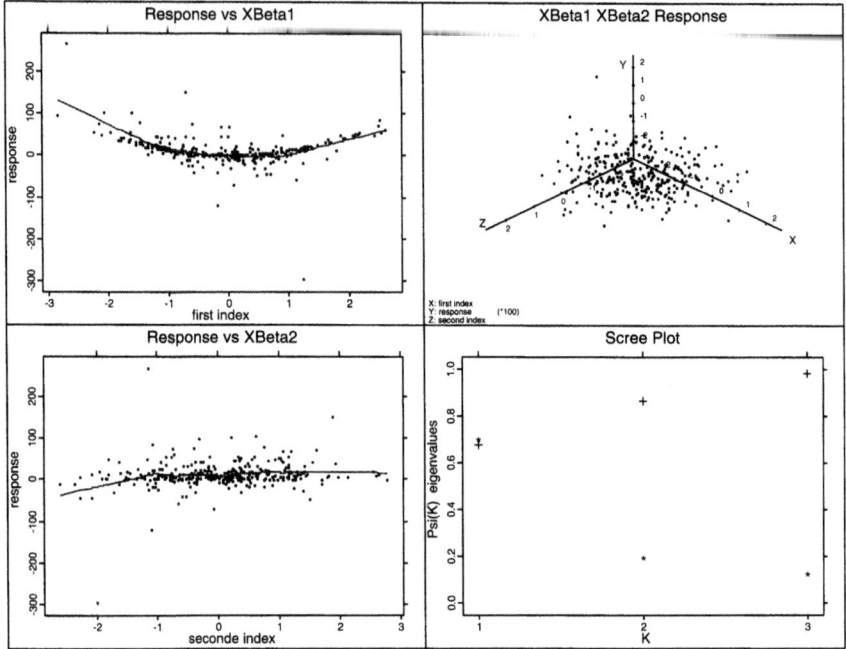

FIGURE 11.7. SIR II sees mainly the direction β_2.

11.4   Average Derivative Estimation

In this section we will develop several routines for *average derivative estimation* (ADE). Given a response variable Y whose expectation is assumed to depend on a d-dimensional variable X via a smooth function m, the aim of ADE is to estimate the average slope of this function. In other words, if $E[Y|X = x] = m(x)$ and ∇ denotes the gradient of partial derivatives with respect to the coordinates of X, the aim is to estimate

$$\delta = E[\nabla m(X)]. \tag{11.6}$$

ADE can be used in many econometric models (see, among others, Stoker, 1992; Härdle, Hildenbrand and Jerison, 1991). As one example, we want to mention *single-index* models (also called *one-term projection pursuit* models). Assume that the unknown function m has the form

$$m(x) = g(x^T \beta),$$

where g is an unknown univariate function and β is a d-dimensional (projection) vector. It is easy to see that in this case we have

$$\nabla m(x) = g'(x^T \beta)\beta,$$

and thus,

$$\delta = \mathrm{E}[g'(X^T\beta)]\beta.$$

This means that ADE allows us to estimate the unknown projection β (respectively, the index $x^T\beta$) up to a scale constant!

Under suitable assumptions on the function m and the marginal density f of the X variable, integration by parts of (11.6) yields to

$$\delta = \mathrm{E}\left[-Y\frac{\nabla f(X)}{f(X)}\right]. \tag{11.7}$$

Various estimation procedures, based on nonparametric estimates of the unknown functions in either (11.6) or (11.7), have been proposed and studied (Stoker, 1991). The most important (and surprising) result is that the proposed estimators are $\sqrt{n}$-consistent (which is typically not the case for nonparametric function estimators). The implementation of these estimators, however, is a little tricky since they involve division by an estimate of the marginal density of X. At the "border" of the observed X-values, these estimates can become so small that the division leads to an erratic behavior of the estimate for δ.

To avoid this problem we will discuss here a variation of ADE called *weighted average derivative estimation*, that is, we modify Equation (11.6) to allow for a nonnegative weight function $w(x)$. Thus, we will try to estimate

$$\delta_w = \mathrm{E}[\nabla m(X)w(X)]. \tag{11.8}$$

With the special choice of $w(x) \equiv f(x)$, we overcome the above problem. This choice is normally called *density-weighted average derivative estimation* and our estimand becomes

$$
\begin{aligned}
\delta_w &= \mathrm{E}[\nabla m(X)f(X)] \\
&= -2\mathrm{E}[Y\nabla f(X)].
\end{aligned} \tag{11.9}
$$

Here, Equation (11.9) follows again by integration by parts. The problem of estimating (11.9) was studied by Powell, Stock and Stoker (1989). They showed that δ_w may be estimated $\sqrt{n}$-consistent too. Their estimator is described below and several possible implementations in XploRe are shown.

Assume that we have observations $(x_i, y_i)_{i=1}^n \in \mathbb{R}^{d+1}$ and we want to estimate δ_w. One possible way to do this is to estimate the gradient of the marginal density of the X variables at each observation point by, say, $\widehat{\nabla f}(x_i)$. An estimator for δ_w would then be

$$\hat{\delta}_w = -\frac{2}{n}\sum_{i=1}^n y_i\widehat{\nabla f}(x_i). \tag{11.10}$$

The easiest method to construct a nonparametric estimate for the gradient of the marginal density is to take the gradient of a nonparametric estimate of the marginal density. Nonparametric density estimation is discussed in another section of this book. In the multivariate case, we need a d-variate kernel $\mathcal{K}$ (think of $\mathcal{K}$ as a d-variate density function) and a $d \times d$ matrix H of smoothing parameters (H should be positive definite). A nonparametric estimate of the marginal density f at a point $z \in \mathbb{R}^d$ would be

$$\hat{f}_H(z) = \frac{1}{n} \sum_{i=1}^{n} \frac{1}{\det(H)} \mathcal{K}\left(H^{-1}(z - x_i)\right). \tag{11.11}$$

For numerical ease, a common choice is to take $\mathcal{K}$ as a product of d univariate kernels K, and to reduce H to a diagonal matrix so that we have only a d-dimensional vector h as smoothing parameters. With these choices, (11.11) simplifies to

$$\hat{f}_h(z) = \frac{1}{n h_1 \ldots h_d} \sum_{i=1}^{n} \prod_{m=1}^{d} K\left(\frac{z_m - x_{im}}{h_m}\right).$$

To estimate the density-weighted average derivative δ_w, Powell et al. (1989) did not use this nonparametric estimator directly, but a *leave-one-out* version of it. To estimate the marginal density f at the observation x_i, they dropped x_i from the sample and calculated $\hat{f}_h(x_i)$ from the remaining sample (of size $n-1$). Plugging this modified density estimate into (11.10) leads to:

$$\hat{\delta}_w = \begin{pmatrix} \frac{\partial}{\partial x_1} \\ \vdots \\ \frac{\partial}{\partial x_d} \end{pmatrix} \left(-\frac{2}{n(n-1)h_1 \ldots h_d} \sum_{i=1}^{n} \sum_{\substack{j=1 \\ j \neq i}}^{n} \prod_{m=1}^{d} K\left(\frac{x_{im} - x_{jm}}{h_m}\right) y_i \right).$$

$$\tag{11.12}$$

As you can see in (11.12), we have two summations and one multiplication sign. This would actually need a three-dimensional array structure for the calculations. Since XploRe can only handle two-dimensional arrays, that is, matrices, the implementation of (11.12) requires at least one loop. This loop can either run over the number of observations (for one of the summation signs in (11.12)), or over the number of dimensions (for the multiplication sign in (11.12)). Both possibilities are described below, where, for further computational ease, K is chosen to be the density function of the standard normal distribution. Note that with this choice of K, we have $K'(u) = -uK(u)$ and $K'(0) = 0$ so it actually makes no difference whether we take in (11.12) the summation over $j \neq i$ or drop this restriction.

11.4.1   How Do We Implement (11.12) with One Loop over the Number of Dimensions?

Assume that our procedure has as input parameters the $n \times p$ matrix x (containing the observed x variable), the $n \times 1$ vector y (containing the observed y variable), and the $d \times 1$ vector h with the smoothing parameters. The strategy is now to calculate the $n \times n$ matrix xs with entries

$$\left(\prod_{m=1}^{d} K \left(\frac{x_{im} - x_{jm}}{h_m} \right) \right)_{\substack{i=1,\ldots,n \\ j=1,\ldots,n}}.$$

After doing this, we can calculate the lth coordinate of $\hat{\delta}_w$ as follows:

- Multiply the above matrix xs pointwise with the $n \times n$ matrix xt with entries

$$\left(\frac{x_{il} - x_{jl}}{h_l} \right)_{\substack{i=1,\ldots,n \\ j=1,\ldots,n}}.$$

  Because of our choice for K, this step takes care of the partial derivative.

- Sum up the columns of the resulting matrix. This is the summation over j in (11.12). Since we sum over the columns and not the rows, we also take into account that $K'(u)$ is equal to $-uK(u)$.

- Take the resulting vector and multiply it pointwise with y and then sum the result. This is the loop over i in (11.12). Since y is also a vector, this operation is equivalent to taking the scalar product of two vectors, which can be done by the function xty.

The final implementation is thus:

```
PROC(del)=DWADE1(x y h)
  d  = cols(x)
  n  = rows(x)
  h  = matrix(d 1 h)
  xs = pdfn( (x[,1]-x[,1]') / h[1,1] )
  i  = 1
  while(i < d)
    i  = i+1
    xs = xs .* pdfn( (x[,i]-x[,i]') / h[i,1] )
  endo
  del = matrix(d)
  i  = 0
  while(i < d)
    i  = i+1
    xt = (x[,i]-x[,i]') / h[i,1]
    del[i,1] = xty( sum(xt .* xs)  y )
  endo
  del = -2*del ./ ( n*(n-1)*prod(h)*h )
ENDP
```

The line

```
h = matrix(d 1 h)
```

ensures that h is a $d \times 1$ matrix. It also allows the user of this procedure to specify the smoothing parameter h as a scalar if he or she wants all of them to have the same value. In the second to last line we multiply the lth component of δ_w by

$$-\frac{2}{n(n-1)h_1 \ldots h_d h_l},$$

which are the normalizing constants in (11.12). Note that the additional h_l comes from the partial derivative.

11.4.2 How Do We Implement (11.12) with One Loop over the Number of Observations?

If you analyze the above routine closely, you will see that it uses large amounts of storage space—several $n \times n$ matrices are created and manipulated. Moreover, we do the "main" calculation, that is,

```
(x[,i]-x[,i]')/h[i,1]
```

twice, once in each loop. This could be avoided by storing the results of this calculation in the first loop (left as an *exercise* for you), but this would use even more storage space. For big data sets, this could lead to memory problems and it would be preferable to have an implementation that loops over the number of observations.

A naive approach for such a routine would just implement a loop over the number of observations for the sum over i in (11.12), and implement the sum over j and the product over m with the matrix-commands of XploRe. In doing this, however, we would perform many unnecessary calculations. Since the kernel K that we have chosen is an even function, its derivative is an odd function. Thus, the matrix xt.*xs, which is calculated in the routine above, is anti-symmetric. In the above routine, there is no simple way to make use of this knowledge, but now we can do better.

To see exactly what is going on, split the sum over j in (11.12) into one sum going from 1 to $i-1$ and another going from $i+1$ to n. (Remember, for $i = j$ we add zero, so it does not matter if we drop this term now.) Now realize that the sum of i from 1 to n and j from 1 to $i-1$ is the same as the sum over j from 1 to n and i from $j+1$ to n. Change these two summation signs and rename the indices. You will see that the double sum in (11.12) is equal to

$$\sum_{i=1}^{n} \sum_{j=i+1}^{n} \prod_{m=1}^{d} K\left(\frac{x_{im} - x_{jm}}{h_m}\right)(y_i - y_j). \tag{11.13}$$

Based on this formula, it is clear how our procedure should work. In a loop over i, we do the following:

- Construct the $(n-i) \times d$ matrix xs with entries

$$\left(\frac{x_{im} - x_{jm}}{h_m}\right)_{\substack{j=i+1,\ldots n \\ m=1,\ldots,d}}.$$

- Apply the kernel K to this matrix and multiply the rows to get the product in (11.13).

- Multiply the result of this operation pointwise with the matrix xs. This takes care of the partial derivatives and results again in a matrix of size $(n-i) \times d$.

- Multiply the last result pointwise with the $(n-i) \times 1$ matrix

$$(y_i - y_j)_{j=i+1,\ldots,n},$$

and sum the resulting $(n-i) \times d$ matrix columnwise. This performs the summation over j in (11.13).

Together with some initializations and the updating of the current value of $\hat{\delta}_w$ in each loop, our algorithm is thus the following:

```
PROC(del)=DWADE2(x y h)
  d    = cols(x)
  n    = rows(x)
  h    = matrix(d 1 h)
  del  = matrix(d 1 0)
  i    = 1
  WHILE(i < n)
    ind = (i+1):n
    xs = (x[i,] - x[ind,]) ./ trn(h)
    xs = xs .* prodr(pdfn(xs))
    del  = del - sum( (y[i,1]-y[ind,1]) .* xs )
    i    = i+1
  ENDO
  del = -2*del ./ (n*(n-1)*prod(h)*h)
ENDP
```

The disadvantage of this procedure is that it loops over the number of observations. If the sample size is already moderately large, this can take several minutes. So if the sample size n and the dimension d are such that you can keep several $n \times d$ matrices in memory, the first routine is preferable. It will be much faster.

The two routines described above are part of the ADDMOD library of XploRe. An example for the application of these methods to real data can be found in Chapter 12.

REFERENCES

Breiman, L. and Friedman, J.H. (1985). Estimating optimal transformations for multiple regression and correlations (with discussion), *Journal of the American Statistical Association* **80**(391): 580–619.

Cook, R.D. and Weisberg, S. (1991). Comment on "sliced inverse regression for dimension reduction", *Journal of the American Statistical Association* **86**(414): 328–332.

Duan, N. and Li, K.C. (1991). Slicing regression: A link-free regression method, *Annals of Statistics* **19**(2): 505–530.

Fan, J. (1992). Design-adaptive nonparametric regression, *Journal of the American Statistical Association* **87**(420): 998–1004.

Fan, J. (1993). Local linear regression smoothers and their minimax efficiency, *Annals of Statistics* **21**(1): 196–216.

Friedman, J.H. (1984). A variable span smoother, *Technical Report No. 5*, Department of Statistics, Stanford University, Stanford, California.

Friedman, J.H. and Stuetzle, W. (1981). Projection pursuit regression, *Journal of the American Statistical Association* **76**(376): 817–823.

Hall, P. and Li, K.C. (1993). On almost linearity of low dimensional projections from high dimensional data, *Annals of Statistics* **21**(2): 867–889.

Härdle, W. (1990). *Applied Nonparametric Regression*, Econometric Society Monographs No. 19, Cambridge University Press, New York.

Härdle, W., Hildenbrand, W. and Jerison, M. (1991). Empirical evidence on the law of demand, *Econometrica* **59**(6): 1525–1549.

Hastie, T.J. and Tibshirani, R.J. (1986). Generalized additive models (with discussion), *Statistical Science* **1**(2): 297–318.

Hastie, T.J. and Tibshirani, R.J. (1990). *Generalized Additive Models*, Vol. 43 of *Monographs on Statistics and Applied Probability*, Chapman and Hall, London.

Hsing, T. and Carroll, R.J. (1992). An asymptotic theory for sliced inverse regression, *Annals of Statistics* **20**(2): 1040–1061.

Huber, P. (1985). Projection pursuit, *Annals of Statistics* **13**(2): 435–475.

Kötter, T.T. (1995). An asymptotic result for sliced inverse regression, *Computational Statistics*. To appear.

Li, K.C. (1991). Sliced inverse regression for dimension reduction (with discussion), *Journal of the American Statistical Association* **86**(414): 316–342.

Li, K.C. (1992). On principal Hessian directions for data visualization and dimension reduction: Another application of Stein's lemma, *Journal of the American Statistical Association* **87**(420): 1025–1039.

Powell, J.L., Stock, J.H. and Stoker, T.M. (1989). Semiparametric estimation of index coefficients, *Econometrica* **57**(6): 1403–1430.

Schimek, M.G., Neubauer, G.P. and Stettner, H. (1994). Backfitting and related procedures for non-parametric smoothing regression: A comparative view, *in* W. Grossmann and R. Dutter (eds), *COMPSTAT '94. Proceedings in Computational Statistics*, Physica, Heidelberg, pp. 64–69.

Schott, J.R. (1994). Determining the dimensionality in sliced inverse regression, *Journal of the American Statistical Association* **89**(425): 141–148.

Sedgewick, R. (1990). *Algorithms in C*, Addison-Wesley, Reading, Massachusetts.

Stoker, T.M. (1991). Equivalence of direct, indirect and slope estimators of average derivatives, *in* W.A. Barnett, J.L. Powell and G. Tauchen (eds), *Nonparametric and Semiparametric Methods in Econometrics and Statistics*, Cambridge University Press, New York.

Stoker, T.M. (1992). *Lectures on Semiparametric Econometrics*, CORE, Université Catholique de Louvain, Louvain-la-Neuve, Belgium.

Stone, C.J. (1985). Additive regression and other nonparametric models, *Annals of Statistics* **13**(2): 689–705.

Stone, C.J. (1986). The dimensionality reduction principle for generalized additive models, *Annals of Statistics* **14**(2): 590–606.

Zhu, L. and Ng, K.W. (1993). On asymptotic theory of sliced inverse regression based on weighted method. Preprint.

12
Comparing Parametric and Semiparametric Binary Response Models

Isabel Proença[1] and Axel Werwatz [2]

12.1 Introduction

Binary response models are frequently applied in economics and other social sciences. Whereas standard parametric models such as Probit and Logit
models still dominate the applied literature, there have been important
theoretical advances in semi- and nonparametric approaches to binary response analysis (see Horowitz, 1993a, for an excellent and up-to-date survey). From the perspective of the applied researcher, the development of
new techniques that go beyond Logit and Probit are important for several
reasons:

1. Economic theory usually does not provide clear guidelines on how
   a parametric model should be specified. Hence, the assumptions underlying Probit and Logit models are rarely justified on theoretical
   grounds. Rather, they are motivated by convenience and by reference
   to "standard practice."

2. Misspecification of parametric models can cause parameter estimates
   and inferences based on these parameters to be inconsistent. Moreover, predictions made from misspecified parametric models can be
   inaccurate and misleading.

Semiparametric models appear to be an alternative to standard techniques that

- do not require the kind of arbitrary distributional assumptions usually invoked in parametric analysis

---

[1]Instituto Superior de Economia e Gestao, Universidade Técnica de Lisboa,
P-1200 Lisboa, Portugal.
[2]Sonderforschungsbereich 373, Humboldt-Universität zu Berlin, D-10178
Berlin, Germany.

- overcome the so-called *curse of dimensionality* that is hampering non-parametric techniques in applications with high-dimensional data and standard sample sizes (Härdle, 1990).

Given that semiparametric models provide a workable alternative to standard parametric methods, the question arises whether standard Probit and Logit models suffice for an adequate treatment of a particular application or whether they should be abandoned in favor of semi- and nonparametric techniques. One way to discriminate between competing parametric and semiparametric models is by means of a formal statistical test. One such test has been developed by Horowitz and Härdle (1994) (in what follows, referred to as the *HH-test*). A different way of using nonparametric and semiparametric techniques to shed light on the adequacy of a parametric model involves using confidence intervals or confidence bands. This approach is less formal than the the HH-test, and it has the distinct advantage that its results can be illustrated graphically.

In this chapter we try to demonstrate how one might use XploRe to estimate and compare parametric and semiparametric binary response models in the context of an application that carries some economic relevance. More specifically, we will estimate both parametric and semiparametric models using data on unemployment following a completed apprenticeship in Germany. The HH-test will be used to test the parametric model against a semiparametric alternative. Moreover, confidence intervals and confidence bands are utilized to analyze the adequacy of the specification of the parametric model.

The chapter is organized as follows. First, we we will give a short description of the application and the data used in this chapter. Then we introduce the statistical binary response models considered in this project and demonstrate how we used XploRe to estimate and compare these models on the basis of specification tests and confidence limits that are described in some detail as we move along. In the final section, we summarize our findings and present some conclusions.

12.2   The Data

In this section we will give a brief description of the data used in this study. All data used in this project refer to the former West Germany. A more detailed outline of how the sample was extracted and how the variables were created form the raw data is available upon request from the authors.

For the purposes of this study we extracted a sample of 462 individuals from the first nine waves (1984 to 1992) of the GSOEP (German socioeconomic panel). For a detailed description of the GSOEP, see Projektgruppe "Das Sozio-ökonomische Panel" (1991). Each year, respondents were asked whether they had completed an apprenticeship in the previous year. Those

Variable	Definition/Comments
AGE	Age of the respondent in the year the apprenticeship was completed
SCHOOLING	Years of schooling
EARNINGS	Gross monthly earnings as an apprentice
CITYSIZE	Size of the city the respondent lives in at the time the apprenticeship was completed
FIRMSIZE	Size of the firm where the respondent was apprenticed
DEGREE	Percentage of people apprenticed in a certain occupation, divided by the percentage of people employed in this occupation in the entire economy
URATE	Unemployment rate in the state the respondent lived in during the year the apprenticeship was completed

TABLE 12.1. Explanatory variables.

who answered "yes" to this question sometime between 1985 and 1992 were included in our sample.

The dependent variable *UNEMP* takes on the value "1" if an individual is registered as unemployed at the time of the survey in the year following the completion of the apprenticeship. It takes on the value "0" if the individual is employed. We assume that the probability that *UNEMP* takes on the value "1" is related to the following set of explanatory variables, summarized in Table 12.1.

SCHOOLING and *AGE* are trying to measure general human capital and are expected to have a negative effect on the probability of being unemployed. *EARNINGS* is supposed to capture "the value of an apprenticeship." Again, we expect a negative coefficient for this variable. The effect of *CITYSIZE* is not clear cut, a priori, but one can make a case that, ceteris paribus, larger cities offer more employment opportunities and therefore a negative coefficient should be expected. It has been observed that for small firms (especially in the artisan sector), the number of apprenticeship positions provided exceeds the number of workers retained after the apprenticeship is completed (see Franz and Soskice (1994) for more on this issue). Hence, *FIRMSIZE* is likely to have a negative effect on the probability of being unemployed. *DEGREE* and *URATE* are variables that are generated from information provided by Germany's federal statistical bureau, the *Statistisches Bundesamt*. *URATE* is supposed to capture the overall employment situation in the state the respondent lived in at the time he or she completed an apprenticeship. Clearly, we expect a positive coefficient for this variable. *DEGREE* derives its name from the fact that by definition, the variable measures the degree to which an occupation is "overapprenticed." That is, suppose that 20% of all apprentices are trained as mechanics but only 10% of all workers in the entire German economy work as mechanics. Then *DEGREE* will take on the value $\frac{0.2}{0.1} = 2$. Hence, if a respondent has completed an apprenticeship that is "overapprenticed" in the sense just described, then *DEGREE* will take on values greater than "1". A significant positive coefficient of this variable would indicate that those who completed an apprenticeship in overapprenticed occupations face a higher probability of becoming unemployed after the apprenticeship is finished.

12.3   Parametric and Semiparametric Binary Response Models

In this section we will introduce the parametric and semiparametric models used in the empirical analysis. We will not aim at a rigorous or complete treatment of binary responses models in general. Rather, the discussion will be largely intuitive and will highlight the aspects of the theoretical model that are relevant for the application at hand.

12.3.1   The Logit Model

The parametric model estimated in this project is the standard Logit model where the conditional probability that the binary random variable Y is assumed to depend on the $k \times 1$ vector of explanatory variables $X = (X_1, X_2, \ldots, X_k)^T$ in the following way:

$$E[Y|X = x] = P(Y = 1|X = x, \beta) = F(x^T \beta). \qquad (12.1)$$

Here, $F(\bullet)$ denotes the cumulative distribution function (*cdf*) of the *Logistic* distribution

$$F(u) = \frac{1}{1 + e^{-u}},$$

and β is a $k \times 1$ vector of unknown coefficients. One may arrive at this specification by using the latent variable model that has dominated economist's thinking about discrete choice, but this is not necessary for our purposes. In particular, specifying $F(\bullet)$ as the cumulative logistic distribution function is neither suggested nor implied by the latent variable model. Rather, we will follow the perspective taken up by McCullagh and Nelder (1989) to view (12.1) as one specific form of the generalized linear model (GLM). In the GLM, the conditional expectation of Y is assumed to be related to X in the following way:

$$E[Y|X = x] = G(x^T \beta), \qquad (12.2)$$

where $G(\bullet)$ is a known function, the *inverse link function* in the terminology of McCullagh and Nelder (1989). In what follows, we will refer to $G(\bullet)$ simply as the *link function*. By comparing (12.1) with (12.2), it is easy to see that the Logit model can be viewed as a special case of the GLM with $G(\bullet)$ chosen to be $F(\bullet)$, the *cdf* of the Logistic distribution. In the following section, we will consider a certain class of semiparametric models, in the econometrics literature referred to as *single index models* (SIM), which generalize the GLM by allowing $G(\bullet)$ to be an arbitrary smooth function that has to be estimated from the data.

12.3.2   The Semiparametric Model

The semiparametric model can be viewed as a generalization of the parametric model. The purpose of the semiparametric approach is to widen the assumptions regarding the link function $G(\bullet)$ while avoiding the *curse of dimensionality* that is hampering fully nonparametric techniques when applied to high-dimensional data. The semiparametric models considered in this project overcome the *curse of dimensionality* by aggregating the multidimensional variable X into the "single (parametric) index" $x^T\beta$, while maintaining the "nonparametric" assumption that the specification of the link in Equation (12.2) is unknown. Hence, the single index model (SIM) considered in this project takes on the following form:

$$E(Y|X = x) = P(Y = 1|X = x) = g(x^T\beta) \qquad (12.3)$$

with $g(\bullet)$ an unknown function.

12.4   Estimation

After introducing parametric and semiparametric binary response models, we will now turn to the task of estimating these models.

12.4.1   Estimating the Logit Model

In the Logit model, we assume that $G(\bullet)$ is known to be of the form

$$F(u) = \frac{1}{1 + e^{-u}}.$$

Hence, estimating the Logit model comes down to estimating the coefficient vector β by the well-known maximum likelihood method. Estimation of the Logit model in XploRe is carried out within the GLM module (see Chapter 10 of this book for a detailed description of this module). In the following section, we present Logit estimates for the unemployment-following-completed-apprenticeship data.

	Coeff.	St. errors
INTERCEPT	-4.27032	2.25014
AGE	0.03718	0.11426
SCHOOLING	-0.01280	0.17347
EARNINGS	-0.00070	0.00098
CITZSIZE	-0.00048	0.00042
FIRMSIZE	0.00025	0.00265
DEGREE	-0.00116	0.00200
URATE	0.23398	0.06504
Degrees of freedom	454	
Deviance	237.394	
Pearson's χ^2	490.030	

TABLE 12.2. Results of the logit fit: Full model.

12.4.2   The Parametric Fit

To start with, we estimated a Logit model including all explanatory variables described above plus an intercept term. We will refer to this model as the "full model".

The results of the parametric fit were obtained by running the following procedure:

```
proc(itres b se t bvar stat) = main()
  capture("on")
  dat = read("xunemp1")  ; reads the data stored in XploRe format
  y   = dat[,1]          ; the dependent variable
  x   = dat[,2:8]        ; the explanatories -- full model
  library(GLM)
  (itres b se t bvar stat) = DOGLM(x y)
endp
```

Here, the data are read from the file **xunemp1.dat** which is a file written in XploRe format. Then the GLM library is called in order to run the DOGLM procedure. This is an interactive procedure that fits a Generalized Linear Model (see Chapter 10 for a detailed description of this procedure). For the Logit fit, the user has to choose the options Binomial (for the exponential family) and then Logit (for the link function).

The results of the fit are shown in Table 12.2. The plot of the Logit regression is shown in Figure 12.1.

The coefficients of all variables, with the exception of *AGE* and *FIRM-SIZE*, have the expected signs. The coefficient of *DEGREE* is negative but highly insignificant. In fact, most coefficients are statistically insignificant at the 5% level. Hence, we decided to drop several variables on the basis of their poor *t*-ratios.

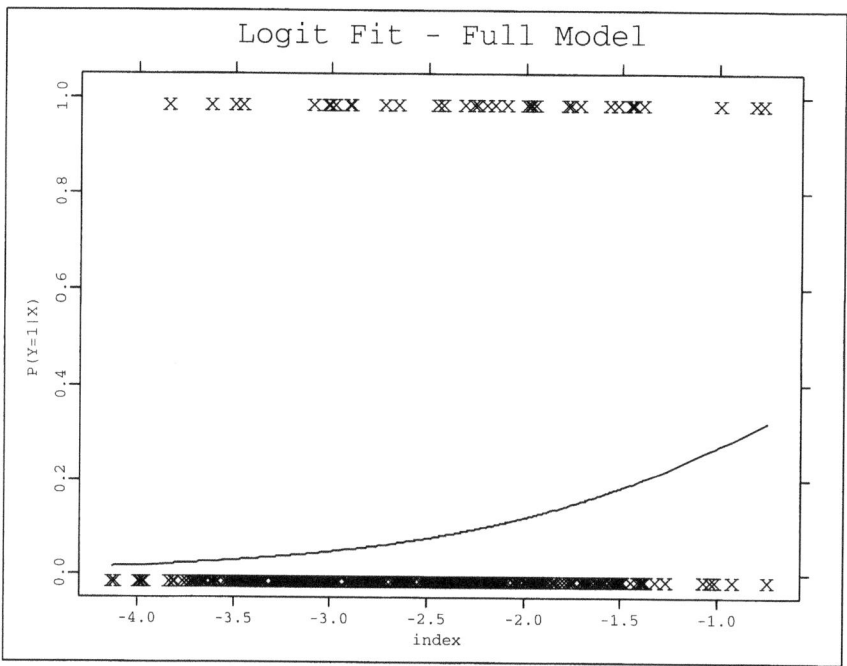

FIGURE 12.1. The Logit fit for the full model. The data points are identified by crosses.

Note that this strategy is open to criticism. First of all, McCullagh and Nelder (1989) have pointed out that in the binary Logit model, the true distribution of the t-statistic may be poorly approximated by the t-distribution for samples of the size encountered in this application. Moreover, t-tests hinge on the assumption that the Logit link is correctly specified. This assumption has not been tested yet.

Still, *AGE*, *SCHOOLING*, and *FIRMSIZE* were eliminated from the model. *EARNINGS* and *DEGREE* were kept in the model despite their low t-values because these variables are particularly interesting from an economic point of view. This leads to the "restricted model" with *DE-GREE*, *EARNINGS*, *CITYSIZE*, and *URATE* as explanatory variables. The results of the Logit fit for the restricted model are given in Table 12.3.

As in the full model, *URATE* has a significant positive effect and appears to be the most important predictor of the probability of being unemployed.

In the following section, we introduce the procedure used in this study to estimate a semiparametric model for the probability of being unemployed following a completed apprenticeship. Following this theoretical discussion, we present estimates for the semiparametric fit of the restricted model and compare these results with the Logit estimates.

	Coeff.	St. errors
INTERCEPT	-3.65076	0.99971
EARNINGS	-0.00065	0.00094
CITYSIZE	-0.00046	0.00041
DEGREE	-0.00116	0.00199
URATE	0.23540	0.06503
Degrees of freedom	457	
Deviance	237.511	
Pearson's χ^2	488.129	

TABLE 12.3. Results of the logit fit: Restricted model.

12.4.3   Estimating the Semiparametric Model

Estimation of the semiparametric model (12.3) proceeds in two steps. First, the coefficient vector β has to be estimated. In this project, we employ the method of *weighted average derivative estimation* (WADE) of Powell, Stock and Stoker (1989) to estimate the coefficients of the index function. This method is a modification of *average derivative estimation* (ADE).

ADE is motivated by the following property of single index models of the form (12.3)

$$E\{\nabla g(X)\} = E\left(\frac{dg}{dX^T\beta}\right)\beta = \gamma\beta,$$

with $\nabla g(X) = \partial g/\partial X$.

Assuming that $g(\bullet)$ is first differentiable in X, β can be estimated up to a constant by estimating the mean of the gradient vector $\nabla g(X)$.

Suppose now that X is continuously distributed with density $p(x)$ which is also first differentiable. Under some suitable regularity conditions, integration by parts allows us to write

$$E\{\nabla g(X)\} = E\left\{-Y\frac{\nabla p(X)}{p(X)}\right\}, \tag{12.4}$$

with $\nabla p(X)$ the gradient vector of $p(X)$. Therefore, estimating $\gamma\beta$ amounts to estimating the density of X and the derivative of this density.

The WADE procedure results from applying a nonnegative weight $w(x)$ to (12.4). For convenience, this weight can be set equal to the density $p(x)$, obtaining (after integration by parts)

$$E\{\nabla g(X)p(X)\} = -2E\{Y\nabla p(X)\}.$$

Finally, the WADE estimator (with weight $p(x)$) for β times a constant is given by

$$\hat{\beta} = -\frac{2}{n}\sum_{i=1}^{n} y_i\widehat{\nabla p}(x_i). \tag{12.5}$$

This estimator is implemented in XploRe macros `DWADE1` and `DWADE2` which are included in the library `ADDMOD`. A detailed description of these procedures is given in Chapter 11.

Note that with WADE, the coefficients in the index function are estimated up to a constant and the intercept of the index is not estimated at all. However, this is a common feature of semiparametric estimators. Scale and intercept will be absorbed in the estimate of the link function.

The advantage of the WADE method is that it provides estimates that have the same asymptotic rate of convergence as parametric estimates. However, it can be applied only to "continuous" explanatory variables.

Secondly, recall that in the semiparametric model, the link function $g(\bullet)$ is assumed to be an unknown, smooth function. Hence, it has to be estimated from the data. We estimate the $g(\bullet)$ by nonparametrically regressing the dependent variable Y on the fitted index $x^T \hat{\beta}$. More specifically, we use the Nadaraya–Watson kernel-estimator given by

$$\hat{F}_h(v) = \frac{\sum\limits_{i}^{n} K\left\{(v - X_i^T \hat{\beta})/h\right\} Y_i}{\sum\limits_{i}^{n} K\left\{(v - X_i^T \hat{\beta})/h\right\}} \tag{12.6}$$

where h is the bandwidth satisfying $h = cn^{-1/5}$ (c being some nonnegative constant) and K is a nonnegative and symmetric kernel function.

12.4.4   The Semiparametric Fit

Recall that in the semiparametric model, the intercept is absorbed in the definition of the link function. To estimate the coefficients of the explanatory variables of the restricted model semiparametrically, we use the following XploRe program:

```
proc(b) = main()
  capture("on")
  dat = read(xunemp1)      ; reads the data
  y   = dat[,1]            ; the dependent variable
  x   = dat[,4:5]~dat[,7:8] ; the explanatories--restricted model
  x   = x .- mean(x)'
  (w v) = eigsm(cov(x))
  mah = v*(sqrt(1./w).*v')
  x   = x*mah              ; Mahalanobis transformation
  library(SMOOTHER)
  library(ADDMOD)
  h   = 1.75*matrix(cols(x)) ; bandwidth for WADE
  b   = DWADE1(x y h)      ; WADE estimation
  b   = mah*b              ; retransforming the coefficents
  b   = b./abs(b[1,])      ; normalization
endp
```

	Logit			WADE		
		$h=1$	$h=1.25$	$h=1.5$	$h=1.75$	$h=2$
INTERCEPT	-5630					
EARNINGS	-1.00	-1.00	-1.00	-1.00	-1.00	-1.00
CITYSIZE	-0.72	-0.23	-0.47	-0.66	-0.81	-0.91
DEGREE	-1.79	-1.06	-1.52	-1.93	-2.25	-2.47
URATE	363.03	169.63	245.75	319.48	384.46	483.31

TABLE 12.4. Results of the semiparametric fit: Restricted model.

For technical reasons, a Mahalanobis transformation is applied to the explanatory variables in order to eliminate correlation and to standardize. After estimating the coefficients of the transformed variables by WADE, one can get coefficient estimates for the untransformed variables by post-multiplying the vector of coefficients of the transformed variables by the transformation matrix (this matrix is labeled mah in the program above). The coefficient estimates of the untransformed variables are reported in Table 12.4.

In semiparametric models of the form (12.3), authors usually subsume the intercept of the index $x^T\beta$ into the definition of the link function. Hence, no estimate for the intercept will be obtained from the semiparametric fit. Moreover, we have normalized the scale of the coefficients by dividing all coefficients by the absolute value of the coefficient of *EARNINGS*. The same scale normalization was applied to the Logit estimates of Table 12.3. The normalized Logit estimates are reported for purposes of comparison.

The signs of the estimated coefficients of all explanatory variables are neither varying with the bandwidth chosen in the second step of the semiparametric estimation nor do they differ between the Logit model and the single index model. Note that for $h = 1.5$, the coefficient estimates of the two models are quite close. Hence, it appears that the Logit link is not grossly misspecified. In the following sections, we will take a closer look at this issue.

12.5 Testing the Adequacy of the Logit Link

In this section we introduce several tools available in XploRe that allow us to test the adequacy of the link function of the Logit model. More specifically, we will cover pointwise confidence intervals, uniform confidence bands, and the specification test of the Logit Link developed by Horowitz and Härdle (1994).

12.5.1 Confidence Intervals and Confidence Bands

Confidence boundaries can be used to check the validity of the link function of the Logit model. Several procedures are available which are either based on pointwise confidence bars or on uniform confidence limits. In general, they involve the following steps:

- First, estimate the index $x^T \beta$ either parametrically or semiparametrically.

- Then regress Y on the fitted index $x^T \hat{\beta}$ nonparametrically.

- Construct pointwise confidence intervals or uniform confidence limits around the nonparametric regression estimate and check whether the parametric link function lies inside the confidence region.

The idea behind this approach is simple and intuitive: If the link function (considered as a function of the index $x^T \beta$) of the parametric model is correctly specified, then it should differ from the nonparametric estimate of the link function only due to sampling error. Hence, if the parametric link lies within the confidence region around the nonparametric estimate, then it is very plausible that the parametric model gives a correct description of the data and consequently should not be rejected. Two questions immediately arise in this context:

1. Should the index $x^T \beta$ be estimated parametrically or semiparametrically?

2. What are the relative merits of uniform confidence bands and pointwise confidence intervals?

Regarding the first question, we may say that the motivation for the confidence procedures outlined above tacitly assumes that the index $x^T \beta$ is known. But the argument doesn't break down when $x^T \beta$ is unknown as long as we estimate β consistently. The parametric estimator of β, however, is consistent only if the Logit model is correctly specified. Since the semiparametric model can be viewed as a generalization of the Logit model, estimating β semiparametrically gives a more robust procedure. Below we will present results based on semiparametric estimates of the index $x^T \beta$, only. The results based on parametrically estimating $x^T \beta$ are quite similar.

Turning to the second question, we may say that uniform confidence bands have the advantage of allowing a truly global evaluation of the parametric model in the sense that the bands cover the entire true regression function with the chosen confidence level. Strictly speaking, pointwise confidence intervals merely allow a local evaluate of the parametric curve at a given value of the index $x^T \beta$. Constructing confidence limits for the entire parametric regression curve via a collection of pointwise confidence intervals provides boundaries that are too optimistic, in the sense that the true, unknown regression curve is not covered by these limits with the level of confidence used to construct the individual confidence intervals. Yet, pointwise intervals are usually easier to calculate. Both approaches share the distinct advantage of giving an easy-to-understand, visual evaluation of the parametric model. Moreover, they may hint at the type of deviation from the correct specification if the link function of the parametric model is misspecified, which may lead to an improved reformulation of the parametric model. In the following sections, we describe both approaches in more detail.

12.5.2   Pointwise Confidence Bars

Härdle (1990) develops pointwise confidence intervals for kernel estimators in nonparametric regression. He uses the fact that asymptotically the Nadaraya–Watson estimator (Härdle, 1990, p. 100) follows a normal distribution to derive the confidence limits. This asymptotic distribution can be used in our case provided that the estimator of β has an asymptotic rate of convergence that is faster than the rate of convergence of the estimator used to nonparametrically regress the binary responses Y_i to the estimated index $x^T \hat{\beta}$. This condition is satisfied for both the parametric estimator of β of the Logit model as well as the WADE estimator of the single index model. Because the Nadaraya–Watson estimator has an asymptotic bias, a bias correction should be used. The bias correction can be incorporated explicitly in the calculation of the confidence limits or can be estimated using bootstrap techniques.

The procedure in XploRe that computes pointwise confidence bars while incorporating an explicit bias correction is called PTWISBAR. It employs the bias correction introduced by Bierens (1987) and relies on the result

$$\sqrt{nh}\{\tilde{F}(v) - F(v)\} \xrightarrow{\mathcal{L}} N\left(0, \frac{\sigma^2(v)}{p(v)} C_K\right) \tag{12.7}$$

with

$$\sigma^2(v) = V(Y|X^T\beta = v), \tag{12.8}$$

$p(v)$ is the density of $X^T\beta$ evaluated at v and estimated by

$$\hat{p}(v) = \frac{1}{nh} \sum_{i}^{n} K\{(v - \hat{v}_i)/h\},$$

$$C_K = \int_{-\infty}^{\infty} K(u)^2 du, \tag{12.9}$$

and

$$\tilde{F}(v) = \left\{\hat{F}_h(v) - \left(\frac{h}{s}\right)^2 \hat{F}_s(v)\right\} \Big/ \left\{1 - \left(\frac{h}{s}\right)^2\right\}. \tag{12.10}$$

Moreover, $h = cn^{-1/5}$, $s = cn^{-\delta/5}$ with $c > 0$ and $0 < \delta < 1$. Equation (12.10) just gives the Bierens' correction.

The procedure BOOTPWCB calculates pointwise confidence bars using the classic Nadaraya–Watson estimator in (12.6) estimating the bias by bootstrap.

12.5.3 Uniform Confidence Limits

There are mainly two ways of constructing uniform confidence limits. One is to take the pointwise confidence bars and correct the level with the Bonferroni correction to have simultaneous coverage. This procedure is implemented in XploRe with the name BOFERBAR.

Another method is based on the asymptotic distribution of

$$\sup_{v} |\hat{F}_h(v) - F(v)|.$$

Horowitz (1993b) derives this distribution when $\hat{F}_h(v)$ is computed semiparametrically according to (12.6), but where $\hat{\beta}$ is the parametric estimate. Incorporating Bierens' bias correction leads to the following result:

$$\lim_{n \to \infty} P\left\{ \sqrt{0.4\delta \log n}\left[\sqrt{nh} \sup_{v} \left\{ \frac{\sigma^2(v)}{\hat{p}(v)} \right\}^{-1/2} |\tilde{F}(v) - F(v)| - d_n \right] < z \right\}$$

$$= \exp\{-2\exp(-z)\}, \qquad (12.11)$$

where $1 < \delta < 5/3$

$$d_n = (0.4\delta \log n)^{1/2} + (0.4\delta \log n)^{-1/2} \log\{C_K^*/(2\pi^2)\}^{1/2},$$

$$C_K^* = \frac{2}{C_K} \int_{\infty}^{-\infty} K'(u)^2 du,$$

and the other variables have the same meaning as before. Uniform confidence bands are computed using the result (12.11). This is the procedure implemented in the XploRe macro UNIFBAND.

12.5.4 Estimated Confidence Limits

In this section we show how to construct confidence intervals when the index $x^T\beta$ is estimated semiparametrically (see Table 12.4 for the coefficient estimates). With β estimated semiparametrically, we believe that the confidence limits are more robust in the case of misspecification. However, comparing graphically the confidence limits with the parametric fit is more complicated in this situation because the semiparametric link function is not defined on the same scale as the Logit link. To overcome this problem, we propose the following procedure:

- Estimate the coefficients by WADE obtaining $\hat\beta$. Using the WADE estimates $\hat\beta$, calculate the fitted index $v_i = x_i\hat\beta$, $i = 1, \ldots, n$.

- Estimate the scale s and intercept c of the Logit with variables y_i and v_i. Store the fitted probabilities as $\hat y_i = (1 + \exp(-\hat c + \hat s v_i))^{-1}$.

- Estimate the link with a kernel regression on y_i and v_i obtaining $\tilde y_i$ and calculate the confidence limits.

- Plot $\hat y_i$, $\tilde y_i$ and the confidence limits against v_i.

This algorithm is implemented in the following XploRe procedure:

```
proc(cb1 cb2 cb3 cb4) = main()
  capture("on")
  dat = read(xunemp1)        ; reads the data
  y   = dat[,1]              ; the dependent variable
  x   = dat[,4:5]~dat[,7:8]  ; the explanatories--restricted model
  x   = x .- mean(x)'
  (w v) = eigsm(cov(x))
  mah = v*(sqrt(1./w).*v')
  x   = x*mah                ; Mahalanobis transformation
  library(SMOOTHER)
  library(ADDMOD)
  h   = 1.75*matrix(cols(x)); bandwidth for WADE
  b   = DWADE1(x y h)        ; WADE estimation
  b   = b./abs(b[1,])        ; normalization
  v   = x*b                  ; the fitted index
  x   = matrix(rows(x))~v
  ;  next parametric estimate of scale and intercept
  library(GLM)
  (itres d bvar stat) = GLMBILO(x y) ;parametric estimation
  yhat= itres[,1]
  z   = y~yhat               ; the data and the fitted probability
  h   = 3
  m   = 1
  ;  plot pointwise bars with Bierens' correction
  cb1 = PTWISBAR(v z h 95 20 m)
  ;  plot pointwise bars with bootstrap bias corr.
  cb2 = BOOTPWCB(v z h)
  ;  plot Bonferroni's uniform pointwise bars
  cb3 = BOFERBAR(v z h 90 20 m)
  ;  plot uniform confidence band with Bierens' corr.
  cb4 = UNIFBAND(v z h)
endp
```

Note that PTWISBAR, BOOTPWCB, BOFERBAR, and UNIFBAND are stored in the library ADDMOD and require as inputs the $n \times 1$ (n the number of observations) vector v containing the fitted index, the $n \times 2$ matrix z with the first column containing the values of the dependent variable and the second column containing the fitted probability of the parametric model, and the bandwidth h for the kernel regression of Y on $x^T\hat{\beta}$. If, as in our application, PTWISBAR and BOFERBAR are applied to data where the dependent variable Y can take on only two values, then the parameter m must be given. For Y Bernoulli, it should be set equal to "1" while for Y Binomial it should be the vector with the binomial coefficients. For PTWISBAR and BOFERBAR, the confidence levels have been set to 95% and 90%, respectively. BOOTPWCB and UNIFBAND use the default values 95% and 90%. The value "20" in PTWISBAR and BOFERBAR specifies the number of points at which pointwise confidence bars are calculated. "20" is also the default for BOOTPWCB.

The location of the bars is determined in a way such that between each bar point there are the same number of data points, following a suggestion of Härdle (1990). This way, the variance of the kernel regression function will not be very different at each bar point and consequently the bars will not be very different in length.

Since PTWISBAR, BOOTPWCB, and BOFERBAR all give roughly the same confidence bars, we included the picture for BOOTPWCB only (see Figure 12.2). The uniform confidence bands are shown in Figure 12.3. In all plots, the parametric fit lies inside the confidence limits, suggesting that the Logit link is correctly specified.

The confidence limits introduced and applied in the previous section equivocally speak in favor of the Logit link. Yet, these procedures are rather informal ways to test the adequacy of $F(u) = 1/[1 + exp(-u)]$. As mentioned earlier, Horowitz and Härdle (1994) have introduced a formal misspecification test for the problem at hand. This test, as well as a modified version of it, will be described and applied in the following sections.

12.5.5   The HH-Test

The HH-test can be used to test the specification of a parametric model like the generalized linear model in (12.2) against the single index model (12.3).

The main idea that inspires the HH-test approach relies on the fact that if the model under the null hypothesis is true, then a nonparametric estimate of $E(Y|x^T\hat{\beta} = v)$ will deviate from the parametric estimate $F(v)$ only due to sampling error. Thus, the specification of the parametric model can be tested by comparing both estimates.

The HH-test statistic is defined as

$$T_h = \sqrt{h} \sum_{i=1}^{n} u(X_i^T\hat{\beta})\{Y_i - F(X_i^T\hat{\beta})\}\{\tilde{F}_i(X_i^T\hat{\beta}) - F(X_i^T\hat{\beta})\}, \quad (12.12)$$

where h is the bandwidth used in the kernel regression, $u(\bullet)$ is a weight function that downweights extreme observations, and $\tilde{F}_i(\bullet)$ is a certain kernel regression of the data. The authors propose to estimate $\hat{\beta}$ under the

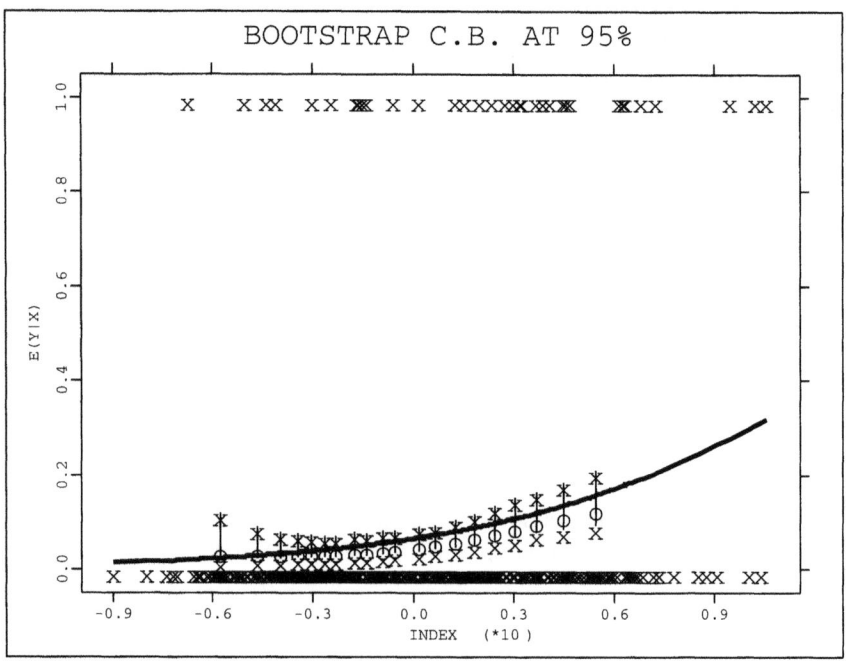

FIGURE 12.2. The Logit fit (thick line) and pointwise confidence bars with bootstrap correction. The index was estimated by WADE.

null. In our example, this corresponds to estimating β by the maximum likelihood estimator of the Logit model. For technical reasons, the estimate $\tilde{F}_i(\bullet)$ has to be asymptotically unbiased and independent of the observations Y_i (for details, see Horowitz and Härdle, 1994). The first property is ensured by using Bierens' correction given by (12.10). Independence is obtained by using the leave-one-out (LOO) kernel regression estimator that eliminates observation i from the calculation of the estimate at v_i. Thus, $\tilde{F}_i$ is the result of combining two regression estimators $\hat{F}_{ih}$ and $\hat{F}_{is}$ according to the Bierens' correction given in Equation (12.10) with

$$\hat{F}_{ti}(v_i) = \frac{\sum\limits_{\substack{j=1 \\ j \neq i}}^{n} Y_j K\{(v - X_j^T \hat{\beta})/t\}}{\sum\limits_{\substack{j=1 \\ j \neq i}}^{n} K\{(v - X_j^T \hat{\beta})/t\}} \qquad t = h, s. \qquad (12.13)$$

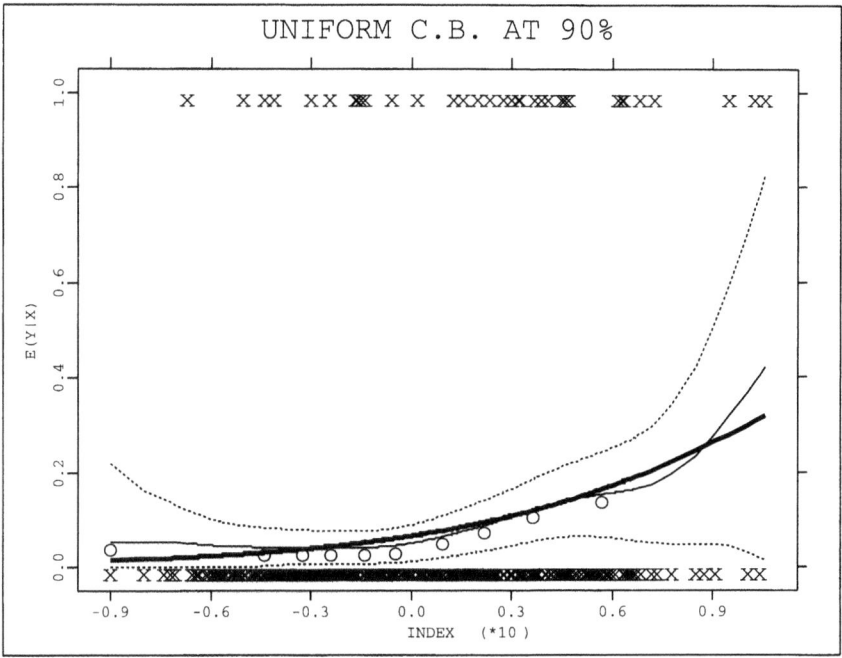

FIGURE 12.3. The Logit fit (thick line), the semiparametric estimate (line with circles), and the uniform confidence bands (broken line). The index was estimated by WADE.

Horowitz and Härdle (1994) have shown that under H_0 and under some suitable regularity conditions, T_n is asymptotically distributed as a $N(0, \sigma_T^2)$ with

$$\sigma_T^2 = 2C_K \int_{-\infty}^{\infty} u(v)^2 \{\sigma^2(v)\}^2 dv \qquad (12.14)$$

with $\sigma^2(v)$ and C_K defined, respectively, in (12.8) and (12.9). A consistent estimator for $\sigma_T^2(v)$ for binary responses is given by

$$\hat{\sigma}_T^2 = \frac{2C_K}{n} \sum_{i=1}^{n} \frac{u(X_i^T \hat{\beta})^2 [\hat{F}_h(X_i^T \hat{\beta})\{1 - \hat{F}_h(X_i^T \hat{\beta})\}]^2}{\hat{p}_h(X_i^T \hat{\beta})}. \qquad (12.15)$$

In practice, $u(v)$ is defined to be identically equal to one for 90% or 95% of the central values of $X_i^T \hat{\beta}$ and zero otherwise.

12.5.6   The Modified HH-Test

Proença and Ritter (1994) have shown with simulations that the HH-test has a negative bias in finite samples that adversely affects the power of the test. The negative bias arises from the fact that the index $v_i = x^T \beta$ is unknown and has to be estimated in order to calculate the test statistic in practice. Moreover, they have detected a dependence of the test statistic on the bandwidth under the null that can induce false rejections for inappropriate bandwidths. They propose a modification of the HH-test that incorporates a bias correction and is robust to the bandwidth under the null hypothesis. The modified statistic has the same limit law as the HH-test and consequently is asymptotically equivalent to the later.

The modified HH-statistic is given by

$$MT_h = \sqrt{h} \sum_{i=1}^{n} u(X_i^T \hat{\beta}) r(X_i^T \hat{\beta}) \hat{r}_h(X_i^T \hat{\beta}), \qquad (12.16)$$

with $r(X_i^T \hat{\beta})$ the residuals of the parametric fit equal to $Y_i - F(X_i^T \hat{\beta})$ and $\hat{r}_h(X_i^T \hat{\beta})$ the residuals smoothed by the LOO Nadaraya–Watson estimator

$$\hat{r}_h(v) = \frac{\sum_{\substack{j=1 \\ j \neq i}}^{n} r(X_j^T \hat{\beta}) K\{(v - X_j^T \hat{\beta}/h\}}{\sum_{\substack{j=1 \\ j \neq i}}^{n} K\{(v - X_j^T \hat{\beta})/h\}}.$$

The crucial difference to the original HH-test is that the modified HH-test statistic smoothes both the data Y_i and the parametric model $F(X_i^T \hat{\beta})$ (by smoothing the residuals), while the HH-test only smoothes the data.

The authors propose an estimator for the first and second conditional moments of MT_h that incorporates a correction for the use of $\hat{\beta}$ instead of the true coefficient β. The expectation conditional on X_i can be estimated by

$$\widehat{E}(MT_h) = -tr(H \, V \, W^*), \qquad (12.17)$$

where $H = VX(X'VX)^{-1}X'$, X is the design matrix with rows X_i, $i = 1, \ldots, n$, V is a diagonal matrix with elements $\sigma^2(X_i^T \hat{\beta})$ on column i and row i, $i = 1, \ldots, n$, and $W^* = (1/2)(UW + W^T U)$ with W the smoothing matrix with elements

$$w_{ij} = \begin{cases} \dfrac{K[(X_j^T \hat{\beta} - X_i^T \hat{\beta})/h]}{\sum_{i \neq j} K[(X_j^T \hat{\beta} - X_i^T \hat{\beta})/h]} & \text{if } i \neq j \\ 0 & \text{if } i = j \end{cases} \qquad (12.18)$$

and U a diagonal matrix with elements $u_i = u(X_i^T \hat{\beta})$. The matrix H adjusts for the distortion resulting from estimating the coefficients β. In general, (12.17) is nonzero and Proença and Ritter (1994) conjecture that it is negative in most cases, supporting the empirical studies about the negative bias.

The variance of MT_h conditional on X_i can be estimated by

$$\hat{\sigma}_{MT}^2 = 2tr(DVDV) + \sum_{i=1}^{n} d_{ii}^2[\sigma^2(X_i^T\hat{\beta}) - 6\sigma^4(X_i^T\hat{\beta})], \qquad (12.19)$$

with $D = (I - H)^T W^*(I - H)$.

The expressions (12.17) and (12.19) can be used to adjust and studentize the statistic by

$$\frac{MT_h - \widehat{E}[MT_h]}{\hat{\sigma}_{MT}}, \qquad (12.20)$$

and the critical values of the test are given by the adequate percentiles of the standard normal distribution.

The modified HH-test provides more reliable results than the HH-test on small and moderate samples because it is bias-corrected and robust to the choice of the bandwidth under the null. However, the estimation of the bias correction and the finite sample variance requires operations with matrices of size $n \times n$ (n the size of the sample) and can be computationally burdensome for moderate to big sample sizes.

12.5.7 Results for the HH-Test and Modified HH-Test

To run the HH-test and the Modified HH-test the following program was implemented in XploRe:

```
proc(b t p) = main()
  capture("on")
  dat = read(xunemp1)      ; reads the data
  y   = dat[,1]            ; the dependent variable
  x   = dat[,4:5]~dat[,7:8] ; the explanatories--restricted model
  x   = matrix(rows(x))~x  ; add intercept
  library(GLM)
  (itres b bvar stat) = GLMBILO(x y)  ; parametric estimation
  yhat= itres[,1]          ; the logit fit
  v   = x*b                ; the fitted index
  h   = 1                  ; bandwidth for kernel regression
  library(SMOOTHER)
  library(ADDMOD)
  (t p) = HHTEST(v y yhat h 0.05 1)  ; HH- test
  (t p) = MODHHTST(x v y yhat h)     ; modified HH-test
  t~p
endp
```

The HH-test and Modified HH-test are stored in library ADDMOD and are named HHTEST and MODHHTST, respectively. The HHTEST needs as inputs the $n \times 1$ (n the number of observations) vector v with the parametrically

h	Statistic	p-value
0.50	-1.0741	0.141
0.75	-1.0210	0.154
1.00	-0.9276	0.177
1.25	-0.8764	0.190
1.50	-0.8449	0.199
1.75	-0.7862	0.216
Modified HH-test		
1.00	-0.4716	0.357

TABLE 12.5. Results for the HH-test and modified HH-test.

estimated index, $\hat{v}_i = x^T \hat{\beta}$, the response in the $n \times 1$ vector y, the $n \times 1$ vector yhat with the logit fit, that is, the fitted $P(Y = 1|X)$, the bandwidth h, and the proportion of points to cut-off on each side of the sample (here, set to 0.05). If the dependent variable is a Bernoulli variable (as in our case), then the last parameter has to be set equal to "1". For Y binomial, it should equal the vector of the binomial coefficients. The HHTEST returns as output the studentized HH-test statistic, the respective p-value, and two plots. The first plot shows the parametric fit along with the leave-one-out estimate used in the HH-test. The second plot puts the logit fit into one picture with the classic Nadaraya–Watson estimate.

The procedure MODHHTST needs as inputs the $n \times k$ design matrix X, v, y, yhat, and h, which have the same meaning as before. By default, the data are assumed to be Bernoulli. For binomial data, the vector with the binomial coefficients has to be given as an additional input. By default, the procedure eliminates a proportion of 0.05 on each "end" of the sample. This parameter may be changed by the user. The procedure gives as output the studentized modified HH-statistic, the respective p-value, a plot of the residuals smoothed by the leave-one-out estimator used in the test, and a plot of the residuals smoothed by the Nadaraya–Watson estimator. Both plots also include the unsmoothed residuals of the Logit fit.

The HH-test was applied to the restricted model for a grid of bandwidths ranging from 0.5 to 1.75. The value of the statistic for each bandwidth along with the respective p-values is shown in Table 12.5. For all bandwidths, the test does not reject the Logit link. Figure 12.4 shows the Logit fit along with the leave-one-out (LOO) estimate used in the HH-test for a bandwidth of $h = 1$. The corresponding plot with the classic Nadaraya–Watson estimate in place of the LOO estimate looks very similar and is not shown here. The parametric and the semiparametric fit are very close, suggesting a correct specification of the logit link.

We also computed the modified HH-test for the restricted model. Because this test is computationally expensive, it was calculated only for one

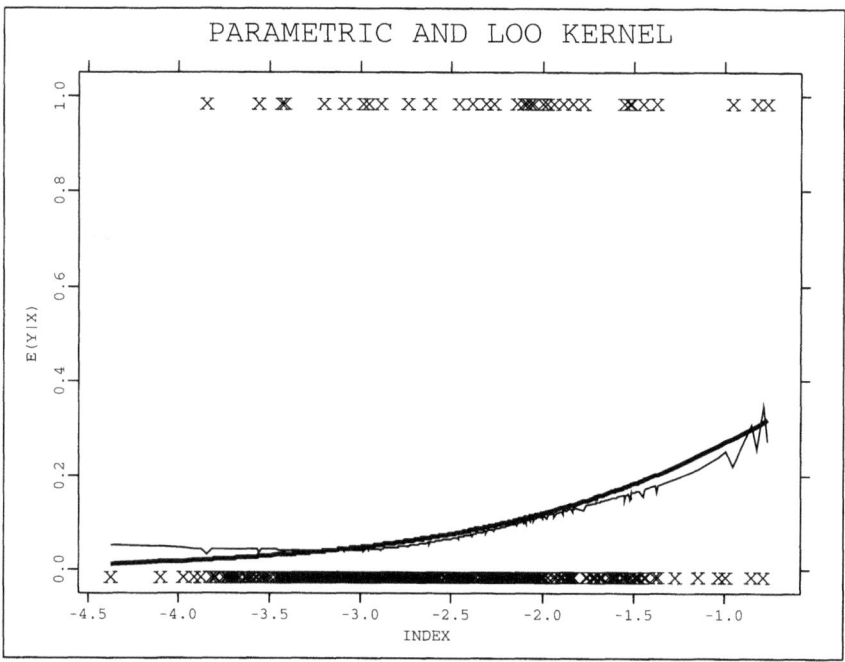

FIGURE 12.4. The Logit fit for the restricted model (thick line) and the
LOO semiparametric fit used in the HH-test (line). The data points are
identified by crosses.

bandwidth. However, the test is bandwidth robust under the null and consequently it is likely that for other bandwidth values, we would obtain very
similar results.

Figure 12.5 shows the residuals of the parametric fit together with their
smoothed counterparts to which the leave-one-out estimator was applied.
The bandwidth was set to $h = 1$. The smoothed residuals are practically
equal to zero, indicating that the Logit link is well specified. The corresponding picture with the residuals smoothed by the Nadaraya–Watson
estimator is very similar and is not included for space reasons. The value of
the test statistic indicates that the Logit link should not be rejected. Note
that the modified statistic has a greater value than the HH-test statistic,
reflecting the bias correction.

12.6   Summary and Conclusions

In this chapter we have illustrated how XploRe can be used to estimate
parametric and semiparametric binary response models using data from the
German Socio Economic Panel. Moreover, we illustrated how the adequacy

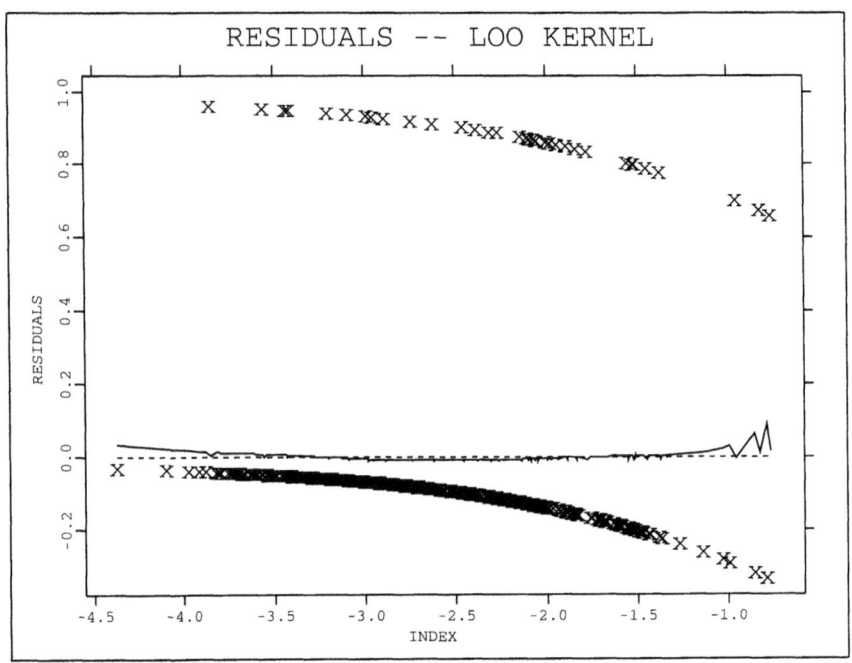

FIGURE 12.5. The residuals of the Logit fit for the restricted model (crosses) and the residuals smoothed by LOO (line).

of the link function of the parametric model can be evaluated using recently developed statistical techniques already implemented in XploRe. In the application at hand, it turned out that the specification of the parametric model could not be rejected.

REFERENCES

Bierens, H.J. (1987). Kernel estimators of regression functions, *in* T.F. Bewley (ed.), *Advances in Econometrics: Fifth World Congress*, Vol. 1, Cambridge University Press, New York, pp. 99–144.

Franz, W. and Soskice, D. (1994). The German apprenticeship system, *Discussionpaper 11*, Center for International Labor Economics, Konstanz, Germany.

Härdle, W. (1990). *Applied Nonparametric Regression*, Econometric Society Monographs No. 19, Cambridge University Press, New York.

Horowitz, J.L. (1993a). Semiparametric and nonparametric estimation of quantal response models, *in* G.S. Maddala, C.R. Rao and H.D. Vinod (eds), *Economics*, Vol. 11 of *Handbook of Statistics*, North-Holland,

Amsterdam, pp. 45–72.

Horowitz, J.L. (1993b). Semiparametric estimation of a work-trip mode choice model, *Journal of Econometrics* **58**(1–2): 49–70.

Horowitz, J.L. and Härdle, W. (1994). Testing a parametric model against a semiparametric alternative, *Econometric Theory*. To appear.

McCullagh, P. and Nelder, J.A. (1989). *Generalized Linear Models*, Vol. 37 of *Monographs on Statistics and Applied Probability*, 2nd edn, Chapman and Hall, London.

Powell, J.L., Stock, J.H. and Stoker, T.M. (1989). Semiparametric estimation of index coefficients, *Econometrica* **57**(6): 1403–1430.

Proença, I. and Ritter, C. (1994). Semiparametric testing of the link function in models for binary outcomes, *SFB 373 Discussion Paper 940017*, Institut für Statistik und Ökonometrie, Humboldt-Universität zu Berlin, Berlin, Germany.

Projektgruppe "Das Sozio-ökonomische Panel" (1991). *Das Sozio-ökonomische Panel (SOEP) im Jahre 1990/91*, Deutsches Institut für Wirtschaftsforschung. Vierteljahreshefte zur Wirtschaftsforschung, pp. 146–155.

13
Approximative Methods for Regression Models with Errors in the Covariates

Raymond J. Carroll[1] and Helmut Küchenhoff[2]

13.1 Introduction

We examine general regression models where some of the covariates are measured with error. The response is denoted by Y and we distinguish between two types of covariates. X are those covariates, that cannot be observed exactly. Instead of X, a surrogate variable W is observed. Z are further covariates measured without error. The example considered in this article is an occupational study on the relationship between dust concentration and chronic bronchitis. In the study, $N = 499$ workers of a cement plant in Heidelberg were observed from 1960 to 1977 (for details, see DFG-Forschungsbericht, 1981). The response Y is the appearance of chronic bronchitis, and the correctly measured covariates Z are smoking and duration of exposure. The effect of the dust concentration in the individual working area X is of primary interest in the study. This concentration was measured several times in a certain time period and averaged, leading to the surrogate W for the concentration. There are two problems, which are typical for this kind of measurement. The first problem concerns the correctness of the single values, that is., the measurement error due to the instrument. The second problem here is the question, "what is the true covariate X?" Here, we have to use an operational definition like the "long-term average of the dust concentration." Obviously, there are many additional sources of measurement error for this quantity: variation of the dust concentration at a special working place over time, change of the working place by an individual, etc.

Failing to take this measurement error (ME) into account leads in most cases to biased estimates of the effect of the true covariate. Typically the effect is underestimated, which is known as attenuation. There is a rich body of literature on the effect of ME and its modeling in the context of

---

[1]Department of Statistics, Texas A&M University, College Station, TX 77843, USA.

[2]Seminar für Ökonometrie und Statistik, Ludwig-Maximilian Universität, München, Germany.

regression models (see monographs by Fuller, 1987; Schneeweiß and Mittag, 1986; Carroll, Ruppert and Stefanski, 1995). In this article we describe two simple generally applicable approximative methods for handling measurement error in estimating general regression models. The representation here is essentially based on Carroll, Ruppert and Stefanski (1995, Chapters 5–6).

The following mean and variance model is considered:

$$
\begin{aligned}
E(Y|Z,X) &= f\{Z,X,\beta\}, & (13.1) \\
\text{Var}(Y|Z,X) &= \sigma^2 g^2\{Z,X,\beta,\theta\}. & (13.2)
\end{aligned}
$$

We use the structural case of the model so that (Y_i, X_i, W_i, Z_i) for $i = 1, \ldots, N$ are assumed to be independently and identically distributed. The (quasi-)likelihood estimator ignoring the ME, that is, regressing X on Z, W, is called the naive estimator $\hat{\beta}_{na}$.

Let us come back to the example mentioned above. Ignoring the ME we conducted a logistic regression with the response chronic bronchitis and the regressors log(1+dust concentration), duration (in years), and smoking. The calculations were conducted by XploRe with the following commands:

```
dat = READ(heid)
y   = dat[,1]
w   = dat[,2]
z   = dat[,3]
LIBRARY(GLM)
DOGLM(w~z y)
LIBRARY(GAM)
DOGAM(w~z y)
```

In interactive modeling, the binomial distribution and the logistic link have to be chosen for the GLM, and the Bernoulli distribution and the logistic link for the GAM. The data matrix were in the file `Heid.dat` with the columns Y, W, Z(smoking and duration). The output table from XploRe for the logistic model is given in Figure 13.1. For the GAM, the estimated regression curve for the variable W is presented in Figure 13.2.

13.2   Regression Calibration

The idea of this method is to replace the unobserved X by its expected value $E(X|W,Z)$ and then perform a standard analysis. This is the *regression calibration* algorithm, suggested as a general approach by Carroll and Stefanski (1990) and Gleser (1990); Rosner, Willett and Spiegelmann (1989) developed the idea for logistic regression into a workable and popular methodology.

$E(X|W,Z)$ can be modeled by a parametric regression function which

```
BETA, S.E.[,1]BETA, S.E.[,2]BETA, S.E.[,3]                          BVAR[,1]        .

-2.536245873        0.4072361222        6.227949180                 0.1658412592
 2.543098411        0.7599982028        3.346190033                -0.100620140
 0.463588640        0.2433014923        1.905408124                -0.050433781
 0.027980142        0.0078925971        3.545112113                -0.002575187
```

```
STAT[,1]        STAT$[,1]

495.0000000        DEGREES OF FREEDOM
495.0000000        N-P
603.9767606        DEVIANCE
496.0752230        PEARSON'S CHI^2
  1.2201551        SCALE
```

FIGURE 13.1. XploRe output logistic model for the Heidelberg data.

is denoted by $m(W, Z, \gamma_{cm})$. The variance function is $V(Z, W, \gamma_{cm})$. Here, γ_{cm} is the calibration model parameter vector describing the relationship between the latent variable X and the observed regressors W and Z. To get to know or to estimate this parameter, respectively, further information is necessary. It can be obtained by a regression using a validation sample, that is, a sample where X and W, Z are known. Other possibilities are to use replicate observations of W on X or to make explicit model assumptions about the ME. The regression calibration algorithm is as follows:

- Using replication, validation, or model assumptions, estimate the mean function $m(X|W, Z, \gamma_{cm})$.

- Replace (or impute) the unobserved X by its estimate $m(X|Z, W, \gamma_{cm})$, and then run a standard analysis to obtain parameter estimates.

- Adjust the resulting standard errors to account for the estimation of γ_{cm}.

There are many methods depending on the given data structure for estimating the mean of the X given (W, Z) (see Carroll, Ruppert and Stefanski, 1995). Here, we restrict ourselves on the classical additive measurement error model.

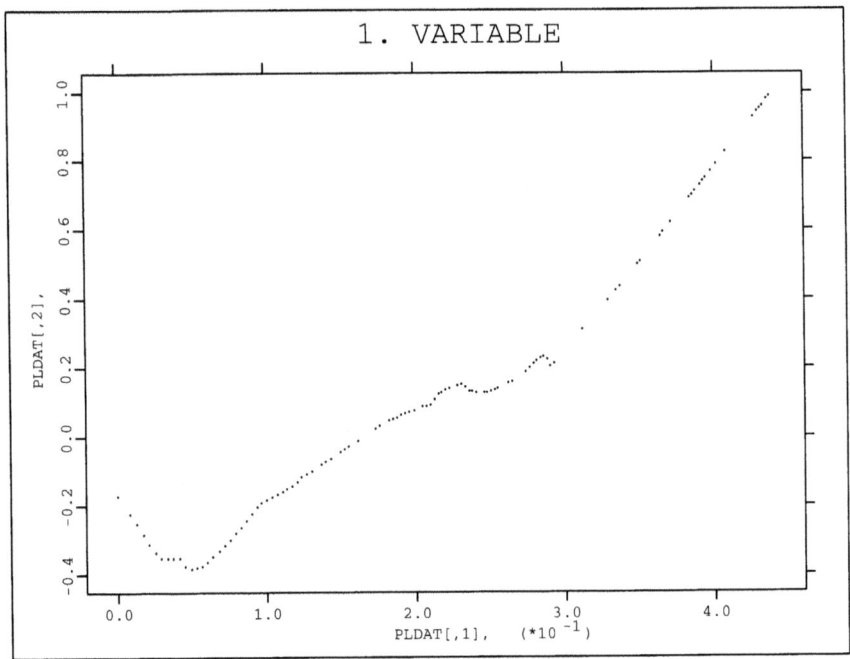

FIGURE 13.2. XploRe output GAM model for the Heidelberg data.

13.2.1   The Additive Measurement Error Model

The model is defined by

$$W \;=\; X + U, \tag{13.3}$$

$$U|(X, Z) \;\sim\; N(0, \Sigma_{uu}). \tag{13.4}$$

If X, U, and Z are multivariate normal, then

$$E(X|W, Z) = \mu_w + (\Sigma_{xx}, \Sigma_{xz}) \begin{bmatrix} \Sigma_{ww} & \Sigma_{wz} \\ \Sigma_{wz}^t & \Sigma_{zz} \end{bmatrix}^{-1} \begin{pmatrix} W - \mu_w \\ Z - \mu_z \end{pmatrix}. \tag{13.5}$$

If we do not assume normality, (13.5) is still the best linear approximation of the conditional expectation of X given W and Z.

In (13.5), $\Sigma_{ww}, \Sigma_{zz}, \mu_w, \mu_z, \Sigma_{wz}$ can be estimated from the observations (W_i, Z_i). From (13.3) and (13.4) we get $\Sigma_{xz} = \Sigma_{wz}$. To find $\Sigma_{xx} = \Sigma_{zz} - \Sigma_{uu}$, additional information is necessary. Σ_{uu} can be assumed to be known. It can also be estimated if there are k_i replicates W_{ij} for the X_i. Then we get, by the well-known components of variance formula:

$$\hat{\Sigma}_{uu} = \frac{\sum_{i=1}^{n} \sum_{j=1}^{k_i} \left(W_{ij} - \overline{W}_{i\cdot} \right) \left(W_{ij} - \overline{W}_{i\cdot} \right)^t}{\sum_{i=1}^{n} (k_i - 1)}. \tag{13.6}$$

The regression calibration model is an approximate working model for the observed data. The exact mean function model for the (Y_i, W_i, Z_i) is

$$E(Y|W, Z) = \int f\{X, Z, \beta\} h_{X|W,Z} dx, \qquad (13.7)$$

where $h_{X|W,Z}$ is the density function of X given W and Z.

In linear regression, the regression calibration model is just the exact model for the mean function. In models with exponential link under the assumption of normality of X, the regression calibration gives a consistent estimator for the slope parameter β_x.

In a logistic regression model, the integral in (13.7) cannot be solved theoretically and it is difficult to compute numerically. The logistic model, however, can be well approximated by the probit model. For $c = 1.7017449$, it is well known that the logistic distribution function $H(t) \approx \Phi(\frac{t}{c})$. Using this approximation, the integral (13.7) can be solved and we get

$$\Pr(Y = 1|Z, W) \approx H\left[\frac{\beta_0 + \beta_x^t m(Z, W, \gamma_{cm}) + \beta_z^t Z}{\{1 + \beta_x^t V(Z, W, \gamma_{cm}) \beta_x / c^2\}^{1/2}}\right]. \qquad (13.8)$$

It turns out that the denominator is, in most practical cases, nearly 1, so regression calibration is a good approximation.

13.2.2   Data Analysis

In the occupational study considered here, there were no replication data or a validation study available. Thus it was assumed that the ME variance is $\sigma_u^2 = 0.25 * \sigma_w^2$. Since the correlation between W and Z is very low, it seems to be reasonable to analyze the data with $E(X|W, Z) = E(X|W)$. Another problem is the occurrence of zero concentrations. Those values are accepted as correctly measured. All other values of W were replaced by $E(X|W)$ using (13.5). This was done using the following XploRe commands:

```
sw2 = VAR(w)
m   = MEAN(w)
su2 = sw2/4
ex  = (w.>0).* (m* (su2/sw2) + w * ((sw2-su2)/sw2))
DOGLM(ex~z y)
```

After that, the same procedure as for the naive model was applied, replacing w by ex. For the slope parameter of the dust concentration β_x, the result was 3.01 (s.e.0.93) compared to the naive estimation 2.54 (s.e.0.76). For the GAM, the result is given in Figure 13.3. Here, the shape of the curve is similar to that obtained by the naive model, although it shows a stronger relationship between the dust concentration and chronic bronchitis.

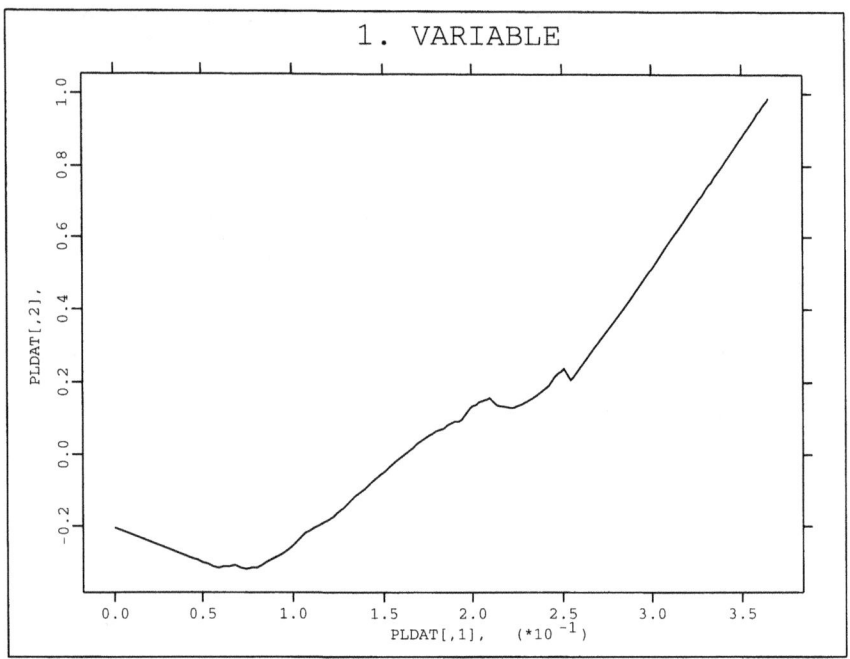

FIGURE 13.3. Regression calibration GAM.

13.3   Simulation and Extrapolation (SIMEX)

We now describe another approximate method for models with additive ME. Simulation extrapolation (SIMEX) is a simulation-based method of estimating and reducing bias due to ME. SIMEX estimates are obtained by adding additional ME to the data in a resampling-like stage, establishing a trend of measurement error-induced bias versus the variance of the added ME, and extrapolating this trend back to the case of no measurement error. The technique was proposed by Cook and Stefanski (1995) and further developed by Carroll, Küchenhoff, Lombard and Stefanski (1995).

The fact that measurement error in a predictor variable induces bias in regression estimates is counter-intuitive to many, even those with training in statistics. An integral component of SIMEX is a self-contained simulation study resulting in graphical displays that illustrate the effect of measurement error on parameter estimates and the need for bias correction. The graphical displays are particularly useful when it is necessary to motivate or explain a measurement error model analysis to nonstatisticians.

13.3.1   *Description of the SIMEX Estimation*

We focus on the case in which X and W are scalars and the measurement error variance, σ_u^2, is known. Furthermore, we assume that the additive measurement error model (13.3) and (13.4) holds:

$$W = X + U, U \sim N(0, \sigma_u^2). \tag{13.9}$$

Typically, the assumption of normality is not critical in practice.

SIMEX is applicable to a broad class of estimation methods, for example, least-squares, maximum likelihood, quasi-likelihood, etc. We denote the estimators as functions of the i.i.d. sample (Y_i, Z_i, W_i)

$$\widehat{\beta}_{\text{naive}} = \mathcal{T}\{(Y_i, Z_i, W_i)_{i=1}^n\}, \qquad \widehat{\beta}_{\text{true}} = \mathcal{T}\{(Y_i, Z_i, X_i)_{i=1}^n\},$$

where $\mathcal{T}$ is the function applied to the data to obtain the estimates. $\widehat{\beta}_{\text{true}}$ is not a real estimator, since the X_i are not observable. In some cases, $\mathcal{T}$ can be determined explicitly, for example, for least squares estimation, although in general it is only implicitly defined by certain estimating equations.

For $\lambda \geq 0$, define

$$W_{b,i}(\lambda) = W_i + \sqrt{\lambda}\, U_{b,i}, \qquad i = 1, \ldots, n, \quad b = 1, \ldots, B,$$

where the generated *pseudo errors*, $\{U_{b,i}\}_{i=1}^n$, are mutually independent, independent of all the observed data and identically distributed, normal random variables with variance σ_u^2. The whole ME variance of the $W_{b,i}$ is given by

$$V(W_{b,i}|X_i) = (1 + \lambda)\sigma_u^2. \tag{13.10}$$

Define

$$\widehat{\beta}_b(\lambda) = \mathcal{T}\left[\{Y_i, Z_i, W_{b,i}(\lambda)\}_{i=1}^n\right], \qquad b = 1, \ldots, B,$$

and

$$\widehat{\beta}(\lambda) = \mathrm{E}\left\{\widehat{\beta}_b(\lambda) \mid (Y_i, Z_i, W_i)_{i=1}^n\right\}. \tag{13.11}$$

The expectation in (13.11) is with respect to the distribution of $(U_{b,i})_{i=1}^n$ only. It can be obtained as the limit as $B \to \infty$ of the average $\{\widehat{\beta}_1(\lambda) + \cdots + \widehat{\beta}_B(\lambda)\}/B$.

Note that $\widehat{\beta}(0) = \widehat{\beta}_b(0) = \widehat{\beta}_{\text{naive}}$. Often it is difficult or even impossible to determine $\widehat{\beta}(\lambda)$ for $\lambda > 0$ analytically; however, it can always be estimated to any desired precision by generating a large number of independent measurement error vectors, $\{(U_{b,i})_{i=1}^n\}_{b=1}^B$, computing $\widehat{\beta}_b(\lambda)$ for $b = 1, \ldots, B$, and approximating $\widehat{\beta}(\lambda)$ by the sample mean of $\{\widehat{\beta}_b(\lambda)\}_{b=1}^B$. This is the simulation component of SIMEX.

The extrapolation step of the proposal entails modeling the components of $\widehat{\beta}(\lambda)$ as functions of λ for $\lambda \geq 0$, and extrapolating the fitted models back to $\lambda = -1$. Cook and Stefanski (1995) propose linear, quadratic, and the following nonlinear extrapolant function:

$$G_{NL}(\lambda, \Gamma) = \gamma_1 + \frac{\gamma_2}{\lambda + \gamma_3}. \qquad (13.12)$$

The vector of extrapolated values yields the simulation-extrapolation estimator denoted $\widehat{\beta}_{\mathrm{SIMEX}}$. The form of the true extrapolation function and the ability to model this function adequately are key factors for the success of SIMEX estimation. In the linear model and some other models, the nonlinear extrapolation function is correct and gives consistent estimates. Cook and Stefanski (1995) show that when the measurement error variance is small, the extrapolation function is approximately quadratic in general. Experience suggests that the quadratic extrapolant or the nonlinear extrapolant defined in (13.12) work well in many situations regardless of the magnitude of the measurement error variance. However, there are exceptions and it is sometimes necessary to study the suitability of a particular extrapolant function, either analytically or by simulation, to ensure its validity in a particular situation.

The estimation of the standard error of the SIMEX-estimator can be done by the delta method (see Carroll, Küchenhoff, Lombard and Stefanski, 1995) or by the following procedure given by Stefanski and Cook (1994), closely related to the jackknife. We denote the sample variances of the SIMEX samples $\beta_b(\lambda), b = 1, \ldots, B$ by $\tau(\lambda)$ and $\tau(0) = 0$. Then by a parametric extrapolation $\hat{\tau}$ (we use a quadratic one), $\mathrm{Var}(\widehat{\beta}_{\mathrm{SIMEX}})$ is estimated by $-\hat{\tau}(-1)$. For a detailed explanation of this procedure, see Carroll, Ruppert and Stefanski (1995).

13.3.2 Application to the Example

As before, we assume that the ME is normal with variance $\sigma_u^2 = 0.25 * \sigma_z^2$. For the grid, the values 0, 0.5, 1, 1.5, and 2 were chosen for λ. The result of the SIMEX is given in Figure 13.4. There, linear, quadratic, and nonlinear extrapolation functions are presented, where the quadratic and the nonlinear curves are nearly identical. The results for $\widehat{\beta}_{\mathrm{SIMEX}}$ were 2.99 (linear), 3.11 (quadratic), and 3.104 (nonlinear). The estimated standard error was calculated with a quadratic extrapolation, yielding 0.368 for all three extrapolants. We give the XploRe procedures for the SIMEX estimation. The first procedure is the simulation step. The input consists of the data vectors Y, W, and Z, the ME variance su2, the vector of λ, and the number B of replications in each simulation. For the parameter estimation function $\mathcal{T}$, the function LOGI is used. The output of LOGI is the slope parameter of the first component of X of a logistic regression of Y on X. It can be replaced by any other estimation procedure (for example, GLMBILO from the

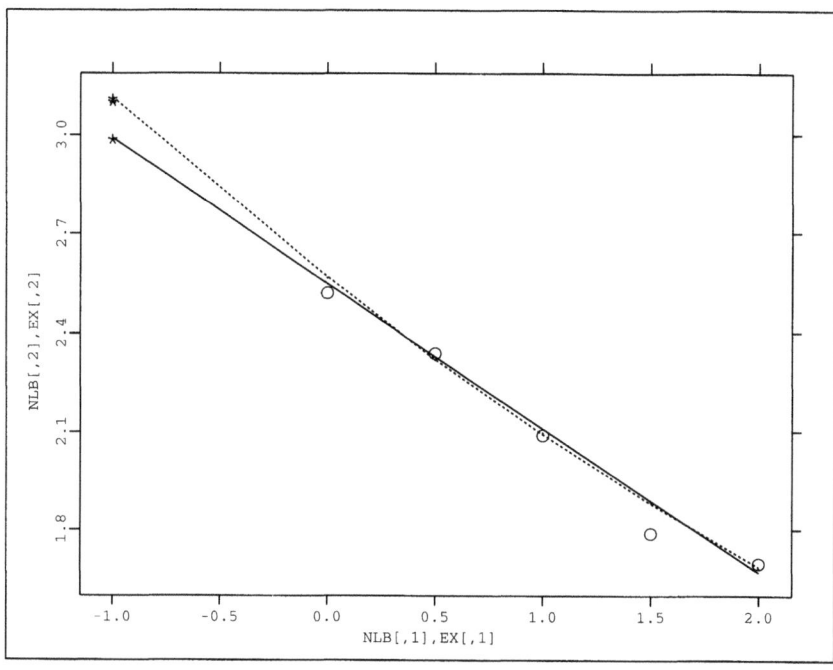

FIGURE 13.4. SIMEX estimation for the Heidelberg data.

GLM library). The output contains the vector of the mean and the variances of the SIMEX samples.

```
PROC (bet1 var1) =  SIMUL(y w z su2 lam B)
   bet1 = LOGI(w~z y)
   var1 = 0
   i=1
   DO
     EXIT(i.>(ROWS(lam)))
     J=1
     bets=ASEQ(0 B 0)
     DO
       EXIT(j.> B)
       wn=w +SQRT(lam*su2)*NORMAL(ROWS(w) 1)
       bets[j,]=LOGI(w~z y)
       j=j+1
     ENDO
     bet1=bet1|MEAN(bets)
     var1=var1|VAR(bets)
   ENDO
ENDP
```

The extrapolation step (linear and quadratic) can be done by the procedure EXTRA. The input is the result of the SIMUL procedure and the grid lam. The output consists of the SIMEX estimates of the linear and the quadratic extrapolation. The nonlinear extrapolant has to be calculated by nonlinear least squares. The XploRe procedure is available from the author. The procedure can also be used for the extrapolation estimation of the variance of the SIMEX estimator.

```
PROC (ergl ergq) =EXTRA(betl lam)
  lxd   = (1x.=1x)~1x
  qxd   = lxd~(1x^2)
  gaml  = INV(lxd'*lxd)*lxd'*betl
  gamq  = INV(qxd'*qxd)*qxd'*betl
  bethl = lxd*gaml
  bethq = qxd*gamq
  ergl  = gaml[1,]-gaml[2,]
  ergq  = gamq[1,]-gamq[2,]+gamq[3,]
ENDP
```

REFERENCES

Carroll, R.J., Küchenhoff, H., Lombard, F. and Stefanski, L.A. (1995). Asymptotics for the SIMEX estimator in structural measurement error models, *Journal of the American Statistical Association*. To appear.

Carroll, R.J., Ruppert, D. and Stefanski, L.A. (1995). *Nonlinear Measurement Error Models*, Chapman and Hall, New York.

Carroll, R.J. and Stefanski, L.A. (1990). Approximate quasilikelihood estimation in models with surrogate predictors, *Journal of the American Statistical Association* **85**(411): 652–663.

Cook, J.R. and Stefanski, L.A. (1995). Simulation-extrapolation estimation in parametric measurement error models, *Journal of the American Statistical Association*. To appear.

DFG-Forschungsbericht (1981). *Chronische Bronchitis, Teil 2*, Harald Boldt, Boppard.

Fuller, W.A. (1987). *Measurement Error Models*, John Wiley & Sons, New York.

Gleser, L.J. (1990). Improvement of the naive approach to estimation in nonlinear errors-in-variables regression models, *in* P.J. Brown and W.A. Fuller (eds), *Statistical analysis of measurement error models and application*, American Mathematical Society.

Rosner, B., Willett, W.C. and Spiegelmann, D. (1989). Correction of logistic regression relative risk estimates and confidence intervals for systematic within-person measurement error, *Statistics in Medicine* **8**(9): 1051–1069.

Schneeweiß, H. and Mittag, H.J. (1986). *Lineare Modelle mit fehlerbehafteten Daten*, Physica Verlag, Heidelberg.

Stefanski, L.A. and Cook, J.R. (1994). Simulation-extrapolation: The measurement error jackknife, *Technical Report 2259*, Institute of Statistics, North Carolina State University, Raleigh.

14
Nonlinear Time Series Analysis

Rong Chen[1] and Christian Hafner[2]

14.1 Introduction

The recent development of nonlinear time series analysis is primarily due to the efforts to overcome the limitations of linear models such as the autoregressive moving-average models of Box and Jenkins (1976) in real applications. It is also attributed to the development of nonlinear/nonparametric regression techniques which provides many useful tools. Advanced computational power and easy-to-use advanced software and graphics such as S-Plus and XploRe make all of these possible.

In order to depart from linearity, it is important and essential to formulate simple nonlinear models that are sufficient in handling most nonlinear phenomena observed in practice, such as limit cycles, amplitude-dependence frequency, modulation efforts, chaos, etc. However, there is no unified theory applicable to all nonlinear models because there are not only many different nonlinear functions, but also many different model structures.

The most general *nonanticipative* time series model is

$$h(x_t, x_{t-1}, \ldots) = \varepsilon_t, \tag{14.1}$$

where h is a prescribed function. This model merely says that x_t is an implicit function of the past values x_{t-j} and a disturbance ε_t. If x_t is invertible, then (14.1) can be rewritten as

$$x_t = g(\varepsilon_t, \varepsilon_{t-1}, \ldots). \tag{14.2}$$

Obviously, without some further restrictions, neither (14.1) nor (14.2) is useful in practical application. In this chapter we focus on nonlinear autoregressive models that assume the form of

$$x_t = f(x_{t-1}, \ldots, x_{t-p}) + \varepsilon_t, \quad t = 1, 2, \ldots \tag{14.3}$$

[1]Department of Statistics, Texas A&M University, College Station, TX 77843, USA.
[2]Graduiertenkolleg "Angewandte Mikroökonomik", Freie Universität Berlin, D-14195 Berlin, Germany.

where (ε_t) is a sequence of i.i.d. random variables. Typically, ε_t is independent of x_s, for $s < t$. There are two ways of applying this model. The first approach is to formulate a certain parametric function $f(\bullet)$ for the model, presumably based on the physical dynamic background and other substantive information of the data. The second approach is to use nonparametric techniques to estimate the function. Then, based on the estimated nonparametric function, one can either make inference directly or formulate a parametric function and hence, build a parameterized nonlinear model for the process.

Section 14.2 discusses the most important existing parametric nonlinear time series models. Section 14.3 focuses on nonparametric approaches to time series analysis, and Section 14.4 surveys some existing nonlinearity tests. Section 14.5 presents basic nonlinear prediction methods.

14.2   Parametric Approaches

14.2.1   Nonlinear Time Series Models

Researchers have developed many successful nonlinear parametric models for time series. For an excellent survey, see Granger and Teräsvirta (1993), Priestley (1988), and Tong (1990).

Threshold Autoregressive Models

Tong (1978; 1983) proposed a *threshold autoregressive* (TAR) model which assumes the form of

$$x_t = \phi_0^{(i)} + \phi_1^{(i)} x_{t-1} + \ldots + \phi_p^{(i)} x_{t-p} + \varepsilon_t^{(i)}$$

if $x_{t-d} \in \Omega_i$, $i = 1, \ldots, k$, where the Ω_i's are nonoverlapping intervals on the real line with $\bigcup_{i=1}^{k} \Omega_i = \mathbb{R}$, and d is called the threshold lag (or delay parameter). Basically, this model can be regarded as a piecewise linear approximation to the general nonlinear AR model of (14.3). One important feature of the threshold models is that they can produce various limit cycles. Hence, if the process under study has an asymmetrically cyclical form, the TAR model might be useful in providing a good description of the series and meaningful forecasts.

Exponential Autoregressive Models

Haggan and Ozaki (1981) proposed an *exponential autoregressive* (EXPAR) model which assumes the form of

$$x_t = [a_1 + b_1 \exp(-c_1 x_{t-d}^2)]x_{t-1} + \ldots + [a_p + b_p \exp(-c_p x_{t-d}^2)]x_{t-p} + \varepsilon_t,$$

where $c_i \geq 0, i = 1, \ldots, p$. Research has shown that this model is capable of reproducing nonlinear phenomena like limit cycle, amplitude-dependent frequency, and jump phenomena.

Bilinear Models

A bilinear time series model assumes the form of

$$x_t - \sum_{i=1}^{p} \phi_i x_{t-i} = \varepsilon_t + \sum_{i=1}^{q} \theta_i \varepsilon_{t-i} + \sum_{i=1}^{r}\sum_{j=1}^{s} \beta_{ij} x_{t-i}\varepsilon_{t-j},$$

where (ε_t) is a sequence of i.i.d. random variables. It was first studied by Granger and Anderson (1978) and followed by Subba Rao (1981) and Subba Rao and Gabr (1980). It can be considered as a second-order nonlinear time series model since it is constructed simply by adding the cross-product terms of x_{t-i} and ε_{t-j} into a linear ARMA model. The bilinear model cannot reproduce the limit cycle phenomenon. However, it has some other nonlinear phenomena. Subba Rao (1981) showed that, with large bilinear coefficients, a bilinear model can have sudden large amplitude bursts and is suitable for some kinds of seismological data sets like earthquakes and underground nuclear explosions. It has a clear indication of non-Gaussian structure and may have unbounded moments of marginal distributions.

Autoregressive Conditional Heteroscedastic Models

The ARCH (autoregressive conditional heteroscedasticity) process can be written as

$$e_t \mid \psi_{t-1} \quad \sim \quad N(0, h_t),$$
$$h_t \quad = \quad \alpha_0 + \sum_{i=1}^{r} \alpha_i e_{t-i}^2, \qquad (14.4)$$

where ψ_{t-1} is the σ-field generated by $\{e_{t-1}, e_{t-2}, \ldots\}$, $\alpha_0 > 0$, and $\alpha_i \geq 0$ for $i = 1, \ldots, r$. Note that the variance of the current disturbance depends on the values of the past disturbances.

A *generalized autoregressive conditional heteroscedastic* (GARCH) process is defined as (14.4) with modification

$$h_t = \alpha_0 + \sum_{i=1}^{r} \alpha_i e_{t-i}^2 + \sum_{i=1}^{s} \gamma_i h_{t-i}.$$

Here a sufficient but not necessary condition for the nonnegativity of h_t is $\alpha_0 > 0$, $\alpha_i \geq 0, \gamma_i \geq 0$ for $i > 0$.

This model is mainly used to handle the following behaviors, which are commonly seen in business and economic time series:

1. marginal distribution has thick tails,

2. large changes tend to be followed by large changes and small changes are followed by small changes, and

3. outliers tend to cluster.

For details, see Engle (1982) and Bollerslev (1986). An excellent survey
of the theory and applications in finance is given by Bollerslev, Chou and
Kroner (1992).

14.2.2   Estimation

In terms of estimation, three methods are used mainly for finite parametric nonlinear time series models. With stationary ergodic Markov chains,
Basawa and Prakasa Rao (1980), and Hall and Heyde (1980) provided consistency and asymptotic normality of the maximum likelihood estimates
under mild conditions. However, in practice it is usually difficult to find
the exact likelihood function for nonlinear time series processes. A conditional least-squares method minimizes the sum of squares of residuals and
is very easy to use for appropriately parameterized nonlinear autoregressive models. Klimko and Nelson (1978) proved that under mild conditions
the least-squares estimates are consistent and asymptotically normal distributed. More references can also be found in Tjøstheim (1986), Robinson
(1977), Hinich and Patterson (1985), and Lai and Wei (1983). Lastly, from a
Bayesian viewpoint, with simple formulation and appropriate prior (usually
conjugate prior), the posterior distribution of the parameters can be found
through numerical integrations or Markov Chain Monte Carlo methods.

14.3   Nonparametric Approach

As we have seen, there are many nonlinear time series models, and each
of them has its own properties and characteristics. In practice, it is important to be able to choose an appropriate model among them. If we
have substantial knowledge about the process under study, that is, the
physical law, the dynamic nature, and the deterministic behavior, etc., we
can choose an appropriate model based on this knowledge. On the other
hand, if we don't have enough information, we may follow the principle of
"letting data speak for themselves." Along with the rapid development of
nonparametric regression techniques, many nonparametric nonlinear time
series model building techniques have been developed.

Local Conditional Mean (Median) Approach

Consider the general nonlinear AR(p) process

$$x_t = f(x_{t-1}, \ldots, x_{t-p}) + \varepsilon_t.$$

Let $Y_t = (X_{t-1}, \ldots, X_{t-p})$, and choose $\delta_n > 0$. For any $y \in \mathbb{R}^p$, let $I_n(y) = \{i : 1 < i < n$ and $||Y_i - y|| < \delta_n\}$ and $N_n(y) = \#I_n(x)$. The conditional mean function estimator is given by

$$\hat{f}_n(y) = \frac{1}{N_n(y)} \sum_{i \in I_n(y)} X_i,$$

and the local conditional median estimator is given by

$$\tilde{f}(y) = median\{X_i,\ i \in I_n(y)\}.$$

Under α-mixing conditions, Truong (1993) provides the optimal rate of convergence.

Nonparametric Kernel Function Estimation Approach

Robinson (1983), Auestad and Tjøstheim (1990), and Härdle and Vieu (1992), etc., used a kernel estimator to estimate the conditional mean and variance under the model

$$x_t = f(x_{t-1}, \ldots, x_{t-p}) + \alpha(x_{t-1}, \ldots, x_{t-p})\varepsilon_t.$$

The function $f(y_1, \ldots, y_p)$ is estimated by the Nadaraya–Watson estimator with productive kernels:

$$\hat{f}(y_1, \ldots, y_p) = \frac{\sum\limits_{t=p+1}^{n} \prod\limits_{i=1}^{p} K\{(y_i - x_{t-i})/h_i\}x_t}{\sum\limits_{t=p+1}^{n} \prod\limits_{i=1}^{p} K\{(y_i - x_{t-i})/h_i\}},$$

and the conditional variance $\alpha(x_{t-1}, \ldots, x_{t-p})$ is estimated by

$$\hat{\alpha}(y_1, \ldots, y_p) = \frac{\sum\limits_{t=p+1}^{n} \prod\limits_{i=1}^{p} K\{(y_i - x_{t-i})/h_i\}x_t^2}{\sum\limits_{t=p+1}^{n} \prod\limits_{i=1}^{p} K\{(y_i - x_{t-i})/h_i\}} - \{\hat{f}(y_1, \ldots, y_p)\}^2, \quad (14.5)$$

where $K(\bullet)$ is a kernel function with bounded support and the h_i's are the bandwidths.

Robinson (1983) proved that under strong mixing conditions, the estimators have the same properties as those in the regression situation.

Härdle and Vieu (1992) apply the algorithm to a gold price series, 1978 to 1986 ($n = 2041$). In Figure 14.1, the returns $r_t = (x_t - x_{t-1})/x_{t-1}$ are plotted against the prices x_{t-1}.

The model

$$r_t = f(x_{t-1}) + \alpha(x_{t-1})\epsilon_t$$

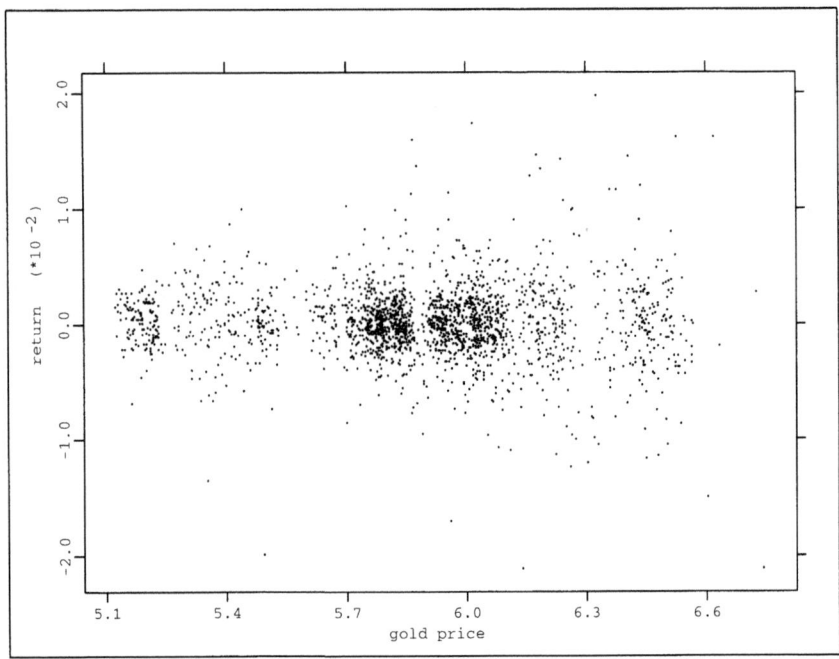

FIGURE 14.1. Gold price returns, 1978 to 1986 ($n = 2040$).

is estimated and the resulting plots for the conditional mean and variance
are shown in Figures 14.2 and 14.4, respectively.

The bandwidths h were selected using the cross-validation technique.
The cross-validation functions in Figures 14.3 and 14.5 correspond to the
cross-validation problems of (14.3) and the first term of (14.5), respectively.
The first function has a minimum at $h = 0.45$, the second one at $h = 0.31$.

The following XploRe code computes the conditional mean, mh, and variance, vh. It requires the minimizers of the cross-validation functions, hcv1
and hcv2, respectively; they have been obtained by the XploRe macro
REGCVL. Z is the data matrix containing in the first column, the gold prices,
and in the second column, the returns.

```
proc(mh vh)=meanvar(z hcv1 hcv2)
  library(smoother)
  mh=regest(z hcv1)              ; conditional mean
  z[,2]=z[,2].*z[,2]
  z=z[,1]~z[,2]
  vh=regest(z hcv2)
  vh[,2]=vh[,2]-mh[,2].*mh[,2]   ; conditional variance
endp
```

It is noted that the pure kernel estimation approach suffers from the curse

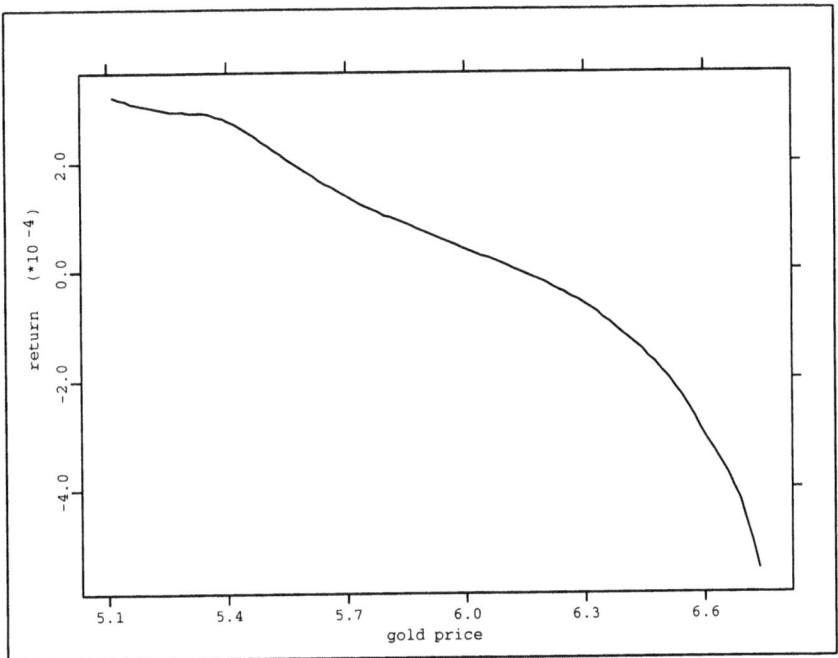

FIGURE 14.2. Nadaraya–Watson estimate for the conditional mean of the gold price returns using the cross-validated bandwidth $\hat{h}_{CV} = 0.45$.

of dimensionality. To overcome this difficulty, researchers have developed more restricted, yet still general enough models.

Functional Coefficient AR Model Approach

A *functional coefficient autoregressive* (FAR) model can be written as

$$x_t = f_1(x_{t-d})x_{t-1} + f_2(x_{t-d})x_{t-2} + \ldots + f_p(x_{t-d})x_{t-p} + \varepsilon_t.$$

The model is general enough to include TAR models (when the coefficient functions are step functions) and EXPAR models (when the coefficient functions are exponential functions), along with many other models (for example, sine function models, logistic function models).

Chen and Tsay (1993a) use an *arranged local regression* procedure (ALR) to roughly identify the nonlinear function forms. For $y \in \mathbb{R}$ and $\delta_n > 0$, let $I_n(y) = \{t : 1 < t < n, |x_{t-d} - y| < \delta_n\}$. If we regress x_t on $x_{t-1}, \ldots, x_{t-p}$ using all the observations x_t such that $t \in I_n(y)$, then the estimated coefficients can be used as estimates of $f_i(y)$. Note that the window width δ_n can be interpreted as the smoothing parameter.

For illustration of the ALR implementation in XploRe we consider the chickenpox data used by Chen and Tsay (1993a) and described by Sugihara

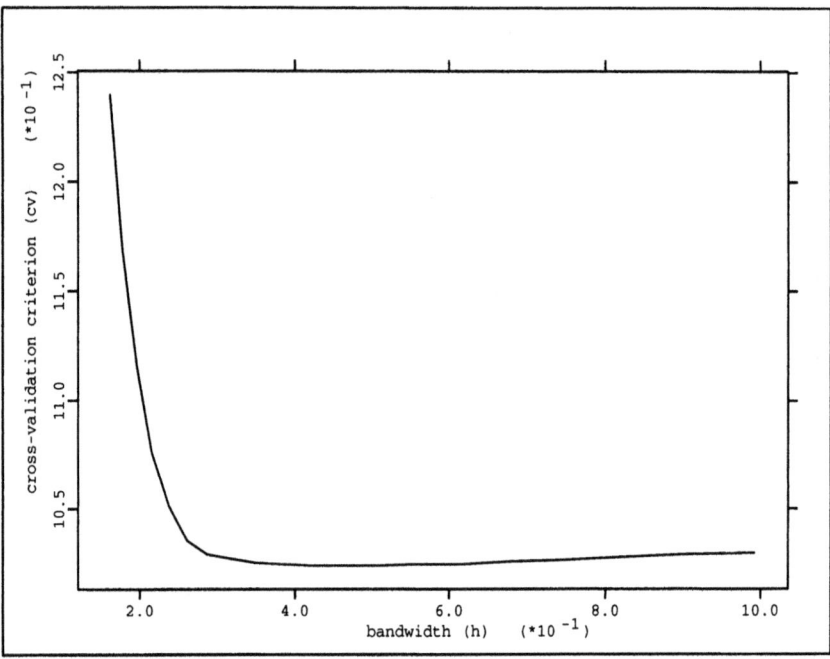

FIGURE 14.3. The cross-validation function for the conditional mean problem of the gold price example has a minimum at $h = 0.45$.

and May (1990) with 533 observations. Natural logarithms are taken for variance stabilization. In the implementation, we require the sample size within each window to be at least k ($> p$) to ensure the accuracy of the coefficient estimates. Lacking an optimal selection criterion, we select the structure parameters heuristically to be $k = 30$ and the window width $c = (x_{max} - x_{min})/10$. Several nonlinearity tests indicate strong nonlinearity for the threshold lag $d = 12$, which is plausible because we have monthly data. The most significant lags are 1 and 24. The resulting model is

$$x_t = f_1(x_{t-12})x_{t-1} + f_2(x_{t-12})x_{t-24} + \varepsilon_t. \qquad (14.6)$$

At first, the values need to be sorted for the window shifting. The number of iterations m is computed such that the whole range $x_{max} - x_{min}$ is covered. At each step, the vector tr gives the time indices of those values that are within the defined window $[r; r + c]$. Finally, beta1 and beta2 give the least-squares estimation of the first and second coefficient, that is, function value, for each window.

```
proc()=main()
  x=read(xchick.dat)
```

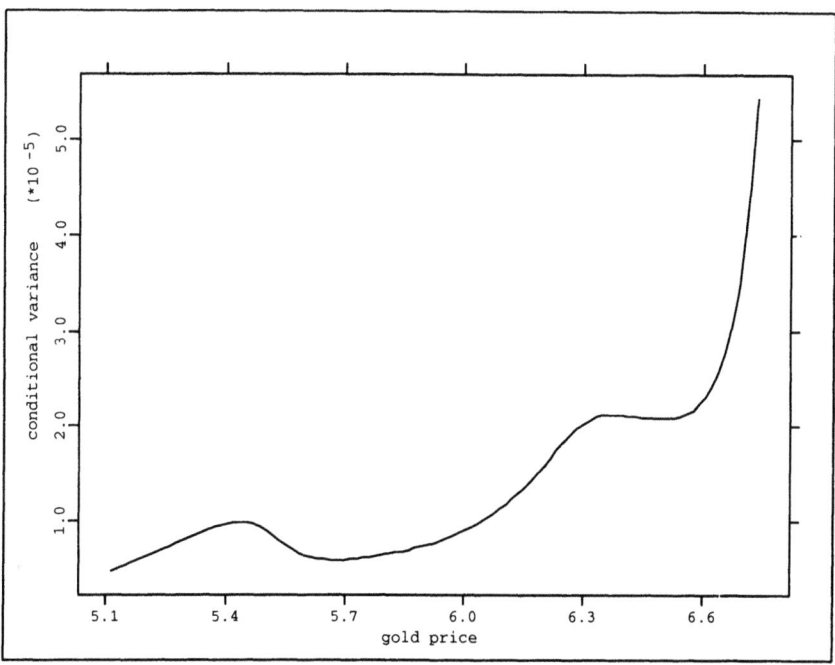

FIGURE 14.4. Conditional variance estimate of gold prices using the cross-validated bandwidth $\hat{h}_{CV}=0.31$.

```
n=rows(x)
t=matrix(n)
t=cumsum(t)                            ; time vector
xs=sort(x~t)
h=10
c=(xs[n,1]-xs[1,1])/h                  ; window width
d=12                                   ; threshold lag
k=30                                   ; minimum # obs
r=xs[1,1]                              ; lower bound
m=floor((xs[(n-d),1]-xs[1,1])*h/c)     ; # iterations
beta=matrix(m 2)
beta1=matrix(m 2)
beta2=matrix(m 2)
l=1
while (1.<=m)
  xn=paf(xs (xs[,1].>r) .& (xs[,1].<=r+c)) ; select entries
  tr=paf((xn[,2]) (xn[,2].<=n-d) .& (xn[,2].>24-d))+d ; indices
  if (rows(tr).>=k)
      y=matrix(rows(tr) 3)
      y[,1]=index(x tr)                ; response variable
      y[,2]=index(x (tr.-1))           ; lag 1
```

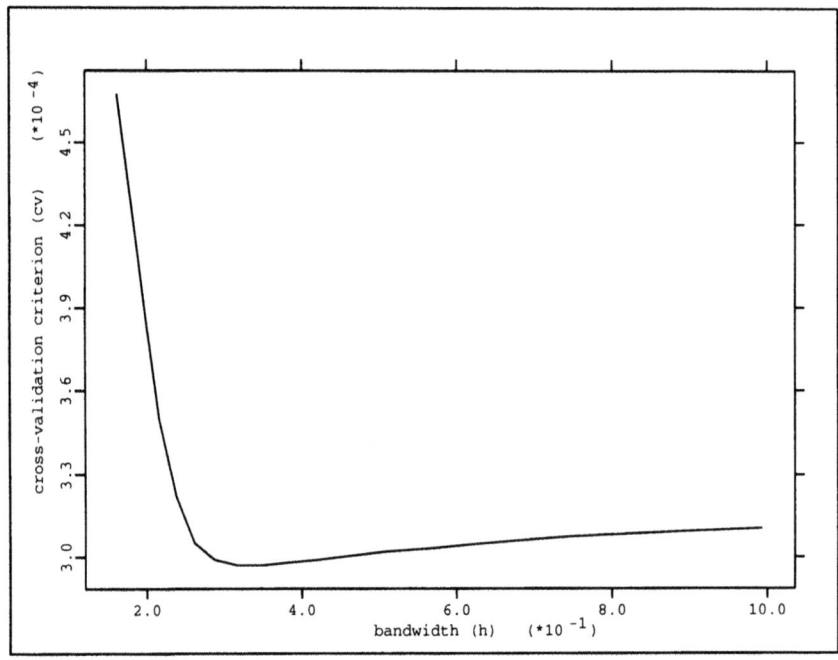

FIGURE 14.5. The cross-validation function for the conditional variance problem of the gold price example has a minimum at $h = 0.31$.

```
      y[,3]=index(x (tr.-24))        ; lag 24
      beta[1,]=gls(y[,2:3] y[,1])'   ; least squares
      beta1[1,]=r+c~beta[1,1]        ; first coeff
      beta2[1,]=r+c~beta[1,2]        ; second coeff
   endif
   r=r+c/h                          ; new lower bound
   l=l+1
  endo
  beta1=paf(beta1 beta1[,2].<>1)
  beta2=paf(beta2 beta2[,2].<>1)
  show (beta1 s2d)
  show (beta2 s2d)
endp
```

The local estimates of $f_1(\bullet)$ and $f_2(\bullet)$, $\beta_1(\bullet)$ and $\beta_2(\bullet)$, are shown in Figures 14.6 and 14.7, respectively.

One can now formulate parametric models based on the forms of the estimated nonlinear functions. There seems to be a level shift around the value $x_{t-12} = 7.2$, so a TAR model with threshold value 7.2 is suggested. For details see Chen and Tsay (1993a). Note that a procedure similar to the local weighted regression of Cleveland and Devlin (1988) can also be used here as well.

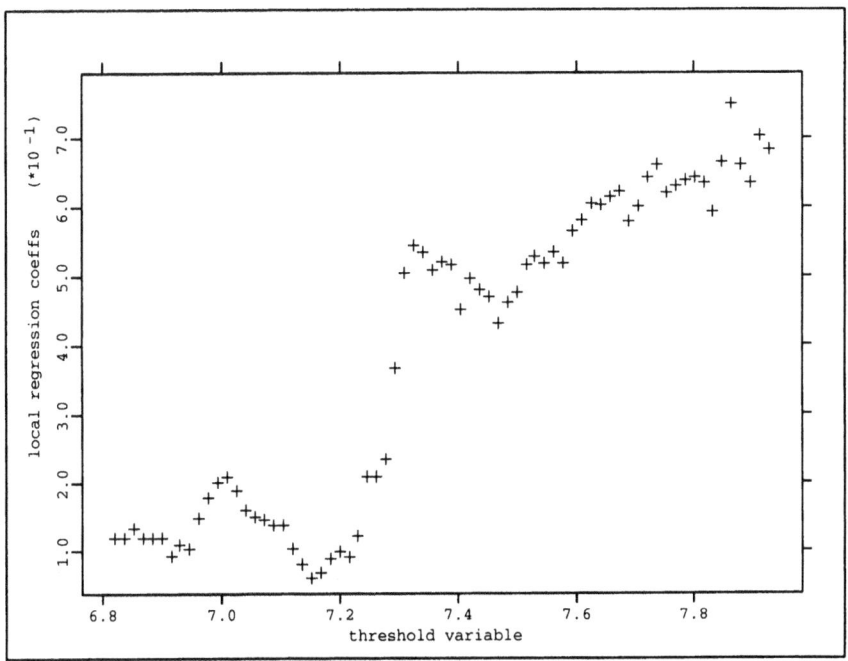

FIGURE 14.6. In the ALR procedure the local regression coefficients $\beta_1(\bullet)$ are used as estimates of $f_1(\bullet)$ for the FAR model (chickenpox example).

Nonlinear Additive AR Model Approach

A *nonlinear additive autoregressive* (NAAR) model can be written as

$$x_t = f_1(x_{t-1}) + f_2(x_{t-2}) + \cdots + f_p(x_{t-p}) + \varepsilon_t.$$

Similar models have been studied extensively in a regression context (see Hastie and Tibshirani, 1990, and the references therein). Chen and Tsay (1993b) used the ACE (Alternating Conditional Expectation) algorithm of Breiman and Friedman (1985) to specify the model. The basic idea is to estimate the transformations $f_1(x_{t-1}), \ldots, f_p(x_{t-p})$ such that the fraction of variance not explained by a regression of x_t on $\sum_{j=1}^{p} f_j(x_{t-j})$ is minimized.

The ACE algorithm has been applied to the riverflow data of the river Jokulsa Eystri in Iceland. This is a multiple time series data set, consisting of daily riverflow (y_t), precipitation (z_t), and temperature (x_t) from January 1, 1972, to December 31, 1974 ($n = 1096$). For further information, see Tong (1990), who used threshold autoregressive models. The time series are plotted in Figure 14.8.

A procedure similar to the best subset regression is suggested by Chen and Tsay (1993b) to select the lag variables in the model. They found $\{y_{t-1}, y_{t-2}, z_t, z_{t-1}, x_{t-1}, x_{t-3}\}$ to be an appropriate explanatory set

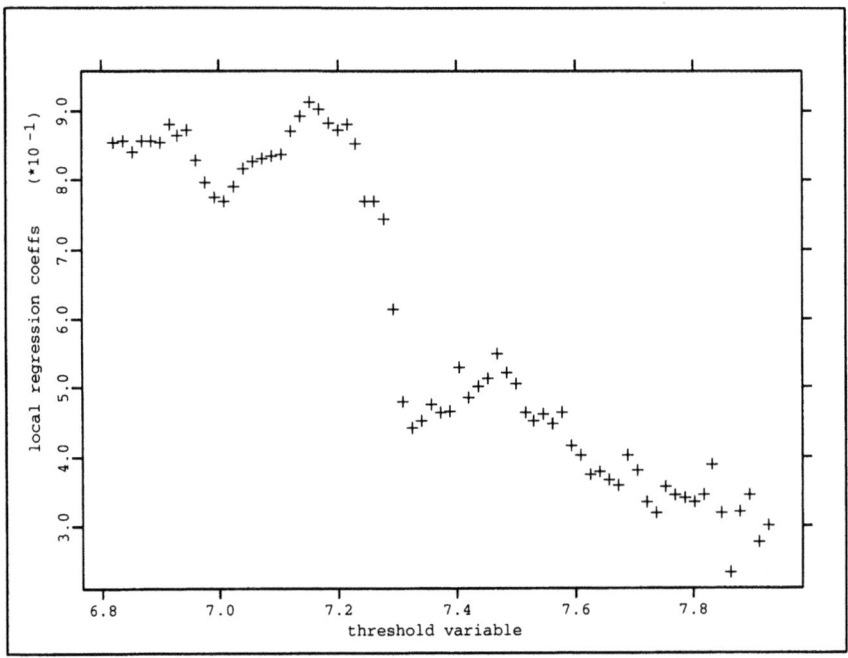

FIGURE 14.7. In the ALR procedure the local regression coefficients $\beta_2(\bullet)$ are used as estimates of $f_2(\bullet)$ for the FAR model (chickenpox example).

for the response variable y_t. The transformations $f_1(y_{t-1})$, $f_2(y_{t-2})$, $f_3(z_t)$, $f_4(z_{t-1})$, $f_5(x_{t-1})$, $f_6(x_{t-3})$ are shown in Figure 14.9.

Linear functions are suggested for the precipitation and piecewise linear functions for the lagged riverflow and temperature variables. In comparison to Tong's threshold model, the obtained model improves out-of-sample forecasts and is preferred by the AIC criterion.

The XploRe code for the riverflow example is:

```
proc()=main()
    x=read(xriver.dat)
    n=rows(x)
    t=matrix(n)
    t=cumsum(t)                    ; the time vector
    createdisplay(disp1, 1 3, s2d) ; for the time plots
    show (t~x[,1] s2d1, t~x[,2] s2d2, t~x[,3] s2d3)
    library(addmod)
    y=x[4:n,1]
    y1=x[3:(n-1),1]
```

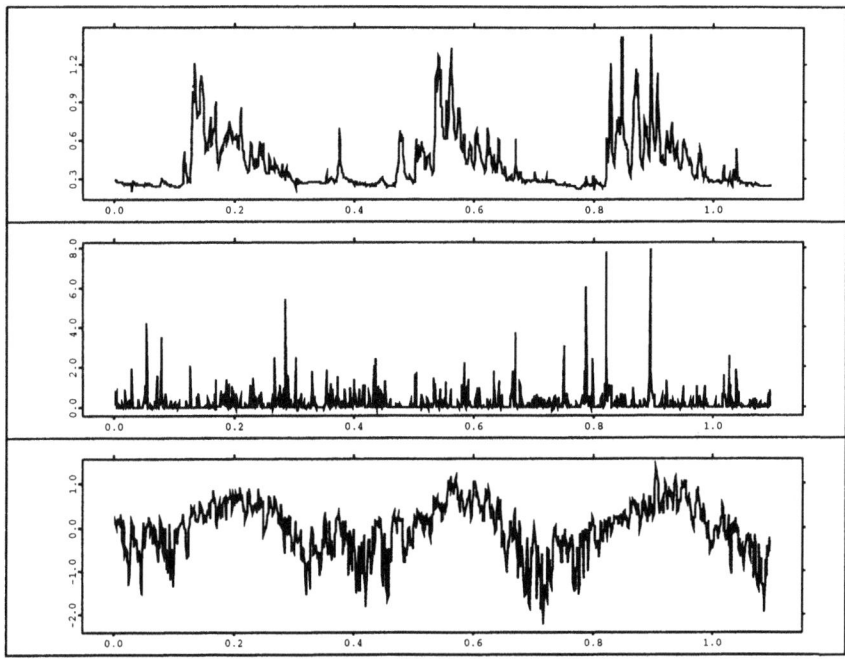

FIGURE 14.8. Time plots of riverflow data: daily riverflow $y_t (m^3/s, *10^2)$, daily precipitation z_t (mm/day, $*10^2$), and daily temperature x_t (C^o, $*10$); time index: $t = 1, \ldots, 1096(*10^3)$.

```
   y2=x[2:(n-2),1]
   z0=x[4:n,2]
   z1=x[3:(n-1),2]
   x1=x[3:(n-1),3]
   x3=x[1:(n-3),3]
   x=y1~y2~z0~z1~x1~x3
   (a b) = ace(x y)          ; a contains the transformations
   createdisplay(disp2, 2 3, s2d)
   d1=y1~a[,1]
   d4=y2~a[,2]
   d2=z0~a[,3]
   d5=z1~a[,4]
   d3=x1~a[,5]
   d6=x3~a[,6]
   show(d1 s2d1, d2 s2d2, d3 s2d3, d4 s2d4, d5 s2d5, d6 s2d6)
endp
```

Note that the *additivity and variance stabilization* (AVAS) algorithm of Tibshirani (1988) can also be used here.

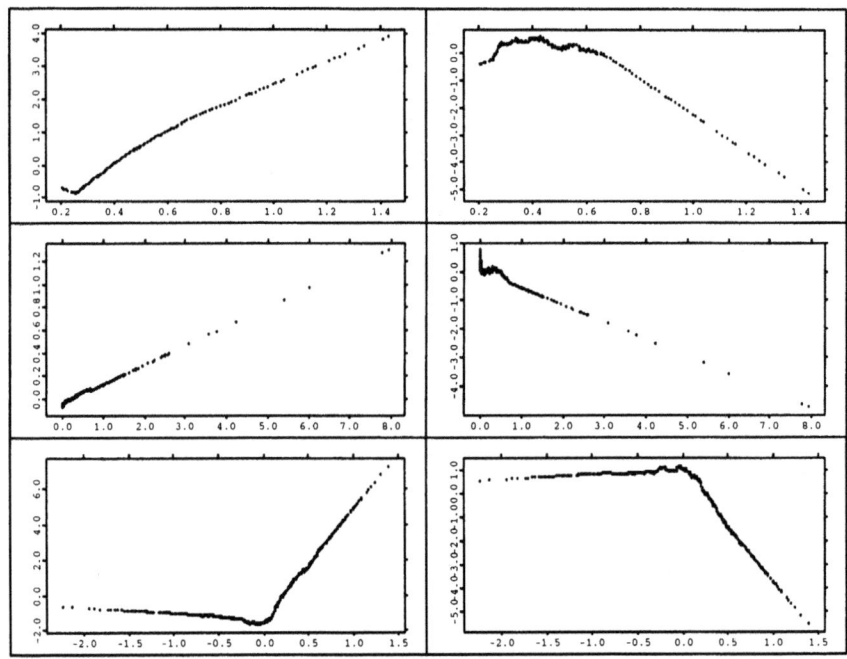

FIGURE 14.9. Results of the ACE algorithm for the riverflow data. The plots show the suggested transformations. First row: y_{t-1} and y_{t-2}; second row: z_t and z_{t-1}; third row: x_{t-1} and x_{t-3}, where y_t indicates the riverflow series, z_t is the precipitation, and x_t is the temperature.

Adaptive Spline Threshold AR Model Approach

Lewis and Stevens (1991) proposed the *adaptive spline threshold autoregressive* (ASTAR) model with the form

$$x_t = \sum_{j=1}^{s} c_j K_j(x) + \varepsilon_t,$$

where $\{K_j(x)\}_{j=1}^s$ are product basis functions of truncated splines $T^-(x) = (t-x)_+$ and $T^+(x) = (x-t)_+$ associated with the subregions $\{R_j\}_{j=1}^s$ in the domain of the predictor variables. For example,

$$x_t = c_1 + c_2 x_{t-1} + c_3(a_1 - x_{t-5})_+ + c_4 x_{t-1}(x_{t-3} - a_2)_+(a_3 - x_{t-4})_+ + \varepsilon_t,$$

where $u_+ = u$ if $u > 0$ and $u_+ = 0$ if $u \leq 0$, is an ASTAR model.

The modeling and estimation procedures follow the *Multivariate Adaptive Regression Splines* (MARS) methodology by Friedman (1991). It is basically a regression tree procedure using truncated regression splines.

14.4   Nonlinearity Tests

Nonlinear modeling is apparently useful in time series analysis. It certainly follows that statistical methods are needed to ascertain when a process is nonlinear. Here we review some of the time domain tests available in the literature. For frequency domain tests, see Subba Rao and Gabr (1980) and Hinich (1982).

14.4.1   Lagrange Multiplier Tests

The F-test of Keenan (1985) consists of three steps:

1. Fit a linear $AR(m)$ model to x_t and calculate the fitted value $\hat{x}_t$ and the residual $\hat{a}_t$ for $t = m+1, \ldots, n$, where n is the sample size.

2. Regress x_t^2 on the set of regressors $\{1, x_{t-1}, \ldots, x_{t-m}\}$ and calculate the residuals $\hat{e}_t$ for $t = m+1, \ldots, n$.

3. Regress $\hat{a}_t$ on $\hat{e}_t$ and compute the associated F-statistic $\hat{F}$.

If x_t is a linear $AR(m)$ process and the sample size n is large, $\hat{F}$ follows approximately an F-distribution with degrees of freedom 1 and $n - 2m - 2$. The Ori-F test of Tsay (1986) improves the F-test by replacing the aggregated quantity $\hat{x}_t^2$ in step 2 by individual quadratic terms

$$\{x_{t-1}^2, x_{t-1}x_{t-2}, \ldots, x_{t-1}x_{t-m}, x_{t-2}^2, x_{t-2}x_{t-3}, \ldots, x_{t-m}^2\},$$

and using multiple linear regression in step 2. The Aug-F test of Luukkonen, Saikkonen and Teräsvirta (1988) modifies the procedure of the Ori-F test by further adding the cubic terms $x_{t-1}^3, x_{t-2}^3, \ldots, x_{t-m}^3$ in step 2.

14.4.2   Tests for Threshold Nonlinearity

The following tests were proposed for testing linear models against threshold models.

CUSUM Test

The CUSUM test estimates a linear AR model sequentially according to the increasing order of the value of a given threshold variable. If the process is linear, the standardized predictive residuals z_i are asymptotically independent normal random variables. Let

$$Z_r = \sum_{i=r_{min}+1}^{r} z_i; \qquad T_n = \max_{r_{min}+1 \le r \le n-p} |Z_r|.$$

Then an invariance principle for random walks (Feller, 1966) can used for the test. For details, see Petruccelli and Davies (1986).

Likelihood Ratio Test

For a threshold model

$$
x_t = \begin{cases} \mu^{(1)} + \sum\limits_{i=1}^{p_1} \phi_i^{(1)} x_{t-i} + \varepsilon_t^{(1)} & x_{t-d} \leq c \\ \mu^{(2)} + \sum\limits_{i=1}^{p_2} \phi_i^{(2)} x_{t-i} + \varepsilon_t^{(2)} & x_{t-d} > c, \end{cases}
$$

the null hypothesis of the likelihood ratio test for testing linearity is H_0: $p_1 = p_2 = p$ and $\phi_i^{(1)} = \phi_i^{(2)}$ for $i = 1, \ldots, p$ and $\mu^{(1)} = \mu^{(2)}$. The asymptotic distribution of the test statistic is given by Chan (1990).

TAR-F Test

The TAR-F test of Tsay (1989) combines the idea of CUSUM and F-tests. The F-statistic is obtained by regressing the predictive residuals of the arranged autoregression (as in CUSUM) on the regressors $\{1, x_{t-1}, \ldots, x_{t-m}\}$. A large F-statistic implies that there are model changes in the series.

14.4.3   New-F Test

The New-F test (Tsay, 1991) modifies the TAR-F test to test linearity against threshold nonlinearity, exponential nonlinearity, and bilinearity. The F-statistic is obtained by regressing the predictive residuals on the regressors $\{1, x_{t-1}, \ldots, x_{t-m}\}$, $\{x_{t-1} \exp(-x_{t-1}^2/\gamma), G(z_{t-d}), x_{t-1}G(z_{t-d})\}$ $\{x_{t-1}\hat{e}_{t-1}, \ldots, x_{t-m}\hat{e}_{t-m}\}$, $\{\hat{e}_{t-1}\hat{e}_{t-2}, \ldots, \hat{e}_{t-m}\hat{e}_{t-m-1}\}$, where $\gamma = \max_t |x_{t-1}|$, and $z_{t-d} = (x_{t-d} - \bar{x}_d)/S_d$ with $\bar{x}_d$ and S_d being the sample mean and standard deviation of x_{t-d}, respectively, and $G(\bullet)$ is the CDF of the standard normal random variable. The F-statistic of this regression follows asymptotically an F-distribution if the process is a linear $\mathrm{AR}(p)$ process.

14.4.4   Nonparametric Test

Hjellvik and Tjøstheim (1994) proposed a nonlinearity test based on the distance between the best linear predictor $\rho_k X_{t-k}$ and the best nonlinear predictor $M_k(X_{t-k}) = \mathrm{E}[X_t \mid X_{t-k}]$ of X_t based on X_{t-k}. The index is defined as

$$
L(M_k) = \mathrm{E}[(M_k(X_{t-k}) - \rho_k X_{t-k})^2 w(X_{t-k})],
$$

where $w(x)$ is a weighting function with compact support and ρ_k is the autocorrelation between X_t and X_{t-k}, assuming X_t has zero mean. The function $M_k(\bullet)$ is estimated using the Nadaraya–Watson estimator.

14.5 Nonlinear Prediction

Nonlinear forecasting can be based on either parametric models or pure nonparametric approaches.

14.5.1 Parametric Approaches

There are two basic parametric approaches in nonlinear time series forecasting, namely, the numerical integration and simulation methods.

Numerical Integration

The basic idea of this method is to use a recursive integral equation to evaluate the conditional density of $[X_{t+k} \mid X_t]$ for $k \geq 1$. Consider the simple nonlinear AR(1) model

$$X_{t+1} = \lambda(X_t) + \varepsilon_{t+1}.$$

To obtain the conditional density $g_k(\bullet)$ of X_{t+k} given X_t, Jones (1978) uses

$$g_k(x|X_t) = \int_{-\infty}^{\infty} f\{x - \hat{\lambda}(y)\} g_{k-1}(y \mid X_t) dy,$$

where $\hat{\lambda}$ denotes the estimated function λ, and $f(\bullet)$ is the density function of the noise ε_t. Pemberton (1987) uses

$$g_k(x|X_t) = \int_{-\infty}^{\infty} f\{y - \hat{\lambda}(X_t)\} g_{k-1}(x \mid y) dy.$$

The methods can be easily extended to the general nonlinear AR(p) case. Note that these two methods can provide exact forecasts, but the computation is intensive. Also, the result heavily depends on the parameter estimation.

Simulation

Assume

$$x_{t+1} = \phi(x_t, \ldots, x_{t-p+1}) + \varepsilon_t,$$

where the distribution of ε_t is known and the nonlinear function ϕ has been estimated by $\hat{\phi}$. Let $\hat{x}_{i,t}(-m) = x_{t-m}$, $m = 0, 1, \ldots, p$. We can get

$$\hat{x}_{i,t}(k) = \hat{\phi}(\hat{x}_{i,t}(k-1), \ldots, \hat{x}_{i,t}(k-p)) + \varepsilon_{i,t+k}, \quad i = 1, \ldots, N,$$

where $\varepsilon_{i,t+k}$ are simulated from the distribution of ε_t. The kth step ahead prediction is then defined as

$$\hat{x}_t(k) = \frac{1}{N} \sum_{i=1}^{N} \hat{x}_{i,t}(k).$$

When there is no distribution assumption of ε_t, the bootstrap method can be used.

14.5.2 Nonparametric Approaches

Consider the nonlinear AR(1) model $X_t = \phi(X_{t-1}) + \varepsilon_t$. Since the conditional mean $E(X_{t+k} \mid X_t = x)$ is the least-squares predictor for k-step ahead prediction, Auestad and Tjøstheim (1990) and Härdle and Vieu (1992) proposed using the ordinary Nadaraya–Watson estimator

$$\hat{m}_{h,k}(x) = \frac{\sum_{t=1}^{n-k} K\{(x - X_t)/h\} X_{t+k}}{\sum_{t=1}^{n-k} K\{(x - X_t)/h\}} \tag{14.7}$$

to estimate $E(X_{t+k} \mid X_t = x)$ directly.

Note that the variables $X_{t+1}, \ldots, X_{t+k-1}$ contain substantial information about the conditional mean function $E(X_{t+k} \mid X_t)$. Chen (1994) proposed a multistage kernel smoother, which utilizes this information. For example, consider two-step ahead forecasting. Due to the Markov property, we have

$$\begin{aligned} m_{h,2}(x) &= E(X_{t+2} \mid X_t = x) \\ &= E(E(X_{t+2} \mid X_{t+1}, X_t) \mid X_t = x) \\ &= E(E(X_{t+2} \mid X_{t+1}) \mid X_t = x). \end{aligned}$$

Define $f(y) = E(X_{t+2} \mid X_{t+1} = y)$. Ideally, if we knew $f(\bullet)$, we would use the pairs $(f(x_{i+1}), x_i)$, $i = 1, \ldots, (n-1)$ to estimate $E(X_{t+2} \mid X_t)$, instead of using the pairs (x_{i+2}, x_i) as the estimator in (14.7). Note that the error between X_{t+2} and $f(X_{t+1})$ is $O(1)$. Hence, if we can estimate the function $f(\bullet)$ with an estimator $\hat{f}(\bullet)$ that has a smaller error rate and use the pairs $(\hat{f}(x_{i+1}), x_i)$ to estimate $E(X_{t+2} \mid X_t)$, we should achieve a smaller error. This observation motivated the following estimator, which is called a "multistage smoother." It is defined as

$$\hat{m}_{h_1,h_2}(x) = \frac{\sum_{t=1}^{n-1} K\{(x - X_t)/h_2\} \hat{f}_{h_1}(X_{t+1})}{\sum_{t=1}^{n-1} K\{(x - X_t)/h_2\}}, \tag{14.8}$$

where

$$\hat{f}_{h_1}(y) = \frac{\sum_{j=1}^{n-1} K\{(y - X_j)/h_1\} X_{j+1}}{\sum_{j=1}^{n-1} K\{(y - X_j)/h_1\}}.$$

The new smoother is proved to have a smaller mean squared error. The estimators in (14.7) and (14.8) are applied to the gold price example. We

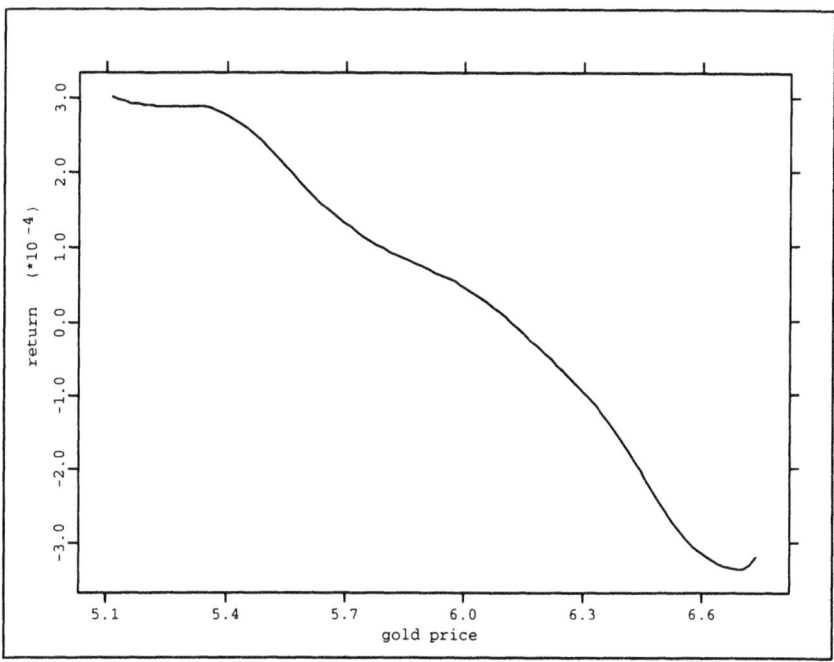

FIGURE 14.10. Ten-step prediction for the gold price example using the direct Nadaraya–Watson estimator, $h = 0.5$.

computed a ten-step prediction with both estimators. For the multistage smoother we need a recursive algorithm that computes at the kth step a smoother of the $(k-1)^{\text{th}}$ smoother, beginning with the simple one-step predictor. The basic element of the algorithm listed below is the XploRe macro sker that computes the Nadaraya–Watson estimator at each data point.

```
proc()=main()
  library(smoother)
  x=read(xgold.dat)
  n=rows(x)
  r=(x[2:n,]-x[1:(n-1),])./x[1:(n-1),]
  x=x[1:(n-1),]
  t=matrix(n-1)
  t=cumsum(t)                 ; the time vector
  i=1
  while (i.<=10)
    cv=regcvl(x~r)            ; cross-validation
    cv=sort(cv 2)
    hcv=cv[1,1]               ; minimizer of cv
```

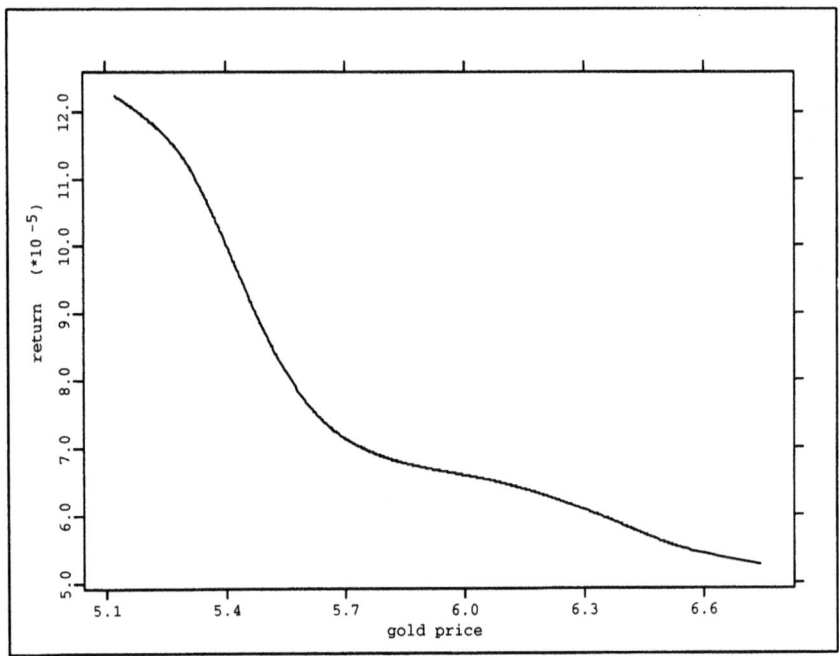

FIGURE 14.11. Ten-step prediction for the gold price example using the multistage smoother, $h = 0.5$.

```
    xs=sort(x~r~t)
    (z f) = sker(xs[,1] xs[,2] hcv)
    mh=z./f                    ; smoother
    s=sort(xs[,1]~mh~xs[,3] 3) ; sort by time index
    x=s[1:(n-1-i),1]
    r=s[2:(n-i),2]             ; the new response variable
    t=t[1:(n-1-i),]
    i=i+1
  endo
  show(xs[,1]~mh s2d)
endp
```

The estimates are shown in Figures 14.10 and 14.11.

REFERENCES

Auestad, B. and Tjøstheim, D. (1990). Identification of nonlinear time series: First order characterization and order estimation, *Biometrika* **77**(4): 669–687.

Basawa, I.V. and Prakasa Rao, B.L.S. (1980). *Statistical Inference for Stochastic Processes*, Academic Press, London.

Bollerslev, T. (1986). Generalized autoregressive conditional heteroscedasticity, *Journal of Econometrics* **31**(3): 307–327.

Bollerslev, T., Chou, R.Y. and Kroner, K.F. (1992). Arch modelling in finance: A review of the theory and empirical evidence, *Journal of Econometrics* **52**(1): 5–59.

Box, G.E.P. and Jenkins, G.M. (1976). *Time Series Analysis: Forecasting and Control*, Holden-Day, San Fransisco.

Breiman, L. and Friedman, J.H. (1985). Estimating optimal transformations for multiple regression and correlations (with discussion), *Journal of the American Statistical Association* **80**(391): 580–619.

Chan, K.S. (1990). Testing for threshold autoregression, *Annals of Statistics* **18**(4): 1886–1894.

Chen, R. (1994). A nonparametric predictor for nonlinear time series, *Technical report*, Department of Statistics, Texas A&M University, College Station.

Chen, R. and Tsay, R.S. (1993a). Functional-coefficient autoregressive models, *Journal of the American Statistical Association* **88**(421): 298–308.

Chen, R. and Tsay, R.S. (1993b). Nonlinear additive ARX models, *Journal of the American Statistical Association* **88**(423): 955–967.

Cleveland, W.S. and Devlin, S.J. (1988). Locally weighted regression: An approach to regression analysis by local fitting, *Journal of the American Statistical Association* **83**(403): 596–610.

Engle, R.F. (1982). Autoregressive conditional heteroscedasticity with estimates of the variance of U.K. inflation, *Econometrica* **50**(4): 987–1007.

Feller, W. (1966). *An Introduction to Probability Theory and Its Application*, Vol. 2, John Wiley & Sons, New York.

Friedman, J.H. (1991). Multivariate adaptive regression splines (with discussion), *Annals of Statistics* **19**(1): 1–141.

Granger, C.W.J. and Anderson, A.P. (1978). *An Introduction to Bilinear Time Series Models*, Vandenhoeck & Ruprecht, Göttingen und Zürich.

Granger, C.W.J. and Teräsvirta, T. (1993). *Modelling Nonlinear Economic Relationships*, Academic Press, Oxford.

Haggan, V. and Ozaki, T. (1981). Modeling nonlinear vibrations using an amplitude-dependent autoregressive time series model, *Biometrika* **68**(1): 189–196.

Hall, P. and Heyde, C.C. (1980). *Martingale Limit Theory and Its Applications*, Academic Press, New York.

Härdle, W. and Vieu, P. (1992). Kernel regression smoothing of time series, *Journal of Time Series Analysis* **13**(3): 209–232.

Hastie, T.J. and Tibshirani, R.J. (1990). *Generalized Additive Models*, Vol. 43 of *Monographs on Statistics and Applied Probability*, Chapman and Hall, London.

Hinich, M.J. (1982). Testing for gaussianity and linearity of a stationary time series, *Journal of Time Series Analysis* **3**(3): 169–176.

Hinich, M.J. and Patterson, D.M. (1985). Identification of the coefficient in a nonlinear time series of the quadratic type, *Journal of Econometrics* **30**(3): 269–288.

Hjellvik, V. and Tjøstheim, D. (1994). Nonparametric tests of linearity for time series, *Biometrika*. To appear.

Jones, D.A. (1978). Nonlinear autoregressive processes, *Proceedings of the Royal Society of London, Series A* **360**: 71–95.

Keenan, D.M. (1985). A Tukey nonadditivity-type test for time series nonlinearity, *Biometrika* **72**(1): 39–44.

Klimko, L.A. and Nelson, P.I. (1978). On conditional least squares estimation for stochastic processes, *Annals of Statistics* **6**(3): 629–642.

Lai, T.L. and Wei, C.Z. (1983). Asymptotic properties of general autoregressive models and strong consistency of the least squares estimates of their parameters, *Journal of Multivariate Analysis* **13**(1): 1–23.

Lewis, P.A.W. and Stevens, G. (1991). Nonlinear modeling of time series using multivariate adaptive regression splines (mars), *Journal of the American Statistical Association* **86**(416): 864–877.

Luukkonen, R., Saikkonen, P. and Teräsvirta, T. (1988). Testing linearity against smooth transition autoregressive models, *Biometrika* **75**(3): 491–499.

Pemberton, J. (1987). Exact least squares multi-step prediction from nonlinear autoregressive models, *Journal of Time Series Analysis* **8**(4): 443–448.

Petruccelli, J. and Davies, N. (1986). A portmanteau test for self-exciting threshold autoregressive-type nonlinearity in time series, *Biometrika* **73**(3): 687–694.

Priestley, M.B. (1988). *Non-linear and Non-stationary Time Series Analysis*, Academic Press, New York.

Robinson, P.M. (1977). The estimation of a nonlinear moving average model, *Stochastic Processes and Their Applications* **5**: 81–90.

Robinson, P.M. (1983). Non-parametric estimation for time series models, *Journal of Time Series Analysis* **4**(3): 185–208.

Subba Rao, T. (1981). On the theory of bilinear time series models, *Journal of the Royal Statistical Society, Series B* **43**(2): 244–255.

Subba Rao, T. and Gabr, M.M. (1980). *An introduction to bispectral analysis and bilinear time series models*, Vol. 24 of *Lecture Notes in Statistics*, Springer-Verlag, New York.

Sugihara, G. and May, R.M. (1990). Nonlinear forecasting as a way of distinguishing chaos from measurement error in time series, *Nature* **344**(6268): 734–741.

Tibshirani, R.J. (1988). Estimating transformations for regression via additivity and variance stabilization, *Journal of the American Statistical Association* **83**(402): 394–405.

Tjøstheim, D. (1986). Estimation in nonlinear time series models, *Stochastic Processes and Their Applications* **21**: 251–273.

Tong, H. (1978). On a threshold model, *in* C.H. Chen (ed.), *Pattern Recognition and Signal Processing*, Sijthoff and Noordholf, The Netherlands.

Tong, H. (1983). *Threshold Models in Nonlinear Time Series Analysis*, Vol. 21 of *Lecture Notes in Statistics*, Springer-Verlag, Heidelberg.

Tong, H. (1990). *Nonlinear Time Series Analysis: A Dynamic Approach*, Oxford University Press, Oxford.

Truong, Y.K. (1993). A nonparametric framework for time series analysis, *in* D. Billinger, P. Caines, J. Geweke, E. Parzen, M. Rosenblatt and M.S. Taqqu (eds), *New Directions in Time Series Analysis*, Springer-Verlag, New York.

Tsay, R.S. (1986). Nonlinearity tests for time series, *Biometrika* **73**(2): 461–466.

Tsay, R.S. (1989). Testing and modeling threshold autoregressive processes, *Journal of the American Statistical Association* **84**(405): 231–240.

Tsay, R.S. (1991). Detecting and modeling nonlinearity in univariate time series analysis, *Statistica Sinica* **1**(2): 431–451.

15
Un Digestif

Wolfgang Härdle[1] and Sigbert Klinke[2]

15.1 A Whole Bunch of Pictures

Un digestif après un bon repas! As a digestif after the XploRe chapters we offer the creation of a dynamic XploRe display. We shall mix six different types of windows into one display and dynamically rotate the **d3d** pictures. We shall do this exercise with columns 4 and 5 of the bank dataset.

We first create a display named **sixpic** and then read only columns 4 and 5 using the format options of **read**. Since we call density estimation routines in one and two dimensions we shall access the **library(smoother)**. Our aim is to show

1. a text window,

2. a 3-D view (of false and genuine banknote observations),

3. an automatic density estimate,

4. a scatterplot,

5. a 3-D view of the joint density, and finally

6. a boxplot of the two variables.

We also want to rotate the 3-D axes in such a way that we have an optimal view on the observations. This is usually done after the display is on the screen but for this digestif we shall use some tricks by writing XploRe internal keystrokes to the console. The XploRe codes for the different keys (for example, **<F6>** to start rotation) can be read off the key table in the XploRe manual. Not always do we have a manual at hand so we write the word **keytable** on the screen and use the *open key* **<F10>** to read off the codes. The picture created by the following XploRe program is shown in Figure 15.1:

---

[1]Institut für Statistik und Ökonometrie, Humboldt-Universität zu Berlin, D-10178 Berlin, Germany.

[2]Institut de Statistique, Université Catholique de Louvain, B-1348 Louvain-la-Neuve, Belgium; and Institut für Statistik und Ökonometrie, Humboldt-Universität zu Berlin, D-10178 Berlin, Germany.

```
proc()=main()
; ** produces picture sixpic
; ** duration 45 sec on 33 MHz
; --
  free(sixpicdisp)
  createdisplay(sixpicdisp, 2 3, text d3d s2d s2d d3d b1d)
  x = read("bank2" "d-d-d-dd")
  library ("smoother")
;  use "keytable" for the XploRe console code !
; text1
  x1 = x
; write ESC on console
  writecon(27)
  show(x1 text1)
; d3d1
  x2 = x~(aseq(1 rows(x)).<101)
; show axes and turn
  writecon(318)
  writecon(317)
  writecon(324)
  writecon(320)
  writecon(316)
  writecon(511)
  writecon(18)
  writecon(328)
  writecon(511)
  writecon(6)
  writecon(331)
  writecon(511)
  writecon(6)
  writecon(328)
  writecon(324)
  writecon(324)
  writecon(13)
  writecon(511)
  writecon(5)
  writecon(317)
  writecon(27)
  writecon(27)
  show(x2 d3d1)
; s2d1
  x31 = denauto(x[1:100,2])
  x32 = denauto(x[101:200,2])
  writecon(324)
  writecon(13)
  writecon(318)
  writecon(336)
  writecon(13)
  writecon(318)
```

```
  writecon(27)
  writecon(27)
  show(x31 x32 s2d1)
; s2d2
  x4 = x
  writecon(324)
  writecon(13)
  writecon(317)
  writecon(27)
  writecon(27)
  show(x4 s2d2)
; d3d2
  d  = (max(x)-min(x))./20
  h  = 5*d
  x5 = denest2(x h d)
  writecon(318)
  writecon(317)
  writecon(324)
  writecon(320)
  writecon(316)
  writecon(511)
  writecon(18)
  writecon(328)
  writecon(511)
  writecon(45)
  writecon(331)
  writecon(511)
  writecon(8)
  writecon(336)
  writecon(324)
  writecon(324)
  writecon(13)
  writecon(318)
  writecon(27)
  writecon(27)
  show(x5 d3d2)
; b1d
  x6 = x
  writecon(324)
  writecon(13)
  writecon(317)
  writecon(317)
  writecon(317)
  writecon(317)
  writecon(27)
  writecon(27)
  show (x6 b1d1)
  display(sixpicdisp)
endp
```

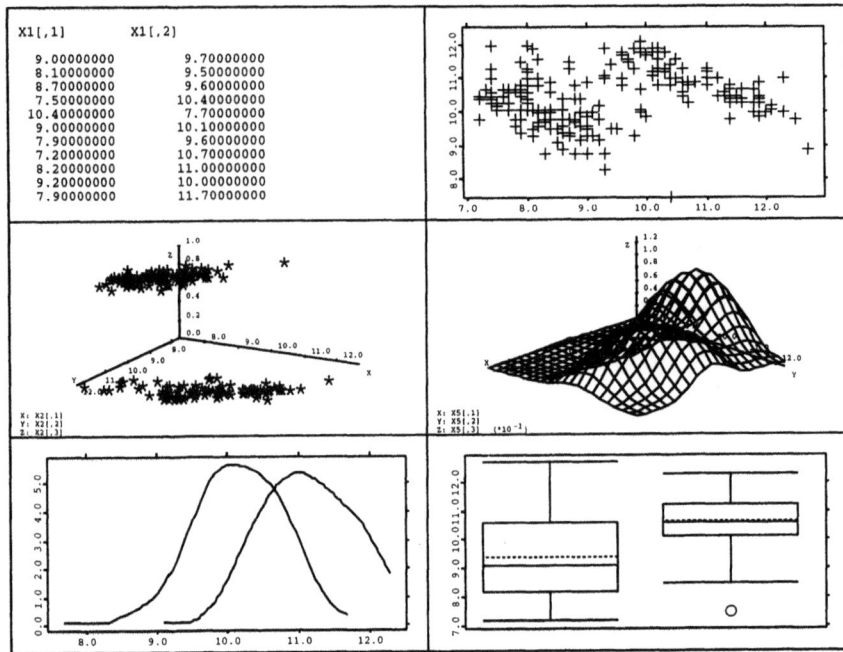

FIGURE 15.1. Swiss banknote dataset shown in various plots.

15.2   Interactive Contouring

As in Chapter 4, already shortly described, XploRe is able to generate contour plots. The macro **contour** of the HIGHDIM library provides an easy tool for interactive contouring for 2-D and 3-D surfaces. The only assumption that is needed is that the data are given in a special form. The data have to be on the vertices of a rectangular mesh. Thus our dataset has the form

$$(x_i, y_j, f_{ij}) \quad \text{or} \quad (x_i, y_j, z_k, f_{ijk})$$

with $i = 1, ..., n_i, j = 1, ..., n_j, k = 1, ..., n_k$. Let us compute as an example the density of three variables of the Swiss banknote data: the width of the banknote, the height of the left side, and the diagonal of the inner box. First we load the libraries HIGHDIM and SMOOTHER and then we construct our dataset:

```
library ("highdim", "smoother")
x = read ("bank2")
x = x[,1]~x[,2]~x[,6]
```

Since we want to make a kernel density estimate at grid points we compute the minima and the gridwidth in each variable:

```
x0 = min(x)
gw = (max(x)-min(x))./15
```

Then we generate a grid. For the kernel estimate we need additionally a bandwidth h and we can easily compute a kernel density estimate for each gridpoint $(15 + 1 = 16)$.

```
nh = 5
h  = nh.*gw
(xb yb) = bindata (x gw)
wx = matrix(cols(x) 1 0)
wy = symweigh (wx 1./nh nh &qua)
wx = grid(wx 1 nh)
(xc yc or) = conv (xb yb wx wy)
xs = (xc.*trn(gw))~(yc./(rows(x).*prod(h)))
xs
```

xs now contains the gridpoints and a density estimate. Looking at xs, you will see how the order of the gridpoints should be so that we can pass them immediately to the **contour** macro. To generate a 2-D surface for a 2-D dataset we have simply to generate an x with two columns. For a 2-D surface we may increase the number of gridpoints from 16 to 50. Please keep in mind that for a 3-D surface we have generated $16^3 = 2048$ gridpoints and for a 2-D surface we would have $50^2 = 2500$ gridpoints. To choose 50 gridpoints for each variable for a 3-D surface would result in $50^3 = 125000$ gridpoints. Depending on your installation this may cause a fatal error in XploRe, because of memory problems; and surely the execution of the program would last a long time.

Now we call the contouring macro with our dataset xs. Since we may want to store some contour plots in files we use the **capture** command. With the parameter on, we force XploRe to ask us before printing in which file we want to save the picture. If we would use another parameter here, for example, **mycont.ps**, the parameter would be used as a name for the graphic file. But if we want to store more than one picture the second picture would overwrite the first, the third the second, and so on.

```
capture ("on")
contour (xs)
```

The result of these commands is shown in Figure 15.2. The right window of the plot shows us some information about our contour lines and our densities. We have computed for each gridpoint (x_i, y_j, z_k) a density estimate f_{ijk}; it is an easy task to rescale the f_{ijk} such that

$$\tilde{f}_{ijk} = \frac{f_{ijk} - \min_{i,j,k} f_{ijk}}{\max_{i,j,k} f_{ijk} - \min_{i,j,k} f_{ijk}}.$$

It holds that $\tilde{f}_{ijk} \in [0, 1]$. Which contours are now plotted? You see a blue,

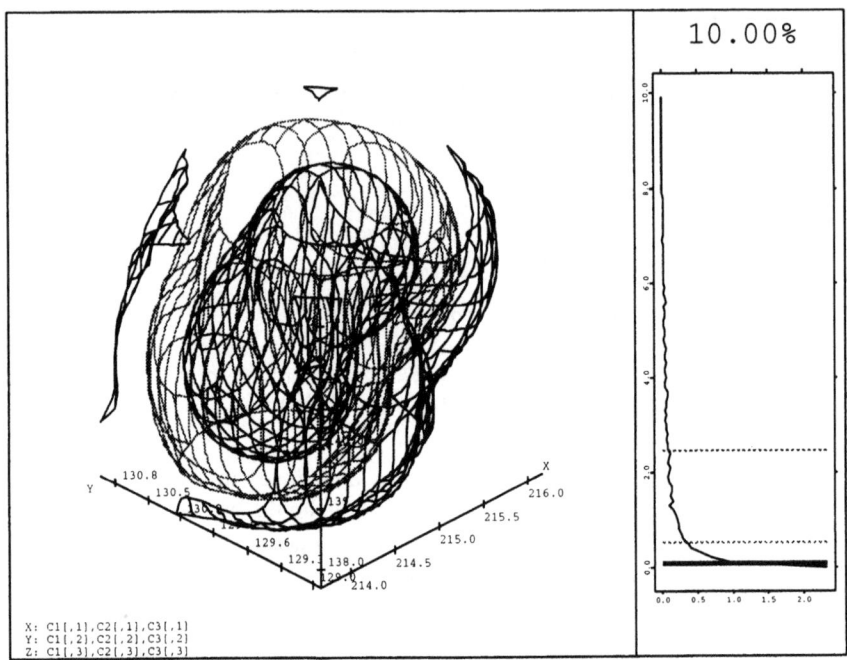

FIGURE 15.2. Kernel density estimate of three variables of the Swiss banknote dataset; automatic choosen contours.

green, and red line in the right window, which corresponds to values near 0.0, 0.05, and 0.25. Thus we have plotted the contours for

$$\tilde{f}(x, y, z) \quad \sim \quad 0.00 \quad \text{blue}$$
$$\sim \quad 0.05 \quad \text{green}$$
$$\sim \quad 0.25 \quad \text{red.}$$

The white (in the graphic black) line gives us information on how many $\tilde{f}_{ijk}$ we have at a specific contouring level. Although we have chosen a big bandwidth we can see that most of the generated gridpoints have a density of 0, which is the smallest value we can get. The "curse of dimensionality" is here again present!

Before we start to change the contour levels, let us have a look how XploRe chooses the contour levels. XploRe takes the 20%, the 50%, and the 80% quantile of the $\tilde{f}_{ijk}$. The blue line represents the 20% quantile, the green line the median, and the red line the 80% quantile. It may happen that they fall together, for example, make the bandwidth smaller (nh = 2), and you will miss one line, but XploRe always computes three contour levels.

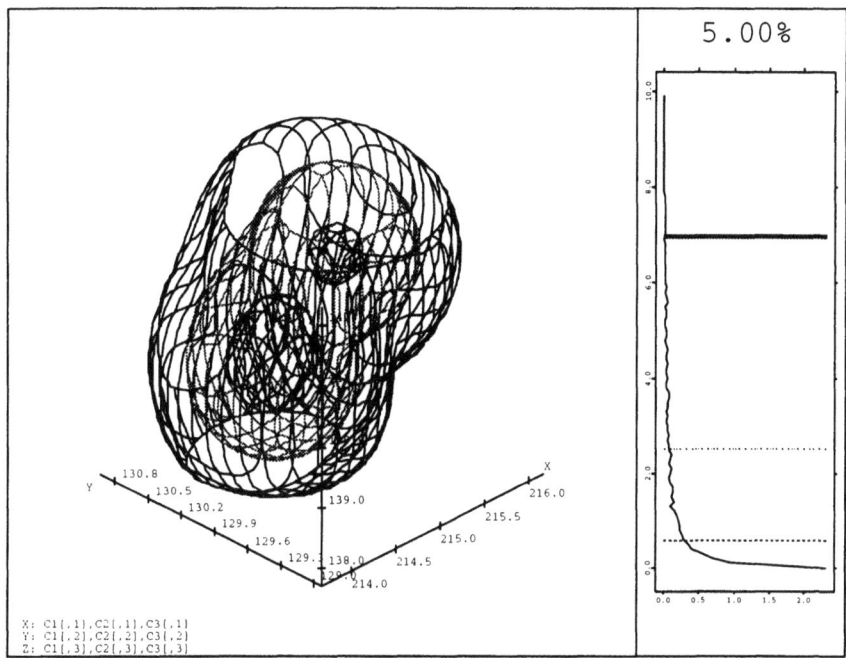

FIGURE 15.3. Kernel density estimate of three variables of the Swiss banknote dataset; user choosen contours.

Now use the keys <↑> and <↓> to change the contour level of the blue line. This is the active line and is plotted as a thick solid line, whereas the others are plotted as thin dashed lines. With <←> and <→> we can change how much a <↑> and a <↓> will increase our active level. The current value is plotted in the right window at the top and is 10%.

The keys <Pg Up> and <Pg Dn> make one of the other lines active. Another picture which you can obtain by changing the contour levels is shown in Figure 15.3.

Maybe you will not be satisfied with the actual perspective you have from the contour plot. Use <ENTER> to enter in the contour plot. Since this a normal 3-D window of XploRe, you can use all facilities of a 3-D picture, like rotating, brushing, and zooming. Leave the plot with <ESC> and the contouring macro with another <ESC>.

We would suggest you to try the following code, which will result in a wild picture.

```
library ("highdim")
x0 = #(0 0)
gw = #(0.02 0.02)
g  = grid(x0 gw 51)
xs = sort(g)~uniform(rows(g))
contour(xs)
```

A
Questions to XploRe

For questions concerning availability of new releases of XploRe, contact `haerdle@wiwi.hu-berlin.de` or

GfKI — Gesellschaft für Kommunikation und Information
Postfach 101248
D-68012 Mannheim
GERMANY
FAX: +49 621 812852

To get more information fill out the form on the next page and fax it to the number given above.

There also exists a mailing list for discussion of software problems. Mail to `sigbert@wiwi.hu-berlin.de` for subscribing or unsubscribing to the mailing list. After subscribing, send your mail to:

`xplmgr@wiwi.hu-berlin.de`

Fax-Message

1 page

To: GfKI — Gesellschaft für Kommunikation und Information
Postfach 101248
D-68012 Mannheim
GERMANY
FAX: ____49 621 812852

I am interested in more information on XploRe. Please send the material
to the following address:

Name   : _____

Address   : _____

                     _____

                     _____

                     _____

                     _____

Date   : _____

Signature : _____

B
Language Reference

This language reference follows the following syntax:

$\Rightarrow$ X means input to XploRe commands
X $\Rightarrow$ means output from XploRe commands

B.1 Operators

=   Assignment for matrices.

$$\boxed{y = x}$$

$\Rightarrow$ X x $n \times k$ matrix
X $\Rightarrow$ y $n \times k$ matrix

$'$ Operator for transposing a matrix.

$$\boxed{y = x'}$$

$\Rightarrow$ X x $n \times k$ matrix
X $\Rightarrow$ y $k \times n$ matrix

This operator is for manual construction of matrices. Float type and text matrices may be constructed.

$$\boxed{x = \#(x_{11} \ldots x_{1p}, \ldots, x_{n1} \ldots x_{np})}$$

$\Rightarrow$ X x_{ij} the matrix elements
X $\Rightarrow$ x $n \times p$ matrix

$\tilde{\ }$ | These two operators are concatenation operators. " $\sim$ " concatenates horizontally. " | " concatenates vertically. The number of rows and columns, respectively, have to match.

$$\boxed{z = x \tilde{\ } y} \qquad \boxed{z = x \mid y}$$

:   This operator creates a vector of values in ascending order.

$$x = s\ :\ n$$

$\Rightarrow$ X	s,n	number
X $\Rightarrow$	x	$(n - s + 1) \times 1$ vector

+   −   *   /   ^   %

These operators are addition (+), subtraction (−), multiplication (∗), division (/), exponentiation (^), elementwise and modulo division (%) of matrices.

$$\boxed{z = x + y}\quad \boxed{z = x - y}\quad \boxed{z = x * y}\quad \boxed{z = x/y}\quad \boxed{z = x\hat{\ }y}$$

$$\boxed{z = x\%y}$$

. >=   . <=   . =

These operators allow elementwise comparison of matrices. The elementwise value 0 is returned if the condition is false. Otherwise, the value of the expression is 1.

$$\boxed{b = x\ .>=\ y}\quad \boxed{b = x\ .<=\ y}\quad \boxed{b = x\ .=\ y}$$

. <>   . >   . <

These operators allow elementwise comparison of matrices. The elementwise value 0 is returned if the condition is false. Otherwise, the value of the expression is 1.

$$\boxed{b = x.<>y}\quad \boxed{b = x.>y}\quad \boxed{b = x.<y}$$

.+   .−   .*   ./   .^

These operators are elementwise addition (.+), subtraction (.−), multiplication (.∗), division (./), and exponentiation (.^). The matrices must be of the same size.

$$\boxed{z = x.+y}\quad \boxed{z = x.-y}\quad \boxed{z = x.*y}\quad \boxed{z = x./y}\quad \boxed{z = x.\hat{\ }y}$$

. * .   This operator perform a Kronecker product of two matrices.

$$z = x.*.y$$

< >   >   <

These operators allow global comparison of matrices. The number 0 is returned if the condition is not satisfied. Otherwise, the value is 1. Note that the value is one only if *all* elements of x and y satisfy the condition.

$$b = x <> y$$ $$b = x < y$$ $$b = x > y$$

= =   > =   < =

These operators allow global comparison of matrices. The number 0 is returned if the condition is not satisfied. Otherwise, the value is 1. Note that the value is one only if *all* elements of x and y satisfy the condition.

$$b = x == y$$ $$b = x >= y$$ $$b = x <= y$$

. <=>   .&   .| .?   .!

These are operators for elementwise logical operations: *equivalence, and, or, xor,* and *not.* Note that in contrast to the global comparison operators, these create matrices.

$$x = a. <=> b$$ $$x = a.\&b$$ $$x = a.|b$$ $$x = a.?b$$ $$x = .!b$$

$\Rightarrow X$ a,b $n \times p$ matrices
$X \Rightarrow$ x $n \times p$ matrix

<=>   && || ??   !!

These are operators for global logical operations: *equivalence, and, or, xor,* and *not.*

$$x = a <=> b$$ $$x = a\&\&b$$ $$x = a||b$$ $$x = a??b$$ $$x = !!b$$

$\Rightarrow X$ a,b $n \times p$ matrix
$X \Rightarrow$ x scalar

!   This operator calculates the factorial.

$$y = x!$$

$\Rightarrow X$ x $n \times p$ matrix
$X \Rightarrow$ y $n \times p$ matrix

& This operator makes a matrix with a list-vector containing the function.

$$\boxed{x = \&f}$$

$\Rightarrow$ X f function or macro
X $\Rightarrow$ x 1×1 matrix

_ This operator is inverse to the "&" operator. If one has defined a matrix with a function by the "&" operator, one calls this function by the underscore operator "_".

$$\boxed{f = \_x}$$

$\Rightarrow$ X x a matrix created by "&"
X $\Rightarrow$ f the result of the function $\_x$

[$[%[This operator creates a submatrix of a float, text, or mask vector.

$$\boxed{y = x[a, b]} \qquad \boxed{y = x\$[a, b]} \qquad \boxed{y = x\%[a, b]}$$

$\Rightarrow$ X x $n \times p$ matrix
            a,b   number or vector
X $\Rightarrow$ y number or submatrix

B.2   Flow Control

; Allows the user to make comments in macros. The rest of a textline after " ; " is ignored.

BREAK serves to break out of XploRe macros.

$$\boxed{\text{BREAK}(x)}$$

$\Rightarrow$ X x $n \times p$ matrix

DO   starts a loop until ENDO or UNTIL is found.

$$\boxed{\text{DO}}$$

ELSE   branches in IF sequences. The sequence is IF ... ELSE ... ENDIF. It is an optional part of this sequence.

$$\boxed{\text{ELSE}}$$

ENDIF

ends an IF sequence. It is an obligatory part of the sequence IF ... ELSE ... ENDIF.

$$\boxed{\text{ENDIF}}$$

ENDO   terminates a DO or a WHILE loop.

$$\boxed{\text{ENDO}}$$

ENDP   terminates XploRe macros.

$$\boxed{\text{ENDP}}$$

ERROR

allows the user to check whether certain conditions are fulfilled and stops the execution of the program if they are not true. In this case, a message (pointing to the error) may be issued to the user. This is a more serious breakpoint than that provided by the command BREAK.

$$\boxed{\text{ERROR}(x \quad mess)}$$

$\Rightarrow$ X x condition
        $mess$ a 1×1 textmatrix

EXIT   leaves a loop if the condition x is true.

$$\boxed{\text{EXIT}(x)}$$

$\Rightarrow$ X x a condition

IF   branches conditionally.

$$\boxed{\text{IF}(cond)}$$

$\Rightarrow$ X   *cond* a condition

PROC starts an XploRe macro. It is possible to write more then one macro in a file.

$$\boxed{\text{PROC } (outputparam) = name \ (inputparam)}$$

$\Rightarrow$ X   *name*   name of macro

  ...*param* list of matrices (could be empty)

UNTIL

  ends a loop if the condition x is true.

$$\boxed{\text{UNTIL } (x)}$$

$\Rightarrow$ X x a condition

WHILE

  begins a loop under condition x.

$$\boxed{\text{WHILE } (x)}$$

$\Rightarrow$ X x a condition

B.3 Commands

ABS   computes the absolute value of the elements of a matrix or a list of matrices.

$$\boxed{y = \text{ABS } (x)}$$

$\Rightarrow$ X x $n \times p$ matrix
X $\Rightarrow$ y $n \times p$ matrix

ACE   computes the alternating conditional expectation functions of Breiman and Friedman (1985). If $X \in I\!\!R^p$ denotes a predictor variable and $y \in I\!\!R$ is the response variable, it finds transformations $\psi, g_1, \ldots, g_p$ such that $E\left(\psi(y) - \sum_{j=1}^{p} g_j(x_j)\right)^2 / E\psi^2(y)$ is minimized.

$$\boxed{(\psi \; \varphi) \; = \; \text{ACE}(x \; y)}$$

$\Rightarrow$ X	x	$n \times p$ matrix
	y	$n \times 1$ matrix
X $\Rightarrow$	ψ	$n \times p$ matrix
	φ	$n \times 1$ matrix

ACOS   ACOSD
returns the arc cosine (in radian/degree) of the elements of a matrix.

$$\boxed{y \; = \; \text{ACOS}(x)}$$

$\Rightarrow$ X	x	$n \times p$ matrix
X $\Rightarrow$	y	$n \times p$ matrix

ACOSH
returns the hyperbolic arc cosine $\ln(x + \sqrt{x^2 - 1})$ of the elements of a matrix.

$$\boxed{y \; = \; \text{ACOSH}(x)}$$

$\Rightarrow$ X	x	$n \times p$ matrix
X $\Rightarrow$	y	$n \times p$ matrix

AGGLOM
performs hierarchical cluster analysis.

$$\boxed{(\text{p tree}) \; = \; \text{AGGLOM}(d \; mtd \; \{b\}\{w\})}$$

$\Rightarrow$ X d $n(n-1)/2 \times 1$ matrix of paired distances between
n objects (for example, d is computed by the command DISTANCE)

mtd string identifying the clustering method. The options (linkage strategies) for mtd are: WARD, SINGLE, COMPLETE, MEAN_LINK, MEDIAN_LINK, AVERAGE, CENTROID, LANCE

b scalar parameter for LANCE&WILLIAMS flexible method

w $n \times 1$ matrix of weights of clustering objects (optional used for the methods AVERAGE, CENTROID, and WARD only)

X $\Rightarrow$ p $n \times 1$ matrix containing the partition of clustering objects into k clusters (k is 20 at most: $k = \min(n, 20)$)

$tree$ 1×2 matrix containing a dendrogram with k clusters as the terminal nodes; show(tree S2d) draws the graph

---

ASEQ   creates an additive sequence of the form $x_i = x_{i-1} + dx$.

$$\boxed{y = \text{ASEQ}(x\ n\ dx)}$$

$\Rightarrow$ X x the starting point
n the number of points
dx the increment (default is 1)
X $\Rightarrow$ y $n \times 1$ the vector containing the additive sequence

---

ASIN   ASIND
returns the arc sine (in radian/degree) of a matrix.

$$\boxed{y = \text{ASIN}(x)}$$

$\Rightarrow$ X x $n \times p$ matrix
X $\Rightarrow$ y $n \times p$ matrix

ASINH

returns the hyperbolic arc sine $\ln(x + \sqrt{x^2 + 1})$ of a matrix or list.

$$y = \text{ASINH}(x)$$

$\Rightarrow$ X	x	$n \times p$ matrix
X $\Rightarrow$	y	$n \times p$ matrix

ATAN   ATAND

returns the arc tangent (in radian/degree) of a matrix.

$$y = \text{ATAN}(x)$$

$\Rightarrow$ X	x	$n \times p$ matrix
X $\Rightarrow$	y	$n \times p$ matrix

ATANH

returns the hyperbolic arc tangent $\frac{1}{2} \ln \left(\frac{1+x}{1-x} \right)$ of a matrix.

$$y = \text{ATANH}(x)$$

$\Rightarrow$ X	x	$n \times p$ matrix
X $\Rightarrow$	y	$n \times p$ matrix

BACKFIT

provides estimates of the functions f_i in the multiple additive regression model $E[Y \mid X] = \alpha + \sum_{i=1}^{d} f_i(x^i)$, $E[f_j(x^j)] = 0$ for all j where $x = (x^1, \ldots, x^d)$ is a d-dimensional variable using either a kernel-weighted local linear smoother, the supersmoother, or a k-NN running line smoother. BACKFIT can also be used within the local scoring algorithm to estimate generalized additive models (see GAMMOD). In this case, the optional parameters w, f, and xs can be used, where xs is a matrix of indices (from 0 to $n - 1$) which indicate how to sort the matrix x (to speed the calculations in the local scoring) and w gives the weights for the weighted fit of f_j.

$$(f \, a \, xs \, ctrl) = \text{BACKFIT}(x \, y \, \{w \, \{f \, \{xs \, \{span \, \{ctrl \, \{mod\}\}\}\}\}\})$$

$\Rightarrow$ X x $n \times d$ matrix with the explanatory variables

y $n \times 1$ matrix with the dependent variable

xs $n \times d$ matrix with integers from 0 to $n-1$ (if not given, by default the x matrix is sorted with respect to each column and the result is given back in xs for further use)

w $n \times 1$ matrix with weights (default is a vector of 1's)

f $n \times d$ matrix with initial guesses for the transformation (default is a matrix of 0's)

$span$ $p \times 3$ matrix, controls the smoothing process.

$ctrl$ 2×1 matrix, controls the iteration within the backfitting algorithm

mod scalar, if unequal to 0 the modified backfitting algorithm is performed (default is 0).

X $\Rightarrow$ f $n \times d$ matrix with the fitted functions

a scalar giving the estimate of the constant term

xs $n \times d$ matrix with integers from 1 to n

$ctrl$ 2×1 matrix, the first element is the number of iterations done in the backfitting procedure, the second element is the maximal (relative) change of the function estimates.

BETAF

returns the beta function $B(z\,w)$.

$$\boxed{bf = \text{BETAF}(z\,w)}$$

$\Rightarrow$ X z $n \times p$ matrix

w $n \times p$ matrix

X $\Rightarrow$ bf $n \times p$ matrix

BFGS   BFGS searches the global minimum of a function.

$$\boxed{(x\,h) = \text{BFGS}\ (x0\,f\,df\,maxit\,\{gam\,h0\,delta\,kappa\,alpha\,\{...\}\})}$$

$\Rightarrow$ X	$x0$	$n \times p$ matrix of starting values
	f	pointer to the procedure where the function is defined
	df	pointer to the procedure where the function derivative is defined
	$maxit$	maximal number of iterations
	gam	startvalue for linesearch (default $= 1$)
	$h0$	startvalue for hesseana matrix (default $=$ unit(n))
	$delta$	control parameter for linesearch (default $= 1/3$)
	$kappa$	control parameter for linesearch (default $= 2/3$)
	$alpha$	control parameter for linesearch (default $= 1/2$)
	$\ldots$	optional parameters that could be needed to calculate f and df
X $\Rightarrow$	x	$n \times p$ matrix with found minimum coordinates
	h	$(np) \times n$ matrix with the hesseana matrices

BINCOF

returns the binomial coefficients n over k.

$$b = \text{BINCOF}(n\ k)$$

$\Rightarrow$ X	x	$n \times p$ matrix
	k	$n \times p$ matrix
X $\Rightarrow$	b	$n \times p$ matrix

BINDATA

bins a p-dimensional data set starting from origin org in steps of d. The binning parameter d can be a scalar or a p-vector which gives a specific step for each column x_j $(j = 1, \ldots, p)$. The output matrix xb contains the index of the nonempty bins. The output yb contains in the first column the absolute frequency of datapoints x_{ij} in each nonempty bin. If the optional parameter y is used, the (elementwise) sum of the k-dimensional row-vectors of y, for which the corresponding row-vectors of x are in the same bin, is given back on the last k columns of yb.

$$(xb\ yb) = \text{BINDATA}(x\ d\ \{org\ \{y\}\})$$

$\Rightarrow$ X	x	$n \times p$ matrix
	d	p-vector or number
	org	p-vector or number (default 0)
	y	$n \times k$ matrix
X $\Rightarrow$	xb	$m \times p$ matrix
	yb	$m \times (k+1)$ matrix

CAPTURE

> redirects the print output from LPTx to a file.

$$\boxed{\text{CAPTURE (x)}}$$

$\Rightarrow$ X x name of the new printfile or one of the strings "off" or "on".

CARTCOOR

> transforms polar coordinates into Cartesian coordinates.

$$\boxed{x \ = \ \text{CARTCOOR}(w\ \{r\})}$$

$\Rightarrow$ X w $n \times p$ matrix
             r $n \times 1$ matrix
X $\Rightarrow$ x $n \times (p+1)$ matrix

CCTOV

> converts a mask vector to a float vector. The outfit and the color of a datapoint are calculated like $x = 16 * outfit + color$ with the following meanings for *outfit* and *color:*

Value	Outfit	Color
0	Unused	Black
1	Point	Blue
2	Plus	Green
3	X	Cyan
4	Y	Red
5	O	Magenta
6	Star	Brown
7	Face	Light Gray
8	Unused	Dark Gray
9	Unused	Light Blue
10	Unused	Light Green
11	Unused	Light Cyan
12	Unused	Light Red
13	Unused	Light Magenta
14	Unused	Yellow
15	Unused	White

$$\boxed{y = \text{CCTOV}(x)}$$

$\Rightarrow$ X x $n \times p$ matrix (p number of mask vectors)
X $\Rightarrow$ y $n \times p$ matrix (p number of float vectors)

CDFB CDFB returns the β-distribution function (with parameters a, b) of a matrix x.

$$y = \text{CDFB}(x\, a\, b)$$

$\Rightarrow$ X x $n \times p$ matrix
          a, b $n \times p$ matrix or a number
X $\Rightarrow$ y $n \times p$ matrix

CDFC returns the χ^2-distribution function with d degrees of freedom of a matrix.

$$y = \text{CDFC}(x\, d)$$

$\Rightarrow$ X x $n \times p$ matrix
          d $n \times p$ matrix or a number
X $\Rightarrow$ y $n \times p$ matrix

CDFF returns the F-distribution function with d_1 and d_2 degrees of freedom of a matrix.

$$y = \text{CDFF}(x\, d_1\, d_2)$$

$\Rightarrow$ X x $n \times p$ matrix
          d_1, d_2 $n \times p$ matrix or a number
X $\Rightarrow$ y $n \times p$ matrix

CDFN returns the standard normal distribution function of a matrix.

$$y = \text{CDFN}(x)$$

$\Rightarrow$ X x $n \times p$ matrix
X $\Rightarrow$ y $n \times p$ matrix

CDFT returns the t-distribution function with d degrees of freedom of a matrix.

$$y = \text{CDFT}(x\, d)$$

$\Rightarrow$ X x $n \times p$ matrix
          d $n \times p$ matrix or a number
X $\Rightarrow$ y $n \times p$ matrix

CEIL   is the ceiling operator and gives the next highest integer value.

$$y = \text{CEIL}(x)$$

$\Rightarrow$ X	x	$n \times p$ matrix
X $\Rightarrow$	y	$n \times p$ matrix

COLS   computes the number of float columns in a matrix.

$$y = \text{COLS}(x)$$

$\Rightarrow$ X	x	$n \times p$ matrix
X $\Rightarrow$	y	the number p

CONTOUR2

CONTOUR2 computes the contour line $f(x_i, y_j) = c$. You have to input the values assumed by $f(.)$ over a two-dimensional equidistant grid (x_i, y_j). If the grid is defined for the n values of x, $x_1, \ldots, x_n$ and for the m values of y, $y_1, \ldots, y_m$, $f(.)$ has to be stored in a vector of dimension $n * m$.

$$ct = \text{CONTOUR2}(x \ y \ f \ c \ \{opt\})$$

$\Rightarrow$ X	x	$n \times 1$ matrix
	y	$m \times 1$ matrix
	f	$n * m \times 1$ matrix
	c	scalar
	opt	0 or 1
X $\Rightarrow$	ct	$p \times 2$ matrix

CONTOUR3

CONTOUR3 computes the contour line $f(x_i, y_j, z_k) = c$. You have to input the values assumed by $f(.)$ over a three-dimensional equidistant grid (x_i, y_j, z_k). If the grid is defined for n values of x, $x_1, \ldots, x_n$, m values of y, $y_1, \ldots, y_m$, and q values of z, $z_1, \ldots, z_q$, $f(.)$ has to be stored in a vector of dimension $n*m*q$.

$$ct = \text{CONTOUR3}(x \ y \ z \ f \ c \ opt)$$

$\Rightarrow$ X	x	$n \times 1$ matrix
	y	$m \times 1$ matrix
	z	$q \times 1$ matrix
	f	$n * m \times 1$ matrix
	c	scalar
	opt	0 or 1

X $\Rightarrow$ ct $p \times 3$ matrix

CONV   performs the convolution of two vectors. It works on grids in the k-dimensional space. Therefore, x and wx are matrices of integer numbers which represent single grid points. The wx are interpreted in the same way. The y and wy vectors represent values at the grid points. CONV works simultaneously on m y-values at every grid point. It computes at the gridpoint

$$xc_w \text{ the value } yc_{w,i} = \sum_{g \in G} y_{w,i} \cdot I(wy_{g-w}),$$

where

$$I(wy_{g-w}) = \begin{cases} wy_{g-w} & \text{if } (g-w) \in wx, \\ 0 & \text{otherwise}, \end{cases}$$

and g runs over all gridpoints of the grid G:

$g = (g_1, \ldots, g_k)$ with $\min_{1 \le q \le n}(x_{j,q}) \le g_j \le \max_{1 \le q \le n}(x_{j,q})$. Thus, CONV computes the whole (!) grid from $\min_{1 \le q \le n}(x_{j,q})$ to $\max_{1 \le q \le n}(x_{j,q})$ in every direction. If the gridpoint xc_w was contained in x, then the output parameter or contains a 1, otherwise a 0. CONV wants to have symmetrical weights. In every direction there are two possibilities for symmetry: symmetry to the y-axis or to the origin. Standard value for sym is zero, which means symmetry on axis in all directions. Suppose you want to say to CONV that in the second and third direction the weights are symmetrically on origin. You build sym like this:

$0 \cdot 2^0$ symmetry on axis in first direction
$+1 \cdot 2^1$ symmetry on origin in second direction
$+1 \cdot 2^2$ symmetry on origin in third direction
$+0 \cdot 2^3$ symmetry on axis in fourth direction

and you get for sym the value 6.

$$\boxed{(xc \ yc \ or) = \text{CONV}(x \ y \ wx \ wy \ \{sym\})}$$

$\Rightarrow$ X x $n \times k$ integer matrix
          y $n \times m$ matrix
          wx $\ell \times k$ integer matrix
          wy $\ell \times 1$ vector
          sym number
X $\Rightarrow$ xc $p \times k$ matrix
          yc $p \times n$ matrix
          or $p \times 1$ vector

COS   COSD

returns the cosine of a matrix given (in radian/degree).

$$y = \text{COS}(x)$$

$\Rightarrow$ X x $n \times p$ matrix
X $\Rightarrow$ y $n \times p$ matrix

COSH

returns the hyperbolic cosine $\dfrac{e^x + e^{-x}}{2}$ of a matrix.

$$y = \text{COSH}(x)$$

$\Rightarrow$ X x $n \times p$ matrix
X $\Rightarrow$ y $n \times p$ matrix

COV

computes the covariance matrix of a matrix x. The covariance y is computed via the formula $x' * x/\text{rows}(x) - \text{mean}(x) * \text{mean}(x)'$.

$$y = \text{COV}(x)$$

$\Rightarrow$ X x $n \times p$ matrix
X $\Rightarrow$ y $p \times p$ matrix

CREATEDISPLAY

creates a new display with different windows.

$$\text{CREATEDISPLAY}(name,\ n\ \{m\},\ w1\ w2\ldots)$$

$\Rightarrow$ X $name$ string
       n integer
       m integer
       w type of the window (b1d, s2d, d3d, fcs or text)

CUMPROD

computes columnwise the cumulative product of a matrix.

$$z = \text{CUMPROD}(x)$$

$\Rightarrow$ X x $n \times p$ matrix
X $\Rightarrow$ z $n \times p$ matrix

CUMSUM

computes columnwise the cumulative sum of a matrix.

$$z \ = \ \text{CUMSUM}(x)$$

$\Rightarrow$ X x $n \times p$ matrix
X $\Rightarrow$ z $n \times p$ matrix

DATE

computes a string of the form $dd.mm.yyyy$ containing the date.

$$y \ = \ \text{DATE}$$

$\Rightarrow$ X y the date string

DET

computes the determinant of a matrix.

$$y \ = \ \text{DET}(x)$$

$\Rightarrow$ X x $n \times n$ matrix
X $\Rightarrow$ y number

DIAG

creates a diagonal matrix from a vector x.

$$y \ = \ \text{DIAG}(x)$$

$\Rightarrow$ X x $p \times 1$ vector
X $\Rightarrow$ y $p \times p$ diagonal matrix

DIFF

computes for each column in a matrix the difference between the elements in row i with the elements in row $i - 1$.

$$z \ = \ \text{DIFF}(x)$$

$\Rightarrow$ X x $n \times p$ matrix
X $\Rightarrow$ z $(n - 1) \times p$ matrix

DISPLAY

pops an existing display on the screen.

$$\text{DISPLAY}(name)$$

$\Rightarrow$ X $name$ name of the display

DOS   allows a temporary exit to the DOS level. A prompt "XploRe>"
is created. A return to XploRe is provided by typing "exit" on
the DOS level.

$$\boxed{\text{DOS}}$$

ECLOCK
measures the time since the last call to SCLOCK.

$$\boxed{y \;=\; \text{ECLOCK}}$$

$\Rightarrow$ X y the time in seconds

EDIT   edits files and matrices.

$$\boxed{\text{EDIT(x)}}$$

$\Rightarrow$ X x $n \times p$ matrix or an external filename

---

EIGGN
computes the eigenvalues and eigenvectors of a general complex
matrix (real part $=$ xr, imaginary part $=$ xi). The resulting
matrices are:
wr the real part of the eigenvalues
wi the imaginary part of the eigenvalues
vr the real part of the eigenvectors
vi the imaginary part of the eigenvectors.

$$\boxed{(wr\ wi\ vr\ vi) = \text{EIGGN}(xr\ \{xi\})}$$

$\Rightarrow$ X xr $p \times p$ matrix
                   xi $p \times p$ matrix
X $\Rightarrow$ wr $p \times 1$ matrix
                   wi $p \times 1$ matrix
                   vr $p \times p$ matrix
                   vi $p \times p$ matrix

EIGSM
computes the eigenvalues and eigenvectors of a symmetrical matrix in unsorted order.

$$\boxed{(w\ v) \;=\; \text{EIGSM}(x)}$$

$\Rightarrow$ X x $p \times p$-matrix

$$X \Rightarrow \quad w \qquad p \times 1\text{-matrix (eigenvalues)}$$
$$v \qquad p \times p\text{-matrix (normalized eigenvectors)}$$

EXIST

EXIST gives a number back, whether an object exists or not. The results are:

0	object does not exist
1	object is a matrix
2	object is a display
3	object is a macro

$$\boxed{r = \text{EXIST}(name)}$$

$$\Rightarrow X \quad name \text{ objectname}$$
$$X \Rightarrow \quad r \qquad 1 \times 1 \text{ matrix}$$

EXP   EXP2   EXP10

produce various forms of exponentiation. EXP is the exponentiation to base $e = 2.7181\ldots$, EXP2 to base 2, and EXP10 to base 10.

$$\boxed{y = \text{EXP}(x)} \quad \boxed{y = \text{EXP2}(x)} \quad \boxed{y = \text{EXP10}(x)}$$

$$\Rightarrow X \quad x \qquad n \times p \text{ matrix}$$
$$X \Rightarrow \quad y \qquad n \times p \text{ matrix}$$

FFT

computes the fast Fourier transformation of a complex vector. The first column is the real part of the vector, and the second column is the imaginary part.

$$\boxed{y = \text{FFT}(x)}$$

$$\Rightarrow X \quad x \qquad p \times k \quad \text{matrix} \qquad (k = 1, 2)$$
$$X \Rightarrow \quad y \qquad p \times 2 \quad \text{matrix}$$

FLOOR

is the operation that gives the next smallest integer value.

$$\boxed{y = \text{FLOOR}(x)}$$

$$\Rightarrow X \quad x \qquad n \times p \quad \text{matrix}$$
$$X \Rightarrow \quad y \qquad n \times p \quad \text{matrix}$$

FREE   FREE is used to dispose memory and delete matrices, displays, and functions. Even the standard displays may be deleted. If x is itself a list of matrices, then FREE kills all elements of x and then x itself.

$$\boxed{\text{FREE}(x)}$$

$\Rightarrow$ X x $n \times p$ matrix or list of matrices

FUNC   This command interprets a list as a function. After this call, the list x can be used as a function x.

$$\boxed{\text{FUNC}(x)}$$

$\Rightarrow$ X x a name of a file

GAMMAF
        returns the logarithm of the gamma function.

$$\boxed{y = \text{GAMMAF}(x)}$$

$\Rightarrow$ X x $n \times p$ matrix
X $\Rightarrow$ y $n \times p$ matrix

GLS   computes the generalized least-squares estimate. If $Y = X\beta + \varepsilon$, then with a weight matrix Ω, the generalized least-squares estimate for β is

$$\widehat{\beta} = (X^T\, \Omega^{-1}\, X)^{-1}\, X^T\, \Omega^{-1}\, Y.$$

$$\boxed{z = \text{GLS}(x\ y\ \{\Omega\})}$$

$\Rightarrow$ X x $n \times p$ matrix
        y $n \times m$ matrix
        Ω $n \times n$ matrix
X $\Rightarrow$ z $p \times m$ matrix

GRID   generates a grid with origin x, stepwidth h, and n steps in each dimension.

$$\boxed{y = \text{GRID}(x\ h\ n)}$$

$\Rightarrow$ X x, h, n vector or number
X $\Rightarrow$ y matrix

HELP   displays all XploRe elements on the standard text display. With the <F10> key, you can get detailed help on the command to which the cursor is pointing. This command is equivalent to <ALT H> and to pressing the <F1> key.

$$\boxed{\text{HELP}}$$

INDEX

generates a new matrix y from an old matrix x by extracting the rows indicated in the index matrix i.

$$\boxed{y = \text{INDEX}(x\ i)}$$

$\Rightarrow$ X x $n \times p$ matrix
              i $m \times 1$ matrix
X $\Rightarrow$ y $m \times p$ matrix

INFO   gives status information on:
- the date, time
- user, license number
- screen, coprocessor type
- available memory
- name of logfile, tracemodus
- mouse parameters
- actual path, number of objects
- printer settings

INFO is equivalent to pressing <F7>.

$$\boxed{\text{INFO}}$$

INTEGRAL

calculates the integral numerically from $b[1,1]$ to $b[2,1]$.

$$\boxed{intval\ =\ \text{INTEGRAL}(b\ n\ func)}$$

$\Rightarrow$ X b 2×1 matrix (integral borders)
              n scalar (exactness of integral)
              $func$ matrix containing a function
X $\Rightarrow$ $intval$ scalar result of the numerical integration

INV computes the inverse of a matrix. The optional parameter gives the dimension of the matrix, which should be inverted. It is expected that m is a divisor of $p = cols(x)$ and $n = rows(x)$. It will be $cols(x)/m \times rows(x)/m$ inversions on the submatrices executed.

$$y = \text{INV}(x \; \{m\})$$

$\Rightarrow$ X	x	$n \times p$	matrix
X $\Rightarrow$	y	$n \times p$	matrix

INVFFT

computes the inverse fast Fourier transformation of a complex vector. The first column is the real part of the vector, and the second column is the imaginary part.

$$y = \text{INVFFT}(x)$$

$\Rightarrow$ X	x	$p \times k$	matrix	$(k = 1, 2)$
X $\Rightarrow$	y	$p \times 2$	matrix	

ISOREG

computes the isotonic regression smoother via the Pool Adjacent Violators algorithm.

Given a data set $\{(X_i, Y_i)\}$ where $X_i \leq X_{i+1}$ $i = 1, \ldots, n$ the algorithm finds the values $\{\hat{X}_i\}$ $i = 1, \ldots, n$, minimizing

$$(1/n) \sum_{i=1}^{n} [Y_i - \hat{X}_i]^2$$

subject to $\hat{X}_i \leq \hat{X}_{i+1}, i = 1, \ldots, n.$

The input matrix must contain in the first column the values of the exogenous variable X, and in the second column the dependent Y. The output matrix contains in the first column the (ordered) values X_i and in the second column the respective $\hat{X}_i$.

$$y = \text{ISOREG}(x)$$

$\Rightarrow$ X	x	$n \times 2$	matrix
X $\Rightarrow$	y	$n \times 2$	matrix

KMEANS

performs cluster analysis, that is, computes a partition of n row points into K clusters.

$$(g\ c\ v\ s) = \text{KMEANS}(x\ b\ \{w\ \{m\}\})$$

$\Rightarrow$ X	x	$n \times p$ matrix: data matrix (n points, p variables)
	b	$n \times 1$ matrix: initial partition (for example, random generated numbers of clusters 1,2,...,K)
	w	$p \times 1$ matrix with the weights of column points
	m	$n \times 1$ matrix of weights (masses) of row points
X $\Rightarrow$	g	$n \times 1$ matrix containing the final partition which gives a minimum sum of within cluster variances
	c	$k \times p$ matrix of means (centroids) of the K clusters
	v	$k \times p$ matrix of within cluster variances divided by the weight (mass) of clusters
	s	$k \times 1$ matrix of the weight (mass) of clusters

KRON   computes the Kronecker (tensor) product of two matrices.

$$z = \text{KRON}(x\ y)$$

$\Rightarrow$ X	x	$n \times p$ matrix
	y	$m \times \ell$ matrix
X $\Rightarrow$	z	$nm \times \ell p$ matrix

L1LINE

computes the least-absolute deviation line from scatterplot data. It gives the estimates $b1$ and $b2$ that minimize

$$\sum_{i=1}^{n} |Y_i - (b1 + b2 * X_i)|.$$

The input matrix contains in the first column the exogenous variable X and in the second column the endogenous Y. The output matrix gives the estimates $b1$ and $b2$.

$$z = \text{L1LINE}(x)$$

$\Rightarrow$ X	x	$n \times 2$ matrix
X $\Rightarrow$	z	2×1 matrix

LIBRARY

> loads an XploRe library.

$$\boxed{\text{LIBRARY}(name)}$$

> $\Rightarrow$ X   *name* name of a library; Standard libraries are: ADD-MOD, COMPLEX, GAM, GLM, HIGHDIM, SMOOTHER, TEACHWAR, and XPLORE

LINK   links dataparts and matrices.

$$\boxed{\text{LINK}(n_1 \; n_2)}$$

> $\Rightarrow$ X n_1 text
> n_2 text

LOG   LOG10   LN   LG   LOG2

> LOG and LN return the natural logarithm, LOG10 and LG return the logarithm of base 10, and LOG2 returns the logarithm of base 2 of a matrix.

$$\boxed{y = \text{LOG}(x)} \quad \boxed{y = \text{LOG2}(x)} \quad \boxed{y = \text{LOG10}(x)}$$

> $\Rightarrow$ X x $n \times p$ matrix
> X $\Rightarrow$ y $n \times p$ matrix

LOGFILE

> allows the definition of a logfile (the development of the session may be seen later by the F6 hotkey).

$$\boxed{\text{LOGFILE}(name)}$$

> $\Rightarrow$ X   *name* name of the new logfile

LOWESS

> computes the smooth of a scatterplot of Y against X using robust locally weighted regression. Fitted values, ys, are computed at each of the values of the horizontal axis in X.

$$\boxed{ys = \text{LOWESS}(x \; y \; \{f\})}$$

> $\Rightarrow$ X x $n \times 1$ matrix
> y $n \times 1$ matrix
> f 1×1 matrix
> X $\Rightarrow$ ys $n \times 1$ matrix

LU   computes the LU decomposition of a matrix.

$$(z\ i) = \mathrm{LU}(x)$$

$\Rightarrow$ X	x	$n \times n$ matrix
X $\Rightarrow$	z	$n \times n$ matrix
	i	$n \times 1$ matrix

MASK creates a mask matrix. This mask matrix can be appended to a data matrix X.

$$y = \mathrm{MASK}(n\ p\ \{c\}\ \{t\})$$

$\Rightarrow$ X	n	length of mask matrix
	p	dimension of mask matrix
	c	color of mask
	t	type of symbols
X $\Rightarrow$	y	mask matrix of size $n \times p$

MATRIX
        generates a matrix with prespecified values.

$$y = \mathrm{MATRIX}(n\ \{p\ \{v\}\})$$

$\Rightarrow$ X	n	the number of rows
	p	the number of columns (default $= 1$)
	v	the entry of the generated matrix
X $\Rightarrow$	y	$n \times p$ matrix

MAX   computes the maximal values of each column for a given matrix.

$$y = \mathrm{MAX}(x)$$

$\Rightarrow$ X	x	$n \times p$	matrix
X $\Rightarrow$	y	$p \times 1$	matrix

MBUTTON
        sets the XploRe codes for the left, middle, and right mousebutton.

$$\mathrm{MBUTTON}(\mathit{left\ mid\ right})$$

$\Rightarrow$ X	left	number
	mid	number
	right	number

MCLICK

sets the time between two mouseclicks. If you set the time to 1 second and press a mousebutton, the next mouseclick is accepted at least 1 second after the last mouseclick.

$$\boxed{\text{MCLICK}(time)}$$

⇒ X *time* scalar between 1 and 1000 expressing time in milliseconds. Default is 200

MEAN

computes the columnwise mean of a data matrix, that is, $y_j = n^{-1}\sum_{i=1}^{n} x_{ij}$.

$$\boxed{y = \text{MEAN}(x)}$$

⇒ X *x* $n \times p$ matrix
X ⇒ *y* $p \times 1$ vector

MEDIAN

calculates the median of each column for a given matrix.

$$\boxed{y = \text{MEDIAN}(x)}$$

⇒ X *x* $n \times p$ matrix
X ⇒ *y* $p \times 1$ vector

MEMAVAIL

returns the amount of available memory measured in units of bytes.

$$\boxed{y = \text{MEMAVAIL}}$$

⇒ X *y* 1×1 matrix

MESSAGE

writes text and numbers to the command line for information. It can also write text- and numbermatrices.

$$\boxed{\text{MESSAGE } (m1 \ \{m2\ldots\}\{WAIT\})}$$

⇒ X *mi* 1×1 text- or numbermatrix

MIN   computes the minimal value of each column of a matrix.

$$y = \text{MIN}(x)$$

$\Rightarrow$ X x $n \times p$ matrix
X $\Rightarrow$ y $p \times 1$ matrix

MIX   mixes the columns of two matrices in order to create a new matrix.

$$z = \text{MIX}(x\ y\ \{m\})$$

$\Rightarrow$ X x $n \times p$ matrix
            y $n \times q$ matrix
            m $l \times 1$ matrix (optional)
X $\Rightarrow$ z $n \times l$ matrix

MOUSE

set the mouse sensitivity.

$$\text{MOUSE}(x)$$

$\Rightarrow$ X x a number between 1 and 1000

MSEQ   creates a multiplicative sequence of the form
$x_i = x_{i-1} \cdot dx$ starting at x.

$$y = \text{MSEQ}(x\ n\ \{dx\})$$

$\Rightarrow$ X x starting point
            n number of point
            dx increment, default is 1
X $\Rightarrow$ y $n \times 1$ vector containing the multiplicative sequence.

MSPEED

sets mouse speed. Changing the speed makes the mouse faster or slower. Low numbers make the mouse fast; big numbers slow it down.

$$\text{MSPEED}\ (x)$$

$\Rightarrow$ X x number between 1 and 1000.

NELMIN

searches a minimum of a function. In each iteration step, the function is evaluated at a simplex consisting of $p+1$ points. The simplex contracts until the variance of the evaluated function values is less than *req* (or the maximal number of iterations is reached).

$$x = \text{NELMIN}(x0 \; f \; req \; \{step \; ic \; \{\dots\}\})$$

$\Rightarrow$ X	$x0$	$n \times p$ matrix with the starting values
	f	pointer to the procedure where the function is defined
	req	scalar
	$step$	scalar with the length of initial simplices sides
	ic	scalar with the maximal number of iteration (default is icount=1000)
X $\Rightarrow$	x	$p \times 1$ matrix with the x-coordinates of a minimum

NORMAL

generates pseudo-random variables with a standard normal distribution. The algorithm by Box–Muller is used.

$$y = \text{NORMAL} \; (n \; \{p\})$$

$\Rightarrow$ X	n	the number of rows to be generated
	p	the number of columns to be generated (default is $p = 1$)
X $\Rightarrow$	y	$n \times p$ matrix or a list of such matrices.

OBJECTS

displays the current XploRe objects. When XploRe is started, one has 6 standard display objects and the numbers EH = 2.718..., PI = 3.1415... and NAN = Not A Number.

Pressing <F10> while the cursor is on one of the objects produces more information. So if the cursor is on top of a matrix object x, the matrix will be displayed. If the cursor is on top of a picture (window) object, this graphical object will be displayed. If the cursor is on top of a macro object, the macro is shown.

$$\boxed{\text{OBJECTS}}$$

ORTHO

generates an orthonormal vector system from x with the Erhard–Schmidt Method. The first vector of the orthonormal system is the normalized first column vector of x. Every other vector is calculated by

$$y_i = \frac{x_i - \sum_{j=1}^{i-1} \langle y_j, x_i \rangle y_j}{\| x_i - \sum_{j=1}^{i-1} \langle y_j, x_i \rangle y_j \|}.$$

$< .,. >$ is standard-scalarproduct of $\mathbb{R}^n$, and $\|.\| = \sqrt{< .,. >}$.

$$\boxed{y = \text{ORTHO}(x)}$$

$\Rightarrow \text{X} \quad x \quad n \times p \quad$ matrix
$\text{X} \Rightarrow \quad y \quad n \times p \quad$ matrix

PAF   deletes entries of a matrix. The entry x_{ij} is deleted if $l_{ij} = 0$.

$$\boxed{y = \text{PAF}(x\ l)}$$

$\Rightarrow \text{X} \quad x \quad n \times p$ matrix
$\qquad\qquad l \quad n \times 1, 1 \times p,$ or $n \times p$ matrix with 0 or 1
$\text{X} \Rightarrow \quad y \quad k \times q$ matrix where $k = max_j \sum_i l_{ij}$ and
$\qquad\qquad\qquad q = max_i \sum_j l_{ij}$

PALETTE

changes the internal color palette to user-specified colors given in the RGB-format. Of course the standard meaning for colors (for example, in VTOCC) is not valid anymore. The columns containing the intensity for Red, Green, and Blue have values from 0 to 63. This also changes the output for color postscript.

$$\boxed{\text{PALETTE (rgb)}}$$

$\Rightarrow \text{X} \quad rgb \quad 16 \times 3$ matrix

PATH   sets the XploRe standard paths for the data, help, library, working, save, print directories.

$$\boxed{\text{PATH } (path\ \ name)}$$

$\Rightarrow$ X   *path*   pathsymbol is one of the following:

     datadir

     helpdir

     libdir

     workdir

     savedir

     printdir

   name string with the name of the path to be set

PAUSE

stops the execution of a macro for some time.

$$\boxed{y = \text{PAUSE}(time)}$$

$\Rightarrow$ X   *time*   number

PDFB   returns the values of the density of the *Beta*-distribution with parameters a, b.

$$\boxed{y = \text{PDFB}(x\ a\ b)}$$

$\Rightarrow$ X x $n \times p$ matrix

   a, b $n \times p$ matrix or a number

X $\Rightarrow$ y $n \times p$ matrix with values of the *Beta*-distribution

---

PDFC   returns the values of the density of the χ^2-distribution with d degrees of freedom.

$$\boxed{y = \text{PDFC}(x\ d)}$$

$\Rightarrow$ X x $n \times p$ matrix

   d $n \times p$ matrix or a number

X $\Rightarrow$ y $n \times p$ matrix with the values of the density

---

PDFF   returns the values of the density of the F-distribution with d_1 and d_2 degrees of freedom.

$$\boxed{y = \text{PDFF}(x\ d_1\ d_2)}$$

$\Rightarrow$ X x $n \times p$ matrix

   d_1, d_2 $n \times p$ matrix or a number

X $\Rightarrow$ y $n \times p$ matrix

PDFN   returns the values of the density of the standard normal distribution.

$$y = \text{PDFN}(x)$$

$\Rightarrow$ X x $n \times p$ matrix
X $\Rightarrow$ y $n \times p$ matrix

PDFT   returns the value of the t-distribution with d degrees of freedom.

$$y = \text{PDFT}(x\ d)$$

$\Rightarrow$ X x $n \times p$ matrix
         d $n \times p$ matrix or a number
X $\Rightarrow$ y $n \times p$ matrix

POLARCOOR
        transforms Cartesian coordinates into polar coordinates.

$$(w\ r) = \text{POLARCOOR}(x)$$

$\Rightarrow$ X x $n \times p$ matrix
X $\Rightarrow$ w $n \times (p-1)$ matrix
         r $n \times 1$ matrix

PRINT
        prints the screen on a connected printer. It is equivalent to
        <Alt P>.

$$\boxed{\text{PRINT}}$$

PROD   performs columnwise multiplication of a matrix.

$$y = \text{PROD}(x)$$

$\Rightarrow$ X x $n \times k$ matrix
X $\Rightarrow$ y $k \times 1$ vector

PRODR
        computes the product of the elements in each row of a matrix.

$$y = \text{PRODR}(x)$$

$\Rightarrow$ X x $n \times k$ matrix

$$X \Rightarrow \quad y \qquad n \times 1 \quad \text{vector}$$

QFC   returns the inverse of a χ^2-distribution function with d degrees of freedom.

$$\boxed{y = \text{QFC}(x \; d)}$$

$\Rightarrow$ X	x	$n \times p$ matrix
	d	$n \times p$ matrix or scalar
X $\Rightarrow$	y	$n \times p$ matrix

QFF   returns the inverse of an F-distribution function with $d1$ and $d2$ degrees of freedom.

$$\boxed{y = \text{QFF}(x \; d1 \; d2)}$$

$\Rightarrow$ X	x	$n \times p$ matrix
	$d1$	$n \times p$ matrix or scalar
	$d2$	$n \times p$ matrix or scalar
X $\Rightarrow$	y	$n \times p$ matrix

QFN   returns the inverse of a Standard Normal distribution function.

$$\boxed{y = \text{QFN}(x)}$$

$\Rightarrow$ X	x	$n \times p$ matrix
X $\Rightarrow$	y	$n \times p$ matrix

QFT   returns the inverse of a t-distribution function with d degrees of freedom.

$$\boxed{y = \text{QFT}(x \; d)}$$

$\Rightarrow$ X	x	$n \times p$ matrix
	d	$n \times p$ matrix or scalar
X $\Rightarrow$	y	$n \times p$ matrix

QUANTILE
     computes the quantiles of the columns of x. Simultaneously, k quantiles are computed.

$$\boxed{y = \text{QUANTILE}(x \; xp)}$$

$\Rightarrow$ X	x	$n \times p$	matrix
	xp	$k \times 1$	matrix

$$\text{X} \Rightarrow \quad y \quad k \times p \quad \text{matrix}$$

QUIT leaves the XploRe system. It is equivalent to <ALT X>.

$$\boxed{\text{QUIT}}$$

RANDOMIZE

sets the seed of the pseudonumber generators. If XploRe starts, it calls **randomize** (0).

$$\boxed{\text{RANDOMIZE}(seed)}$$

$\Rightarrow$ X   *seed*   the seed number.

READ   is a multipurpose command to read data from a file. Each column of the file will be interpreted as a vector with a given type (float, text, list, or mask). You can have vectors of different types in the same file.

$$\boxed{y = \text{READ}(file \ \{fmt \ \{lno\}\})}$$

$\Rightarrow$ X   *file*   name of file
          fmt   text (formatstring)
          lno   vector (max. linenumbers)
X $\Rightarrow$ y matrix

READCON

reads one character from the keybuffer and computes the ASCII value of it. The parameter t gives the time in seconds READCON waits for a keypress. If $t < 0$, then READCON will wait until a key is pressed.

$$\boxed{x = \text{READCON } (t)}$$

$\Rightarrow$ X t scalar, maximum time to wait
X $\Rightarrow$ x scalar, the ASCII value of the read character

---

READSTR

reads text from a box. A message in the box is shown to the user and a string input is expected. After writing the string, if <ENTER> is pressed, a text vector is created containing the string; if <ESC> is pressed, it creates an empty text vector.

$$\boxed{s = \text{READSTR}(t)}$$

$\Rightarrow$ X t $n \times 1$ textmatrix or text
X $\Rightarrow$ s 1×1 textmatrix with the string

READVAL

reads a number (integer or float) from a box. The input is terminated with \<ENTER\> or \<ESC\>. If the \<ESC\> key is pressed, READVAL returns zero.

$$v = \mathrm{READVAL}(t)$$

$\Rightarrow$ X t $n \times 1$ textmatrix or text
X $\Rightarrow$ v 1×1 matrix with the value

RMED

computes the running median from an input vector x. The optimal median smoothing algorithm of Härdle and Steiger (1995) is used.

$$y = \mathrm{RMED}(xk)$$

$\Rightarrow$ X x $n \times p$ vector
       k scalar smoothing parameter
X $\Rightarrow$ y $n \times p$ vector
            containing the running medians of each column of x
            the $k/2$ boundary values are set constant.

ROWS computes the number of rows in a matrix.

$$y = \mathrm{ROWS}(x)$$

$\Rightarrow$ X x $n \times p$ matrix
X $\Rightarrow$ y the number n

RPR builds up a recursive partitioning regression tree. Several options are possible, among them to pursue the tree by cross-validating the complexity. An output protocol may be created and the minimal bucket size for terminal nodes may be selected. For details, see Breiman, Friedman, Olshen and Stone (1984).

$$\mathrm{pred} = \mathrm{RPR}\ (x\ y\ \{pvec\}\ \{z\})$$

$\Rightarrow$ X x $n \times p$ matrix
         y $n \times 1$ matrix
         $pvec$ 5×1 matrix
         z $m \times p$ matrix
X $\Rightarrow$ $pred$ $m \times 1$ matrix, if z is given

SCLOCK

sets the internal clock to start. Time of operations can be measured by a call to ECLOCK.

$$\boxed{y = \text{SCLOCK}}$$

$\Rightarrow$ X
X $\Rightarrow$ y the time in seconds

SETPRINTER

changes the options for the chosen printer-driver, that has been defined via the configuration program `xconfig.exe`.

$$\boxed{\text{SETPRINTER } (\{r\}\ \{h\}\ \{v\}\ \{f\})}$$

$\Rightarrow$ X r resolution
         h horizontal expansion
         v vertical expansion
         f format

SHOW

is a multipurpose function to display text and data. With SHOW you can obtain BoxPlot pictures, two-dimensional graphics, three-dimensional graphics, and Flury Faces.

$$\boxed{\text{SHOW } (x_1\ \{x_2\{\ldots\}\}\ w)}$$

$\Rightarrow$ X x_i $n \times p$ matrix
         w string with the window type. Possible types are **b1d**, **s2d**, **d3d**, **fcs**, and **text**. If we have more than one window in a display, a number has to be concatenated to the type (for example, **b1d1**).

SIGN   calculates the sign function. If $b_{ij} \geq 0$, then $z_{ij} = |x_{ij}|$ else $z_{ij} = -|x_{ij}|$.

$$\boxed{z = \text{SIGN}(x\ b)}$$

$\Rightarrow$ X x $n \times p$ matrix
         b $n \times p$ or $n \times 1$ or $1 \times p$ matrix

$$X \Rightarrow \quad z \quad n \times p \quad \text{matrix}$$

SIN   SIND
returns the sine of a matrix (given in radian/degree).

$$\boxed{y = \text{SIN}(x)}$$

$\Rightarrow X \quad x \quad n \times p \quad$ matrix
$X \Rightarrow \quad y \quad n \times p \quad$ matrix

SINH   returns the hyperbolic sine $\dfrac{e^x - e^{-x}}{2}$ of a matrix.

$$\boxed{y = \text{SINH}(x)}$$

$\Rightarrow X \quad x \quad n \times p \quad$ matrix
$X \Rightarrow \quad y \quad n \times p \quad$ matrix

SKNN   calculates the symmetric k-nearest-neighbor smooth from scatter plot data. As inputs, you have to specify the explanatory variable x, the dependent variable y, and the smoothing parameter k.

$$\boxed{z = \text{SKNN}(x \ y \ k)}$$

$\Rightarrow X \quad x, y \quad n \times 1 \quad$ matrix
$\quad\quad\quad k \quad\quad n \times 1 \quad$ matrix or a number
$X \Rightarrow \quad z \quad\quad n \times 1 \quad$ matrix

SORT   sorts the rows of a matrix. If column c_j is specified, the matrix will be sorted with respect to column c_j. That is, the rows of the matrix will be arranged in order such that elements of column c_j are in ascending (descending) order.

$$\boxed{y = \text{SORT}(x \ \{c_1 \ \{c_2\{\ldots\}\}\})}$$

$\Rightarrow X \quad x \quad\; n \times p \quad$ matrix
$\quad\quad\quad c_j \quad\; \text{scalar}$
$X \Rightarrow \quad y \quad\; n \times p \quad$ matrix

SPLINE

computes a cubic spline from data $\{(x_i, y_i)\}_{i=1}^n$, that is, solves the problem to minimize with respect to m:

$$\sum_{i=1}^{n}(y_i - m(x_i))^2 + \lambda \int (m'')^2.$$

The spline s is computed at the gridpoints x_i.

$$\boxed{s = \text{SPLINE}(x \; y \; \{\lambda\})}$$

$\Rightarrow$ X	x	$n \times 1$	vector
	y	$n \times 1$	vector
	λ	number	
X $\Rightarrow$	s	$n \times 1$	vector

SPLIT splits a matrix and glues the parts together. A matrix x with n rows and p columns may be split every qth row $(n = \ell \cdot q)$ and glued together to give a matrix with q rows and $\ell \cdot p$ columns.

$$\boxed{y = \text{SPLIT}(x \; \{q\})}$$

$\Rightarrow$ X	x	$n \times p$	matrix
	q	number	
X $\Rightarrow$	y	$q \times (\ell \cdot p)$	matrix

SQRT computes the square root of a matrix.

$$\boxed{y = \text{SQRT}(x)}$$

$\Rightarrow$ X	x	$n \times p$	matrix
X $\Rightarrow$	y	$n \times p$	matrix

SSD computes the simplicial depth function of Liu (1990). The simplicial depth is a way of defining a multivariate median.

$$\boxed{y = \text{SSD}(x \; \{k\})}$$

$\Rightarrow$ X	x	$n \times p$ matrix or list of matrices
	k	number determining the precision of the procedure $(4 \le k \le 10)$
X $\Rightarrow$	y	the simplicial depth

STRCMP

compares elementwise two strings.

$$\boxed{s = \text{STRCMP}(t1\ t2)}$$

$\Rightarrow$ X	$t1$	$n \times p$ textmatrix or text
	$t2$	$n \times p$ textmatrix or text
X $\Rightarrow$	s	scalar (result of comparison)

SUM   computes the columnwise sum of a matrix.

$$\boxed{y = \text{SUM}(x)}$$

$\Rightarrow$ X	x	$n \times p$	matrix
X $\Rightarrow$	y	$p \times 1$	column vector

SUMR   computes the rowwise sum of a matrix.

$$\boxed{y = \text{SUMR}(x)}$$

$\Rightarrow$ X	x	$n \times p$	matrix
X $\Rightarrow$	y	$n \times 1$	column vector

SUPSMO

computes the supersmoother for a two-dimensional data set (x, y).

$$\boxed{s = \text{SUPSMO}(x)}$$

$\Rightarrow$ X	x	$n \times 2$	matrix
X $\Rightarrow$	s	$n \times 1$	matrix

SVD   computes the singular value decomposition of an $n \times p$ matrix. The singular value decomposition finds matrices u, ℓ, and v such that $u * \text{diag}(\ell) * v' = x$.

$$\boxed{(u\ \ell\ v) = \text{SVD}(x)}$$

$\Rightarrow$ X	x	$n \times p$	matrix of rank
X $\Rightarrow$	u	$n \times p$	matrix
	ℓ	$p \times 1$	vector
	v	$p \times p$	matrix

TAN   TAND

returns the tangent of a matrix (in radian/degree).

$$y = \text{TAN}(x)$$

$\Rightarrow$ X	x	$n \times p$	matrix
X $\Rightarrow$	y	$n \times p$	matrix

TANH   returns the hyperbolic tangent $\dfrac{e^x - e^{-x}}{e^x + e^{-x}}$ of a matrix.

$$y = \text{TANH}(x)$$

$\Rightarrow$ X	x	$n \times p$	matrix
X $\Rightarrow$	y	$n \times p$	matrix

TIME   computes a string with the current time in the form *hh:mm:ss*.

$$y = \text{TIME}$$

$\Rightarrow$ X		
X $\Rightarrow$	y	the time string

TRACE

is designed to trace execution of commands.

$$\text{TRACE } (n)$$

$\Rightarrow$ X	n	scalar

TRN   computes the transpose of a matrix x. It is equivalent to the " ' " operator.

$$y = \text{TRN}(x)$$

$\Rightarrow$ X	x	$n \times p$	matrix or list of matrices
X $\Rightarrow$	y	$p \times n$	matrix or list of matrices

UNIFORM

generates pseudo-random variables with a standard uniform distribution.

$$y = \text{UNIFORM}(n\{p\})$$

$\Rightarrow$ X n the number of rows to be generated

          p the number of columns to be generated (default is $p = 1$)

X $\Rightarrow$ y $n \times p$ matrix

UNIT   generates a unit matrix of given dimension.

$$y = \text{UNIT}(p)$$

$\Rightarrow$ X p the dimension of the unit matrix

X $\Rightarrow$ y $p \times p$ unit matrix

UPDATE

allows multiple drawings in the active picture object. With UP-DATE you can replace or add a datapart in a window of the actual display and you can manipulate the style matrix of this datapart.

$$z = \text{UPDATE}(x\ n\ \{s_1\ \{s_2\{\ldots\}\}\})$$

$\Rightarrow$ X x $n \times p$ matrix with the datapart to be manipulated

          n datapart number of x in the window where x will be inserted. If n is bigger than the data displayed in the window, then x will be added; if not, it will replace the nth dataset

          si styleparameter ($i = 1, \ldots$) or textmatrix with parameters; these parameters specify in which window x will be inserted, the drawstyle of the data x, the title, and the text of the axis.

X $\Rightarrow$ z datapart number of x in the window ($z = n$).

VAR   computes the variance for each column of a data matrix.

$$y = \text{VAR}(x)$$

$\Rightarrow$ X x $n \times p$ matrix

X $\Rightarrow$ y $p \times 1$ matrix

VTOCC

converts a float vector to a mask vector. The input value m represents the color and the outfit of a datapoint which is calculated according to:

$$x = 16 * outfit + color$$

with the following meanings for *outfit* and *color:*

Value	Outfit	Color
0	Unused	Black
1	Point	Blue
2	Plus	Green
3	X	Cyan
4	Y	Red
5	O	Magenta
6	Star	Brown
7	Face	Light Gray
8	Unused	Dark Gray
9	Unused	Light Blue
10	Unused	Light Green
11	Unused	Light Cyan
12	Unused	Light Red
13	Unused	Light Magenta
14	Unused	Yellow
15	Unused	White

$$\boxed{y = \text{VTOCC}(x)}$$

⇒ X	x	$n \times p$	matrix (p number of float vectors)
X ⇒	y	$n \times p$	matrix

WRITE

writes matrices or lists to a file.

$$\boxed{\text{WRITE}(x\ \{file\{fmt\}\})}$$

⇒ X	x	$n \times p$	matrix
	$file$	filename	
	fmt	XploRe format	

WRITECON

puts an XploRe code into the key buffer. The codes can be accessed by typing KEYTABLE and pressing F10.

$$\boxed{\text{WRITECON}(x)}$$

$\Rightarrow$ X x $n \times 1$ vector or number

XTY   computes the formula $X' * Y$.

$$z = \text{XTY}(x\ y)$$

$\Rightarrow$ X x $n \times k$ matrix
       y $n \times l$ matrix
X $\Rightarrow$ z $k \times l$ matrix

References

Aitkin, M., Anderson, D., Francis, B. and Hinde, J. (1989). *Statistical Modelling in GLIM*, Vol. 4 of *Oxford Statistical Science Series*, Oxford University Press, Oxford.

Andrews, D.F. and Herzberg, A.M. (1985). *Data: A Collection of Problems from Many Fields for the Student and Research Worker*, Springer Series in Statistics, Springer-Verlag, New York.

Anscombe, F.J. (1953). Contribution to the discussion of H. Hotelling's paper, *Journal of the Royal Statistical Society, Series B* **15**(2): 229–230.

Auestad, B. and Tjøstheim, D. (1990). Identification of nonlinear time series: First order characterization and order estimation, *Biometrika* **77**(4): 669–687.

Balke, N.S. (1993). Detecting level shifts in time series, *Journal of Business & Economic Statistics* **11**(1): 61–85.

Basawa, I.V. and Prakasa Rao, B.L.S. (1980). *Statistical Inference for Stochastic Processes*, Academic Press, London.

Baxter, L.A., Coutts, S.M. and Ross, G.A.F. (1980). Applications of linear models in motor insurance, *Proceedings of the 21st International Congress of Actuaries*, Zürich, pp. 11–29.

Becker, R.A., Chambers, J.M. and Wilks, A.R. (eds) (1988). *The new S Language*, Wadsworth & Brooks/Cole Advanced Books & Software, Pacific Grove, California.

Bierens, H.J. (1987). Kernel estimators of regression functions, *in* T.F. Bewley (ed.), *Advances in Econometrics: Fifth World Congress*, Vol. 1, Cambridge University Press, New York, pp. 99–144.

Bollerslev, T. (1986). Generalized autoregressive conditional heteroscedasticity, *Journal of Econometrics* **31**(3): 307–327.

Bollerslev, T., Chou, R.Y. and Kroner, K.F. (1992). Arch modelling in finance: A review of the theory and empirical evidence, *Journal of Econometrics* **52**(1): 5–59.

Bond, J. and Tabazadeh, A. (1994). Super Smoothing with Xlisp-Stat. Available as PostScript file `supersmu.ps` via anonymous ftp from `ftp.stat.ucla.edu` in the directory `pub/lisp/xlisp/xlisp-stat/code/homegrown/smoothers/supersmoother`.

Box, G.E.P. and Jenkins, G.M. (1976). *Time Series Analysis: Forecasting and Control*, Holden-Day, San Fransisco.

Breiman, L. and Friedman, J.H. (1985). Estimating optimal transformations for multiple regression and correlations (with discussion), *Journal of the American Statistical Association* **80**(391): 580–619.

Breiman, L., Friedman, J.H., Olshen, R. and Stone, C.J. (1984). *Classification and Regression Trees*, Wadsworth, Belmont.

Carroll, R.J., Küchenhoff, H., Lombard, F. and Stefanski, L.A. (1995). Asymptotics for the SIMEX estimator in structural measurement error models, *Journal of the American Statistical Association*. To appear.

Carroll, R.J., Ruppert, D. and Stefanski, L.A. (1995). *Nonlinear Measurement Error Models*, Chapman and Hall, New York.

Carroll, R.J. and Stefanski, L.A. (1990). Approximate quasilikelihood estimation in models with surrogate predictors, *Journal of the American Statistical Association* **85**(411): 652–663.

Chambers, J.M., Cleveland, W.S., Kleiner, B. and Tukey, P.A. (1983). *Graphical Methods for Data Analysis*, Duxbury Press, Boston.

Chan, K.S. (1990). Testing for threshold autoregression, *Annals of Statistics* **18**(4): 1886–1894.

Chen, R. (1994). A nonparametric predictor for nonlinear time series, *Technical report*, Department of Statistics, Texas A&M University, College Station.

Chen, R. and Tsay, R.S. (1993a). Functional-coefficient autoregressive models, *Journal of the American Statistical Association* **88**(421): 298–308.

Chen, R. and Tsay, R.S. (1993b). Nonlinear additive ARX models, *Journal of the American Statistical Association* **88**(423): 955–967.

Chernoff, H. (1973). Using faces to represent points in k-dimensional space graphically, *Journal of the American Statistical Association* **68**(342): 361–368.

Cleveland, W.S. (1979). Robust locally-weighted regression and smoothing scatterplots, *Journal of the American Statistical Association* **74**(368): 829–836.

Cleveland, W.S. and Devlin, S.J. (1988). Locally weighted regression: An approach to regression analysis by local fitting, *Journal of the American Statistical Association* **83**(403): 596–610.

Cook, D., Buja, A. and Cabrera, J. (1993). Projection pursuit indexes based on orthonormal function expansions, *Journal of Computational and Graphical Statistics* **2**(3): 225–250.

Cook, J.R. and Stefanski, L.A. (1995). Simulation-extrapolation estimation in parametric measurement error models, *Journal of the American Statistical Association*. To appear.

Cook, R.D. and Weisberg, S. (1991). Comment on "sliced inverse regression for dimension reduction", *Journal of the American Statistical Association* **86**(414): 328–332.

Cox, D.R. and Snell, E.J. (1968). A general definition of residuals (with discussion), *Journal of the Royal Statistical Society, Series B* **30**(2): 248–275.

DFG-Forschungsbericht (1981). *Chronische Bronchitis, Teil 2*, Harald Boldt, Boppard.

Dillon, W.R. and Goldstein, M. (1984). *Multivariate Analysis*, John Wiley & Sons, New York.

Dobson, A.J. (1990). *An Introduction to Generalized Linear Models*, Chapman and Hall, London.

Donoho, D.L. and Johnstone, I.M. (1989). Projection based approximation and a duality with kernel methods, *Annals of Statistics* **17**(1): 58–106.

Duan, N. and Li, K.C. (1991). Slicing regression: A link-free regression method, *Annals of Statistics* **19**(2): 505–530.

Emerson, J.D. and Strenio, J. (1983). Boxplots and batch comparison, *in* D.C. Hoaglin, F. Mosteller and J.W. Tukey (eds), *Understanding Robust and Explanatory Data Analysis*, John Wiley & Sons, New York, chapter 3, pp. 58–96.

Engle, R.F. (1982). Autoregressive conditional heteroscedasticity with estimates of the variance of U.K. inflation, *Econometrica* **50**(4): 987–1007.

Fan, J. (1992). Design-adaptive nonparametric regression, *Journal of the American Statistical Association* **87**(420): 998–1004.

Fan, J. (1993). Local linear regression smoothers and their minimax efficiency, *Annals of Statistics* **21**(1): 196–216.

Fan, J. and Gijbels, I. (1995a). Data-driven bandwidth selection in local polynomial fitting: Variable bandwidth and spatial adaptation, *Journal of the Royal Statistical Society, Series B*. To appear.

Fan, J. and Gijbels, I. (1995b). *Local Polynomial Modeling and Its Application—Theory and Methodologies*, Chapman and Hall, New York.

Fan, J. and Marron, J.S. (1994). Fast implementations of nonparametric curve estimators, *Journal of Computational and Graphical Statistics* **3**(1): 35–56.

Feller, W. (1966). *An Introduction to Probability Theory and Its Application*, Vol. 2, John Wiley & Sons, New York.

Flury, B. and Riedwyl, H. (1981). Graphical representation of multivariate data by means of asymmetrical faces, *Journal of the American Statistical Association* **76**(376): 757–765.

Flury, B. and Riedwyl, H. (1988). *Multivariate Statistics, A Practical Approach*, Cambridge University Press, New York.

Franz, W. and Soskice, D. (1994). The German apprenticeship system, *Discussionpaper 11*, Center for International Labor Economics, Konstanz, Germany.

Friedman, J.H. (1984). A variable span smoother, *Technical Report No. 5*, Department of Statistics, Stanford University, Stanford, California.

Friedman, J.H. (1987). Exploratory projection pursuit, *Journal of the American Statistical Association* **82**(397): 249–266.

Friedman, J.H. (1991). Multivariate adaptive regression splines (with discussion), *Annals of Statistics* **19**(1): 1–141.

Friedman, J.H. and Stuetzle, W. (1981a). Projection pursuit classification. unpublished manuscript.

Friedman, J.H. and Stuetzle, W. (1981b). Projection pursuit regression, *Journal of the American Statistical Association* **76**(376): 817–823.

Friedman, J.H., Stuetzle, W. and Schroeder, A. (1984). Projection pursuit density estimation, *Journal of the American Statistical Association* **79**(387): 599–607.

Friedman, J.H. and Tukey, J.W. (1974). A projection pursuit algorithm for exploratory data analysis, *IEEE Transactions on Computers* **C-23**(9): 881–890.

Fuller, W.A. (1987). *Measurement Error Models*, John Wiley & Sons, New York.

Gleser, L.J. (1990). Improvement of the naive approach to estimation in nonlinear errors-in-variables regression models, *in* P.J. Brown and W.A. Fuller (eds), *Statistical analysis of measurement error models and application*, American Mathematical Society.

Gordon, A.D. (1981). *Classification*, Vol. 16 of *Monographs on Statistics and Applied Probability*, Chapman and Hall, London.

Granger, C.W.J. and Anderson, A.P. (1978). *An Introduction to Bilinear Time Series Models*, Vandenhoeck & Ruprecht, Göttingen und Zürich.

Granger, C.W.J. and Teräsvirta, T. (1993). *Modelling Nonlinear Economic Relationships*, Academic Press, Oxford.

Greenacre, M.J. (1988). Clustering the rows and columns of a contingency table, *Journal of Classification* **5**: 39–52.

Haggan, V. and Ozaki, T. (1981). Modeling nonlinear vibrations using an amplitude-dependent autoregressive time series model, *Biometrika* **68**(1): 189–196.

Hall, P. (1989a). On polynomial-based projection indices for exploratory projection pursuit, *Annals of Statistics* **17**(2): 589–605.

Hall, P. (1989b). On projection pursuit regression, *Annals of Statistics* **17**(2): 573–588.

Hall, P. and Heyde, C.C. (1980). *Martingale Limit Theory and Its Applications*, Academic Press, New York.

Hall, P. and Li, K.C. (1993). On almost linearity of low dimensional projections from high dimensional data, *Annals of Statistics* **21**(2): 867–889.

Hall, P., Marron, J.S. and Park, B.U. (1992). Smoothed cross-validation, *Probability Theory and Related Fields* **92**: 1–20.

Härdle, W. (1990). *Applied Nonparametric Regression*, Econometric Society Monographs No. 19, Cambridge University Press, New York.

Härdle, W. (1991). *Smoothing Techniques, With Implementations in S*, Springer-Verlag, New York.

Härdle, W., Hildenbrand, W. and Jerison, M. (1991). Empirical evidence on the law of demand, *Econometrica* **59**(6): 1525 1549.

Härdle, W. and Simar, L. (1994). Applied multivariate statistical analysis. Lecture script.

Härdle, W. and Steiger, M. (1995). Optimal median smoothing, *Applied Statistics*. To appear.

Härdle, W. and Turlach, B.A. (1992). Nonparametric approaches to generalized linear models, *in* L. Fahrmeier, B. Francis, R. Gilchrist and G. Tutz (eds), *Advances in GLIM and Statistical Modelling*, Vol. 78 of *Lecture Notes in Statistics*, Springer-Verlag, New York, pp. 213–225.

Härdle, W. and Vieu, P. (1992). Kernel regression smoothing of time series, *Journal of Time Series Analysis* **13**(3): 209–232.

Harvey, A.C. (1989). *Forecasting, structural time series models and the Kalman filter*, Cambridge University Press, New York.

Hastie, T.J. and Loader, C.R. (1993). Local regression: Automatic kernel carpentry (with discussion), *Statistical Science* **8**(2): 120–143.

Hastie, T.J. and Tibshirani, R.J. (1986). Generalized additive models (with discussion), *Statistical Science* **1**(2): 297–318.

Hastie, T.J. and Tibshirani, R.J. (1990). *Generalized Additive Models*, Vol. 43 of *Monographs on Statistics and Applied Probability*, Chapman and Hall, London.

Hilbe, J.M. (1993). Generalized additive models software, *The American Statistician* **47**(1): 59–64.

Hilbe, J.M. (1994a). Generalized linear models, *The American Statistician*. To appear.

Hilbe, J.M. (1994b). Log negative binomial regression as a generalized linear model, *Technical Report 26*, Graduate College Commitee on Statistics, Arizona State University, Tempe.

Hinich, M.J. (1982). Testing for gaussianity and linearity of a stationary time series, *Journal of Time Series Analysis* **3**(3): 169–176.

Hinich, M.J. and Patterson, D.M. (1985). Identification of the coefficient in a nonlinear time series of the quadratic type, *Journal of Econometrics* **30**(3): 269–288.

Hjellvik, V. and Tjøstheim, D. (1994). Nonparametric tests of linearity for time series, *Biometrika*. To appear.

Horowitz, J.L. (1993a). Semiparametric and nonparametric estimation of quantal response models, *in* G.S. Maddala, C.R. Rao and H.D. Vinod (eds), *Economics*, Vol. 11 of *Handbook of Statistics*, North-Holland, Amsterdam, pp. 45–72.

Horowitz, J.L. (1993b). Semiparametric estimation of a work-trip mode choice model, *Journal of Econometrics* **58**(1–2): 49–70.

Horowitz, J.L. and Härdle, W. (1994). Testing a parametric model against a semiparametric alternative, *Econometric Theory*. To appear.

Hsing, T. and Carroll, R.J. (1992). An asymptotic theory for sliced inverse regression, *Annals of Statistics* **20**(2): 1040–1061.

Huber, P. (1985). Projection pursuit, *Annals of Statistics* **13**(2): 435–475.

Hubert, L.J. and Arabie, P. (1985). Comparing partitions, *Journal of Classification* **2**: 193–218.

Jee, J.R. (1985). Exploratory projection pursuit using nonparametric density estimation, *Proceedings of the Statistical Computing Section*, American Statistical Association, Statistical Computing Section, Washington, D.C., pp. 335–339.

Jones, D.A. (1978). Nonlinear autoregressive processes, *Proceedings of the Royal Society of London, Series A* **360**: 71–95.

Jones, M.C., Marron, J.S. and Park, B.U. (1991). A simple root n bandwidth selector, *Annals of Statistics* **19**(4): 1919–1932.

Jones, M.C. and Sibson, R. (1987). What is projection pursuit? (with discussion), *Journal of the Royal Statistical Society, Series A* **150**(1): 1–36.

Keenan, D.M. (1985). A Tukey nonadditivity-type test for time series nonlinearity, *Biometrika* **72**(1): 39–44.

Klimko, L.A. and Nelson, P.I. (1978). On conditional least squares estimation for stochastic processes, *Annals of Statistics* **6**(3): 629–642.

Klinke, S. (1993). A fast implementation of projection pursuit indices, *Discussion Paper 9320*, Institut für Statistik und Ökonometrie, Humboldt-Universität zu Berlin, Berlin, Germany.

Klinke, S. and Polzehl, J. (1994). Experiences with bivariate projection pursuit indices, *Discussion paper 9416*, Institut de Statistique, Université Catholique de Louvain, Louvain-la-Neuve, Belgium.

Kötter, T.T. (1995). An asymptotic result for sliced inverse regression, *Computational Statistics*. To appear.

Kruskal, J.B. (1969). Toward a practical method which helps uncover the structure of a set of observations by finding the line tranformation which optimizes a new "index of condensation", *in* R.C. Milton and J.A. Nelder (eds), *Statistical Computation*, Academic Press, New York, pp. 427–440.

Kruskal, J.B. (1972). Linear transformation of multivariate data to reveal clustering, *in* R.N. Shepard, A.K. Romney and S.B. Nerlove (eds), *Multidimensional Scaling: Theory and Applications in the Behavioural Sciences*, Vol. 1, Seminar Press, London, pp. 179–191.

Lai, T.L. and Wei, C.Z. (1983). Asymptotic properties of general autoregressive models and strong consistency of the least squares estimates of their parameters, *Journal of Multivariate Analysis* **13**(1): 1–23.

Lee, D.K.C. (1992). N-Kernel and XploRe, *Journal of Economic Surveys* **6**(1): 89–104.

Lee, Y. and Nelder, J.A. (1994). Double generalized linear models. Manuscript.

Lewis, P.A.W. and Stevens, G. (1991). Nonlinear modeling of time series using multivariate adaptive regression splines (mars), *Journal of the American Statistical Association* **86**(416): 864–877.

Li, K.C. (1991). Sliced inverse regression for dimension reduction (with discussion), *Journal of the American Statistical Association* **86**(414): 316–342.

Li, K.C. (1992). On principal Hessian directions for data visualization and dimension reduction: Another application of Stein's lemma, *Journal of the American Statistical Association* **87**(420): 1025–1039.

Liang, K.Y. and Zeger, S.L. (1986). Longitudinal data analysis using generalized linear models, *Biometrika* **73**(1): 13–22.

Liu, R.Y. (1990). On a notion of data depth based upon random simplices, *Annals of Statistics* **18**(1): 405–414.

Lubischew, A.A. (1962). On the use of discriminant functions in taxonomy, *Biometrics* **18**: 455–477.

Luukkonen, R., Saikkonen, P. and Teräsvirta, T. (1988). Testing linearity against smooth transition autoregressive models, *Biometrika* **75**(3): 491–499.

Mardia, K.V., Kent, J.T. and Bibby, J.M. (1979). *Multivariate Analysis*, Academic Press, Duluth, London.

Marron, J.S. and Nolan, D. (1988). Canonical kernels for density estimation, *Statistics & Probability Letters* **7**(3): 195–199.

McCullagh, P. and Nelder, J.A. (1989). *Generalized Linear Models*, Vol. 37 of *Monographs on Statistics and Applied Probability*, 2nd edn, Chapman and Hall, London.

Mucha, H.J. (1992). *Clusteranalyse mit Mikrocomputern*, Akademie Verlag, Berlin.

Mucha, H.J. and Klinke, S. (1993). Clustering techniques in the interactive statistical computing environment XploRe, *Discussion Paper 9318*, Institut de Statistique, Université Catholique de Louvain, Louvain-la-Neuve, Belgium.

Nelder, J.A. and Wedderburn, R.W.M. (1972). Generalized linear models, *Journal of the Royal Statistical Society, Series A* **135**(3): 370–384.

Ng, F.T. and Sickles, R.C. (1990). 'XploRe'-ing the world of nonparametric analysis, *Journal of Applied Econometrics* **5**(3): 293–298.

Nishisato, S. (1980). *Analysis of Categorical Data: Dual Scaling and Its Applications*, University of Toronto Press, Toronto.

Park, B.U. and Marron, J.S. (1992). On the use of pilot estimators in bandwidth selection, *Nonparametric Statistics* **1**: 231–240.

Park, B.U. and Turlach, B.A. (1992). Practical performance of several data driven bandwidth selectors, *Computational Statistics* **7**(3): 251–270.

Pemberton, J. (1987). Exact least squares multi-step prediction from nonlinear autoregressive models, *Journal of Time Series Analysis* **8**(4): 443–448.

Petruccelli, J. and Davies, N. (1986). A portmanteau test for self-exciting threshold autoregressive-type nonlinearity in time series, *Biometrika* **73**(3): 687–694.

Polzehl, J. (1993). Projection pursuit discriminant analysis, *Discussion Paper 9320*, CORE, Université Catholique de Louvain, Louvain-la-Neuve, Belgium.

Powell, J.L., Stock, J.H. and Stoker, T.M. (1989). Semiparametric estimation of index coefficients, *Econometrica* **57**(6): 1403–1430.

Priestley, M.B. (1988). *Non-linear and Non-stationary Time Series Analysis*, Academic Press, New York.

Proença, I. and Ritter, C. (1994). Semiparametric testing of the link function in models for binary outcomes, *SFB 373 Discussion Paper 940017*, Institut für Statistik und Ökonometrie, Humboldt-Universität zu Berlin, Berlin, Germany.

Projektgruppe "Das Sozio-ökonomische Panel" (1991). *Das Sozio-ökonomische Panel (SOEP) im Jahre 1990/91*, Deutsches Institut für Wirtschaftsforschung. Vierteljahreshefte zur Wirtschaftsforschung, pp. 146–155.

Rand, W.R. (1971). Objective criteria for the evaluation of clustering methods, *Journal of the American Statistical Association* **66**(336): 846–850.

Ritter, C. and Bouckaert, A. (1993). Modeling the morbidity of newborns. Submitted for publication. Available as manuscript from authors.

Robinson, P.M. (1977). The estimation of a nonlinear moving average model, *Stochastic Processes and Their Applications* **5**: 81–90.

Robinson, P.M. (1983). Non-parametric estimation for time series models, *Journal of Time Series Analysis* **4**(3): 185–208.

Rosner, B., Willett, W.C. and Spiegelmann, D. (1989). Correction of logistic regression relative risk estimates and confidence intervals for systematic within-person measurement error, *Statistics in Medicine* **8**(9): 1051–1069.

Rudemo, M. (1982). Empirical choice of histograms and kernel density estimators, *Scandinavian Journal of Statistics* **9**: 65–78.

Ruppert, D., Sheather, S.J. and Wand, M.P. (1996). An effective bandwidth selector for local least squares regression, *Journal of the American Statistical Association*. To appear.

Ruppert, D. and Wand, M.P. (1994). Multivariate weighted least squares regression, *Annals of Statistics*. To appear.

Schimek, M.G. (1991). Non-parametric regression techniques for biometric problems: Concepts and software, *in* K.P. Adlassnig, G. Grabner, S. Bengtsson and R. Hansen (eds), *Medical Informatics Europe 1991*, Springer-Verlag, Berlin, pp. 562–566.

Schimek, M.G. and Kubik, W. (1992). Möglichkeiten und Grenzen des Werkzeuges XploRe aus der Sicht des Biometrikers, *in* H. Enke, J. Gölles, R. Haux and K.D. Wernecke (eds), *Methoden und Werkzeuge für die exploratorische Datenanalyse in den Biowissenschaften*, G. Fisher, Stuttgart, pp. 129–139.

Schimek, M.G., Neubauer, G.P. and Stettner, H. (1994). Backfitting and related procedures for non-parametric smoothing regression: A comparative view, *in* W. Grossmann and R. Dutter (eds), *COMPSTAT '94. Proceedings in Computational Statistics*, Physica, Heidelberg, pp. 64–69.

Schimek, M.G. and Schmaranz, K.G. (1993). The statistical computing environment XploRe and state-of-art density and regression smoothing, *Statistics and Computing* **3**(1): 23–26.

Schneeweiß, H. and Mittag, H.J. (1986). *Lineare Modelle mit fehlerbehafteten Daten*, Physica Verlag, Heidelberg.

Schott, J.R. (1994). Determining the dimensionality in sliced inverse regression, *Journal of the American Statistical Association* **89**(425): 141–148.

Scott, D.W. (1992). *Multivariate Density Estimation: Theory, Practice, and Visualization*, John Wiley & Sons, New York, Chichester.

Scott, D.W. and Terrell, G.R. (1987). Biased and unbiased cross-validation in density estimation, *Journal of the American Statistical Association* **82**(400): 1131–1146.

Sedgewick, R. (1990). *Algorithms in C*, Addison-Wesley, Reading, Massachusetts.

Sheather, S.J. (1992). The performance of six popular bandwidth selection methods on some real data sets, *Computational Statistics* **7**(3): 225–250.

Sheather, S.J. and Jones, M.C. (1991). A reliable data-based bandwidth selection method for kernel density estimation, *Journal of the Royal Statistical Society, Series B* **53**(3): 683–690.

Silverman, B.W. (1985). Some aspects of the spline smoothing approach to nonparametric curve fitting (with discussion), *Journal of the Royal Statistical Society, Series B* **47**(1): 1–52.

Silverman, B.W. (1986). *Density Estimation for Statistics and Data Analysis*, Vol. 26 of *Monographs on Statistics and Applied Probability*, Chapman and Hall, London.

Späth, H. (1985). *Cluster Dissection and Analysis. Theory, FORTRAN Programs, Examples*, Ellis Horwood Limited, Chichester.

Stefanski, L.A. and Cook, J.R. (1994). Simulation-extrapolation: The measurement error jackknife, *Technical Report 2259*, Institute of Statistics, North Carolina State University, Raleigh.

Stoker, T.M. (1991). Equivalence of direct, indirect and slope estimators of average derivatives, *in* W.A. Barnett, J.L. Powell and G. Tauchen (eds), *Nonparametric and Semiparametric Methods in Econometrics and Statistics*, Cambridge University Press, New York.

Stoker, T.M. (1992). *Lectures on Semiparametric Econometrics*, CORE, Université Catholique de Louvain, Louvain-la-Neuve, Belgium.

Stone, C.J. (1985). Additive regression and other nonparametric models, *Annals of Statistics* **13**(2): 689–705.

Stone, C.J. (1986). The dimensionality reduction principle for generalized additive models, *Annals of Statistics* **14**(2): 590–606.

Stone, G. (1990). XploRe 2.0—A computing environment for exploratory regression and data analysis. T. BROICH, W. HÄRDLE AND A. KRAUSE, *The Economic Journal* **100**: 1401–1403.

Stuetzle, W. (1987). Plot windows, *Journal of the American Statistical Association* **82**(398): 466–475.

Subba Rao, T. (1981). On the theory of bilinear time series models, *Journal of the Royal Statistical Society, Series B* **43**(2): 244–255.

Subba Rao, T. and Gabr, M.M. (1980). *An introduction to bispectral analysis and bilinear time series models*, Vol. 24 of *Lecture Notes in Statistics*, Springer-Verlag, New York.

Sugihara, G. and May, R.M. (1990). Nonlinear forecasting as a way of distinguishing chaos from measurement error in time series, *Nature* **344**(6268): 734–741.

Tibshirani, R.J. (1988). Estimating transformations for regression via additivity and variance stabilization, *Journal of the American Statistical Association* **83**(402): 394–405.

Tjøstheim, D. (1986). Estimation in nonlinear time series models, *Stochastic Processes and Their Applications* **21**: 251–273.

Tong, H. (1978). On a threshold model, *in* C.H. Chen (ed.), *Pattern Recognition and Signal Processing*, Sijthoff and Noordholf, The Netherlands.

Tong, H. (1983). *Threshold Models in Nonlinear Time Series Analysis*, Vol. 21 of *Lecture Notes in Statistics*, Springer-Verlag, Heidelberg.

Tong, H. (1990). *Nonlinear Time Series Analysis: A Dynamic Approach*, Oxford University Press, Oxford.

Truong, Y.K. (1993). A nonparametric framework for time series analysis, *in* D. Billinger, P. Caines, J. Geweke, E. Parzen, M. Rosenblatt and M.S. Taqqu (eds), *New Directions in Time Series Analysis*, Springer-Verlag, New York.

Tsay, R.S. (1986). Nonlinearity tests for time series, *Biometrika* **73**(2): 461–466.

Tsay, R.S. (1989). Testing and modeling threshold autoregressive processes, *Journal of the American Statistical Association* **84**(405): 231–240.

Tsay, R.S. (1991). Detecting and modeling nonlinearity in univariate time series analysis, *Statistica Sinica* **1**(2): 431–451.

Turlach, B.A. (1993). Bandwidth selection in kernel density estimation: A review, *Discussion Paper 9317*, Institut de Statistique, Université Catholique de Louvain, Louvain-la-Neuve, Belgium.

Wand, M.P. (1994). Fast computation of multivariate kernel estimators, *Journal of Computational and Graphical Statistics.* in press.

Wand, M.P., Marron, J.S. and Ruppert, D. (1991). Transformations in density estimation (with discussion), *Journal of the American Statistical Association* **86**(414): 343–361.

Ward, J.H. (1963). Hierarchical grouping methods to optimize an objective function, *Journal of the American Statistical Association* **58**(301): 236–244.

Zhu, L. and Ng, K.W. (1993). On asymptotic theory of sliced inverse regression based on weighted method. Preprint.

Author Index

Subject Index